The Upper Cambrian *Rehbachiella* and the phylogeny of Branchiopoda and Crustacea

DIETER WALOSSEK

Walossek, D. 1993 07 08: The Upper Cambrian *Rehbachiella* and the phylogeny of Branchiopoda and Crustacea. *Fossils and Strata*, No. 32, 1–202, Oslo. ISSN 0300-9491. ISBN 82-00-37487-4.

More than 130 specimens representing various growth stages of *Rehbachiella kinnekullensis* Müller, 1983, have permitted a detailed description of its ontogeny. It begins with a nauplius already able to swim and feed actively. The 30th stage is about 1.7 mm long, but still immature. Because the type specimens belong to earlier instars, the original diagnosis of Müller (1983) is emended. Details of the limb apparatus of late instars suggest that the animals were able to filter-feed by this stage, possibly while swimming close to the bottom. Two larval series are distinguished by size and morphology in their early stages, but their structural differences become almost balanced subsequently. This is interpreted as intraspecific differentiation rather than as existence of two species. The entire postnaupliar feeding apparatus of Branchiopoda, which is basically adapted to filtration, is recognized here as an apomorphic character of this group. Branchiopoda comprise the two monophyletic units Anostraca and Phyllopoda (Calmanostraca, with Notostraca and Kazacharthra, and Onychura). *Rehbachiella* shares all major aspects of the branchiopod filter apparatus, which led to identify it as an ancestral marine branchiopod. Moreover, there are indications that *Rehbachiella* is a representative of the anostracan lineage, i.e. a representative of the stem-group of Sarsostraca, which include the Devonian Lipostraca and the extant Euanostraca. The long larval sequence of *Rehbachiella* and selective external features, including the locomotory and feeding apparatus, are evaluated for their bearing upon the phylogeny of Branchiopoda and Crustacea in general. This study on *Rehbachiella* supports the monophyly of the crown-group Crustacea (sensu Walossek & Müller 1990). It also has revealed that only the first maxilla was morphologically and functionally included into the crustacean head, while subsequent limbs were addted to the head in a stepwise manner and became modified separately within the different crustacean lineages, which is of great relevance when evaluating the relationships between these. □ *Crustacea, Branchiopoda*, Rehbachiella, *functional morphology, filter apparatus, life habits, ontogeny, phylogeny.*

Dieter Walossek, Institut für Paläontologie, Rheinische Friedrich-Wilhelms-Universität, Nußallee 8, D-53115 Bonn 1, Germany; 2nd February, 1991; revised version 3rd September, 1991.

T0224314

Contents

List of figures

Introduction

Stratigraphy and taphonomy

The *orsten* arthropods, etched from anthraconitic stink-stones, have been recovered from various localities in the southern part of Sweden (Fig. 1). They are from two time intervals within the Upper Cambrian sequence, the majority of forms coming from nodules of zone 1 (*Agnostus pisiformis*), in some cases extending up to zone 2, subzone 2a (*Olenus gibbosus* with *Homagnostus obesus*). Another series of collections was made in zone 5 (*Peltura* sp.); zones 3 and 4 did not yield any such material. The geological range of the arthropods appears to be restricted to one or the other set of zones; so far the only exception is the so-called 'type-A larvae' (Müller & Walossek 1986b, Walossek & Müller 1989; see Addendum).

The original integument of the fossils was impregnated by phosphate and prevented from compaction by being embedded in limestone matrix. This resulted in a three-dimensional preservation largely retaining most of the delicate cuticular details. Details of taphonomy are still not well understood. Possibly the animals sank (alive or dead) into the anoxic zone below the still aerated surface layer (see Fig. 2), but the phosphate source is unclear, because the surrounding rock contains no significant amount of phosphate. It is unlikely that the *orsten* fossils represent exuviae, or that they were mummified, as claimed by Chen & Erdtmann 1991). Impregnation must have occurred rapidly, for if they had sunk a long way through the water column down to the anoxic zone where phosphatization took place, more extensive decay would have resulted than if they had only travelled a short distance (indicated on right side of Fig. 2). Hence, the relative degree of decomposition might perhaps help in estimating the preferred life zone of the faunal components (vertical stratification). Again, the distribution of developmental stages of an animal could point to special life strategies, because forms can be represented by either larval stages up to adults (A in Fig. 2), only larval and immature stages (B), only early larvae (C), or only adults (D).

Remarkably, no specimen in the *orsten* material, whether complete or fragmented, exceeds two millimetres. This may be explained by a very selective mode of phosphatization which affected only small-sized fossils with chitinous or chitin-like cuticular components. Such a chemically controlled preservation mechanism may be confirmed by the poor record of phosphatized trilobite remains in the etched material (only one clear find so far), although their calcareous exoskeletal remains are common in the rock. A remarkable exception is *Agnostus pisiformis* (cf. Müller & Walossek 1987), but its relationships with trilobites are not unequivocal (cf. Walossek & Müller 1990).

Palaeoecology and environmental conditions

Orsten arthropods document a wide range of life form types (Müller & Walossek 1985a, Fig. 5), the majority of them seeming to have been adapted to a life at or near the bottom, presumably on or within a soft surface layer, a flocculent zone, rich in detrital matter (Müller & Walossek 1986c). In other words, they may not have ventured greatly above the sediment–water interface. Flocculent layers exist today in all regimes from deep sea to shallow water and are preferentially inhabited by the meiofauna. The assumption of a flocculent layer at the bottom of the alum shale sea carries significance for ecological as well as environmental interpretations, on account of the special nature of such a layer, for example:

- high availability of nutrients,

- a water column that is oxygenated down to the benthic boundary layer,

- rapid decrease of oxygen immediately below the flocculent layer in accordance with rapid formation of sulphides (Ott & Novak 1989) and enrichment of

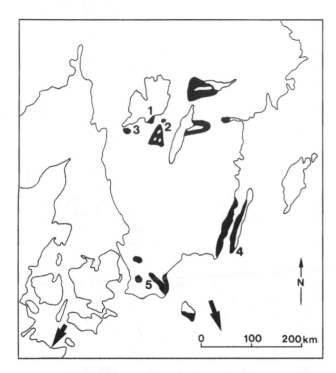

Fig. 1. Map of Southern Sweden, including localities that yielded phosphatized arthropods with preserved cuticle (black areas = exposed Cambrian). 1 Kinnekulle, Västergötland; 2 Falbygden–Billingen, Västergötland; 3 Hunneberg; 4 Öland; 5 Skåne. Arrows point to further discoveries outside Sweden – at lower left: in drift boulders from Northern Germany, at right bottom: in a borehole in Poland (Walossek & Szaniawski 1991). Modified from Bergström & Gee (1985, Fig. 1).

phosphate in the upper part at low rates of sedimentation (U. Pfretschner, Bonn, personal communication, 1992), or

• vertical stratification by gradual compaction, providing niches for animals of different sizes to live and to escape from predators.

Very small animals, and especially the meiofauna, are adapted to conditions of the viscous regime at low Reynolds numbers, which necessitates quite different life strategies from those of larger animals. Primarily, they feed on detritus and small-sized algae and bacteria (Coull 1988), if they have not otherwise evolved different life styles. Several *orsten* arthropods were most likely bottom dwellers and encounter-feeders on similar particulate matter. Examples are the Skaracarida and *Martinssonia elongata*, but these differ in details of their feeding strategies: the latter had only rigid spines with which to push food toward an exposed sucking mouth (Müller & Walossek 1986a), while the former possessed delicate setulate setae on their cephalic appendages for sweeping or brushing particles into the atrium oris underneath the labrum (Müller & Walossek 1985b).

The two species of *Skara* differ mainly in size and details of the feeding apparatus: *S. minuta*, about 0.7 mm long,

may have lived below the sediment–water interface, while *S. anulata*, about 1.2 mm long, lived at the interface or slightly above it in the benthic boundary layer. *Bredocaris admirabilis* must have been fairly mobile, as is apparent from its set of swimming postmaxillulary limbs, but its size, about 0.85 mm, and the effacement of segmentation of its various body parts point rather to a life below or at the sediment–water interface (Müller & Walossek 1988b; type A in Fig. 2).

While some forms were without a distinct head shield, such as the newly discovered *Cambropachycope clarksoni*, or even lacked external body segmentation completely, such as *Goticaris longispinosa* (cf. Walossek & Müller 1990), the Phosphatocopina (Müller 1964, 1979, 1982) and *Agnostus* had their body entirely enclosed in two valves. Undoubted predators, infaunal organisms, or crawling forms have not been discovered as yet, with the possible exception of *Henningsmoenicaris scutula* with its bowl-shaped shield covering most of the body (Walossek & Müller 1990, 1991). Yet, predation must have occurred, as is indicated by the anterior–posterior compaction of complete phosphatocopines, feces pellets which contain larval phosphatocopines and setae, or specimens with lost legs in the same fashion as produced by predators (such as a specimen of *Bredocaris*, illustrated in Müller & Walossek 1988b, Pl. 1:2; after Strickler, personal communication, 1989). With the chelicerate larva (Müller & Walossek 1986b, 1988a) and several larvae with remarkable resemblance to the extant Pentastomida (Müller & Walossek, in preparation) also ectoparasites existed in the *orsten* assemblages.

The size range, morphotypes, life styles and cycles of *orsten* arthropods accord well with a typically minute meiofauna. These are to be separated into (1) typical meiofaunal elements that never exceed the upper size limit of preservation, and (2) forms of the 'transitory meiobenthos'. Examples of the former type are *Bredocaris*, found with the complete set of developmental stages and the adult, and the Skaracarida, known only from adults (types a, A and D, d in Fig. 2). The transitory type is represented by larval stages of forms whose later stages do exceed the size limit and, since departing from the flocculent layer, are not preserved; examples are *Agnostus* and the Phosphatocopina, of which empty shells of larger stages can be found, but also *Rehbachiella* and possibly *Martinssonia* (b, B, c, C in Fig. 2), known from five growth stages (three egg- to spindle-shaped early instars and two stages with a segmented tail). Whereas twice as much material has been recovered of *Martinssonia* since, no specimen has been found larger than those already known. Other forms are represented only by early larvae (e.g., type-A larvae; type C in Fig. 2).

The presence of meiofaunal components in the fossil material also indicates that the special nature of the environment limited the size range of candidates for preserva-

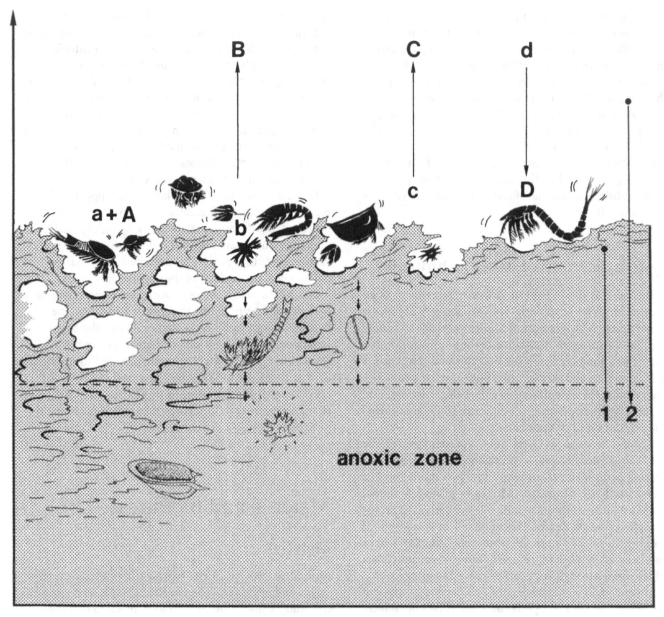

Fig. 2. Scheme showing supposed flocculent layer at the bottom of the Alum Shale sea; lower-case letters = larvae; upper-case letters = adult stages; 1 = short distance of sinking into zone of preservation; 2 = long distance (explanations see text).

tion, but this can be stated only for a part of the fauna. With this, the fossil record has an important bearing on the reconstruction of the presence of life at the bottom of the alum shale sea. It may also have an impact on conceptions not only of the palaeo-environment and genesis of the Upper Cambrian shale sequence but of deposits of similar type elsewhere from the Cambrian and from other geological periods. The *orsten* fauna points to the long existence of the flocculent zone as an environment preferably for small-sized organisms. Indications of the even longer existence of this regime may be seen in finds of faecal-pellet microfossils in rocks up to 1.9 billion years old (Robbins *et al.* 1985), because faecal pellets are typical components of Recent soft-bottom layers (e.g., Watling 1988).

Size and developmental stages

Considering the restricted size range, whether controlled by fossilization or environment, it is not surprising that the bulk of the material comprises larval stages, even down to 100 μm in body length. Larger animals are more rarely preserved, are mostly fragmented, and likewise do not exceed the upper size limits. Müller has commented several times (e.g., 1979, 1982, 1983) that the material embraces a mixture of immature stages and adults. It has been frequently contended that the *orsten* assemblages consist exclusively of larvae, and size by itself has been used as an argument against the adult state of certain forms (e.g., Schram 1986, pp. 522–524). As a matter of fact, crustaceans are, with the exception of most Malacostraca, commonly rather small, often in the same size range as the *orsten* arthropods or even much smaller (e.g., cephalocarids, branchiopod and, in particular, maxillopod taxa).

Similarly, considerable misinterpretation may result if the material is treated as though it were comprised exclusively of adults (Lauterbach 1988, for *Walossekia* and *Rehbachiella*). In *Walossekia quinquespinosa*, for example, which is known as yet only from larval specimens (cf. Müller 1983), its immature status is evident from the few trunk segments and rudimentary shape of the posterior limbs. Additional material of younger as well as later stages confirms this.

Systematic status of *orsten* arthropods

Since the first discoveries of cuticular remains in open shells of Phosphatocopina in 1975 (Müller 1979, 1985), the major research programme of Müller has yielded a variety of minute arthropods in addition to the phosphatocopines (the most abundant non-trilobite arthropod components in the nodules) and various other phosphatic microfossils. They document not only remarkable ecological adaptation but also distinctive body plans indicating different systematic positions and evolutionary levels.

Besides *Agnostus* as a possible representative of the arachnate–trilobite line, a probably ectoparasitic larva bearing prominent cheliphores and two more pairs of limbs has been recovered (Müller & Walossek 1986b, 1988a). This larva shows remarkable similarities to protonymph larvae of Recent Pantopoda. The small pair of outgrowths located near the frontal mouth of this larva may be interpreted to represent the reduced first antennae, which would strongly support the general assumption that Chelicerata s. str. (= crown-group chelicerates, including Euchelicerata and Pantopoda) have lost these appendages early in their evolution (e.g., Pross 1977). It also gives further evidence, together with the finds of *Sanctacaris* in the Burgess shale fauna (cf. Briggs & Collins 1988), that the roots of Chelicerata s. str. reach down well into the Cambrian.

Some *orsten* forms can be definitely assigned to particular crustacean taxa. Confirmation of the presumed maxillopod relationships of Skaracarida has been given in the description of *Bredocaris admirabilis*, now known with its complete larval series from a metanauplius of about 0.2 mm to the 0.85 mm long adult (Müller & Walossek 1988b). Evidence for the adult state of the largest specimens is seen in their tagmosis and segmentation which agree with the basic plan of Maxillopoda, in the full development of seven pairs of swimming thoracopods appearing after a metamorphosis-like jump from the last metanaupliar stage (with four pairs of thoracopod rudiments), and the effaced segmentation on various body parts (tho-

rax, abdomen, thoracopods, articulation of furca), which is recognized here as a special adaptation to a meiobenthic life style, possibly below the sediment–water interface.

Others resemble crustaceans but do not exactly fit within this group as characterized. The relationships of four forms, *Henningsmoenicaris scutula*, *Cambropachycope clarksoni*, *Goticaris longispinosa* and *Martinssonia elongata* as representatives of the stem group of Crustacea (short: stem-group crustaceans in the following text) have been worked out recently (Walossek & Müller 1990, 1991, 1992; for the stem-lineage concept, see Ax 1985).

Given the diversity of body plans in the Upper Cambrian *orsten* material and especially with the availability of ontogenetic stages, the potential for studying the external morphology in full detail is of outstanding value for the understanding of the evolution and early life history of Arthropoda, particularly of the Crustacea.

For *Rehbachiella kinnekullensis*, recognized here as a branchiopod crustacean, this restudy includes the first description of the life cycle, and evaluates also aspects of functional morphology and life habits. The enhanced information on the ontogenetic sequence and morphology of this fossil permits detailed comparisons with other crustaceans as well as a discussion of the status of particular characters and of evolutionary processes among Branchiopoda and Crustacea in general.

Material and methods

Material

More than 140 specimens representing different growth stages, and initially assigned to *Rehbachiella*, were used for this study. Closer examination soon revealed that the material was less homogeneous than had first been assumed. Dissimilarities were apparent, for example, in the position of a large dorsal spine in front rather than at the end of the last segment. Eventually, a total of 134 specimens remained clearly identified as *Rehbachiella kinnekullensis* 'in the strict sense' (Tables 1, 2). Of these, 117 specimens could more or less be definitely grouped into growth stages, while the rest are fragments without clear assignment.

The 25 samples with specimens are from four localities in the Kinnekulle area and Billingen–Falbygden, Västergötland, Sweden (Fig. 1); at three of these localities the material comes from zone 1 and at one from zone 2a. From the single sample of zone 2a, 6404 (road cut between Häggården and Marieberg, Kinnekulle) only three specimens have been recovered: the holotype, one early larval specimen, and one unassignable. The majority comes from three samples from Gum: 6409 with 27 specimens, 6761 with 21, and 6783 with 17 (6409 is the most productive

A

Series A	N	6364	6404	6409	6410	6411	6414	6415	6417	6729	6730	6748	6750	6752	6753	6754	6755	6760	6761	6763	6765	6773	6776	6783	6784	6787
		1	2a	1	1	1	1	1	1	1	1	1	1	1	1	1	1	1	1	1	1	1	1	1	1	1
L 1	1					1																				
L 2	7				2		2	1											2							
L 3	20	1		9							1								4	1				4		
L 4	10	1		2		1	1	1											3						1	
TS 1i	12	1	1	2	2				1									1	1					1	2	
TS 1	2			1																						1
TS 2i	1													1												
TS 2	3				1														1							1
TS 3i	2			1																						1
TS 3	1																		1							
TS 4i																										
TS 4	1																							1		
TS 5i																										
TS 5	4			1														1	1						1	
TS 6i	2			1											1											
TS 6	2																1								1	
TS 7i	1																			1						
TS 7	2				1																				1	
TS 8i	1																1									
TS 8	1												1													
TS 9i																										
TS 9																										
TS 10i																										
TS 10	2															1								1		
TS 11i																										
TS 11																										
TS 12i																										
TS 12																										
TS 13i	1											1														
TS 13	1																									1
ΣA	77	3	1	17	3		4		5	3	1	1	1	1		1	1	2	14	2				2	11	3	1

B

Series B	N	6364	6404	6409	6410	6411	6414	6415	6417	6729	6730	6748	6750	6752	6753	6754	6755	6760	6761	6763	6765	6773	6776	6783	6784	6787
		1	2a	1	1	1	1	1	1	1	1	1	1	1	1	1	1	1	1	1	1	1	1	1	1	1
L 4	2																	1						1		
TS 2	2			1					1															1		
TS 3	3			1			1																	1		
TS 4i	2					1																		1		
TS 4	10			3	1		1													1	3			1		
TS 5i	3			1																1				1		
TS 5	6				1		1												1				1	2		
TS 7i	1																		1							
TS 8i	3				1	1																				
TS 9i				1																						
TS 10	1																		1							
TS 11	3																	2	1							
TS 12	1	1																								
TS 13	3														1						1		1			1
ΣB	40	1		7	2	1	3		2	1					1			5	7		1		1	5	3	
unass.	17	1	1	3	5		1	1										2	1	1			1			
Σtotal	134	4	3	27	10	1	8	1	7	3	2	1	1	1	1	1	1	2	21	10	1	1	3	17	6	1

Table 1. Sample productivity. Sample 6364 from locality Stolan, 6404 from Haggården-Marieberg, E of Kinnekulle, 6729, 6730 from Backeborg SW of Kinnekulle, all other samples from Gum south of Kinnekulle (unass. = unassignable). □A. Larval series A. □B. Larval series B.

sample; the paratype is from sample 6411, also from Gum; see Table 1).

In general, the material is more distorted than, for example, that of Skaracarida or Orstenocarida. In particular the larger specimens are rare and in most cases rather fragmentary. This may indicate that *Rehbachiella* was less well sclerotized than the other forms. Complete preservation of setation is rare, but is occasionally found (e.g., Pls. 4:5; 6:9; 14:5, 6; 15:3–5; 16:1–7; 19:6; 22:7; 25:4; 28:7). In the majority of specimens, the setae or spines are broken off at their insertions, leaving tubercles, small holes, or rings on the surface (e.g., Pls. 6:1; 9:1; 17:2 25:3, 5, 6; 29:3). Remains of thinner setules or denticles mainly appear as tiny pustules (e.g., Pls. 6:3; 15:5; 29:2; 33:4; 34:2), but these also may be preserved in some cases (e.g., Pls. 11:4; 13:4; 14:5, 6; 16:1–7; 22:8; 25:4, 6; 28:7). Because the total lengths of setae and spines are in most cases unknown, they are mainly illustrated either cut short or by dots demarcating their insertions. The setae may have been even more numerous and, in various cases, longer originally than could be included in the reconstructions.

As in other *orsten* arthropods described, the arthrodial membranes are often collapsed, probably due to loss of turgor pressure after death of the animal or to osmotic changes. This shrinkage in particular at joints may repeatedly lead to a similar orientation of body parts. As an example, Pls. 4:3 and 5:2 show the posterior flexure of the exopod in the same fashion as in *Bredocaris* (Müller & Walossek 1988b, Fig. 13:2) or in dead Recent crustaceans

(e.g., Perryman 1961). Preservation of the membrane covering the anal region is also rare, obviously due to its softness (Pls. 3:2; 4:1; 6:3; 9:6; 10:3; 12:3; 14:2, 3; 18:6; 19:1; 22:2; 24:8). In some instances it seems as if internal effects (gas production by decay?) have caused extrusion of the caecum (e.g., Pls. 7:8; 8:2; 20:1; 34:3).

Methods

Processing and measurements

Techniques of preparation have been described earlier (e.g., Müller 1985; Müller & Walossek 1985b). SEM micrographs were taken with a CamScan series II and an Asahi Pentax K1000. A few specimens were lost, partly due to drying out and cracking of the double adhesive tape on which the specimens are mounted (see Table 2). Reconstructions were based on actual specimens representing particular stages as far as possible.

Measurements were made in the same way as was described earlier (e.g., Müller & Walossek 1985b for Skaracarida). In most cases the data were slightly adjusted according to the degree of distortion of the individuals. Hence the resulting means are not statistical values but approximations in order to give an impression of the growth of *Rehbachiella.* These values are incomplete because of preservation, and several important data, such as those of the head shield, the trunk, and the total length, could not definitely be established for the later growth stages (Table 3). To compensate for this, the distance between 1st antenna and 2nd maxilla was taken as a measure of head length, because this value could be obtained even when appendages were not preserved.

Two larval series

The measurements and the morphological analysis revealed two sets of larvae, series A with 77 specimens, B with 40. The specimens were grouped into 35 stages, 21 of series A and 14 of series B. The early developmental stages of both series could be quite readily distinguished due to differences in size and various morphological features (e.g., morphogenesis of appendages, head structures, furcal rami, 'dorsocaudal spine'). After a number of instars the major distinctions are, however, almost balanced between the two series. Thus, advanced larvae could be ascribed with certainty to a particular developmental stage of the one or the other series only when sufficient data from sizes and appendages were available. Hence, in the light of uneven representation (especially since larger stages are often known only from single specimens) it cannot be entirely excluded that the occasional individual may be still misplaced.

Since only external features are recognizable in *Rehbachiella*, the ontogenetic sequence is described along with the progressive formation of body segments. This method follows that of, in particular, Weisz (1946, 1947) who argues strongly against the use of moulting stages to describe sequences due to relative growth of individuals ('biochronism') and inconsistencies when stages are 'lost' either by non-recognition or by abbreviated development (even Weisz missed stages). Again, moult intervals may also vary on account of environmental influences, such as temperature or salinity (e.g., Hentschel 1967, 1968, for euanostracan Branchiopoda). As to the existence of two series, the working hypothesis is made that:

- both series belong to the same species,

- both were equal in consisting of 30 stages up to an instar with 13 trunk segments (12 limb-bearing ones),

- missing stages are caused by preservation failure,

- larger stages existed, beyond the 30th instar,

- probably the unsegmented abdomen becomes segmented in the subsequent developmental phase, and

- 13 is the final number of thoracomeres, and 12 of the thoracopods.

Terminology

The terminology is in general accordance with that of Kaestner (1967), Moore & McCormick (1969), and Mc-

Laughlin (1980). The classification of setal types proposed by Watling (1989) is applicable only to the large-sized Eumalacostraca and is not adopted here. Some principal terms used for *Rehbachiella* are included in Fig. 3 for an early larva (a), a later instar (b), and special parts (c–d). Other terms that are in different use are explained below (additional notes in the text, when necessary).

Appendages. – Discussion of the terminology of crustacean appendages has had a long history (cf. McLaughlin 1982, p. 200, for compilation of references on this subject), which is also true for branchiopod limbs (cf. Eriksson 1934, pp. 30–50 for historical overview). Difficulties arose in particular because terminology from other crustaceans with different segmentation and even other arthropods (e.g., trilobites, insects) was applied, although the homology of parts was at least not unequivocal. Additional problems resulted from the distinctiveness of the naupliar from the subsequent limbs: limb stems or corms of the 2nd antenna and mandible always show a clear subdivision, while in postmandibular limbs such a distinctive bipartition is not the rule. In phyllopodous limbs or similar types, the corms may, for example, be more or less completely devoid of any such division (often named 'sympodite' accordingly).

The homology of the subdivisions of the limb stems and rami in Crustacea has never been sufficiently clarified. Herein, the terminology of Walossek & Müller (1990) is adopted. This expands Sanders (1963b) convincing homologization of the coxal portions of the 2nd antenna and mandible with the 'proximal endite' of the 1st maxilla of Cephalocarida onto a separate endite at the medioproximal edge of the limb basis (corm) of recently discovered stem-group crustaceans. Accordingly, all proximal endites or portions of postmandibular limbs of Crustacea in the strict sense, whether termed 'arthrite', 'gnathite', 'gnathobase' or 'median endite', are homologized with the 'proximal endite' retained from the limb at the stem-group level, and, furthermore, with the coxae of the naupliar limbs (see also Fig. 54).

In particular in these two naupliar mouthparts, the endite has enlarged to form a distinct coxal portion. Coxa and basipod of these appendages carry a single enditic outgrowth each. Basically the 'proximal endite' (pe; = coxa) of subsequent limbs is a single outgrowth too, but may also be subdivided in certain crustaceans and/or enlarge to form a distinct portion similar to that of the anterior two limbs. The limb basis or basipod, regardless of its size, represents the primordial basis of the euarthropod limb, which carries two rami, as can be observed in the limbs of stem-group crustaceans as well as virtually all trilobitoid-type limbs of the various early Palaeozoic fossils (e.g., Cisne 1975, Fig. 3; Whittington 1979; Briggs & Whittington 1985; Müller & Walossek 1987; Chen *et al.* 1991, Fig. 6 of a *Naraoia* leg; inner ramus often termed 'telopod[ite]'; Fig. 54A herein). At the ground-plan level of

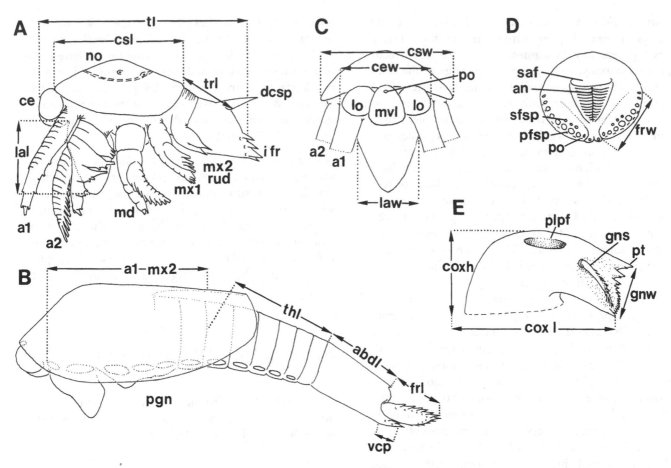

Fig. 3. Measurements and gross morphology of *Rehbachiella kinnekullensis* at different stages. For abbreviations in this and subsequent figures, see list at the end of paper. □A. Lateral view of early larva. □B. Advanced larva, appendages omitted (insertions indicated). □C. Anterior view of head. □D. Posterior view of abdomen with furcal rami and anus. □E. Coxal body of mandible from anterior; palp, comprising basipod and rami, omitted.

Crustacea s. str. the inner edge of the basipod was most probably still uniform, but within the different lineages it became subdivided into up to 7–8 spinose or setiferous enditic lobes (e.g., *Rehbachiella* herein). Again, the basipod may also subdivide entirely due to functional needs and attain new joints.

Some confusion arose because this basipod portion has variously been understood as the 'protopod[ite]', while in those limbs with a distinctive or coxal portion, the 'protopod' included both parts. Hence, herein the term 'corm' is used when it is referred to the entire limb stem, in accordance with Cannon (1933) and Fryer (1983). Since there are only two portions in the crustacean limb corm, the 'old' basis and the 'new' coxa (primarily a small endite), hypotheses that the basipod originated from the fusion of proximal endopodal and exopodal segments, as proposed by Ito (1989a), or of the existence of an additional precoxal portion, are rejected.

For major aspects of the branchiopod type of limbs the terminology used herein follows Eriksson (1934), McLaughlin (1982), Fryer (e.g., 1983, 1988), and Schram (1986), because this terminology is largely compatible with

observations on the limbs of *Rehbachiella* as well as the Devonian *Lepidocaris rhyniensis* Scourfield, 1926. Although at least some of Eriksson's interpretations concerning the segmentation of euanostracan phyllopodia may not conform with the concept of the crustacean limb, as accepted here, this author convincingly explained the nature of rami and exites (epipods/pre-epipods) by their shape, morphogenesis, serial modification, and function.

In accordance with Fryer (1988, also his Fig. 121) the term 'palp' is used for the distal part of the mandible, comprising the basipod and the two rami. Its area of articulation with the coxal body is termed 'palp foramen' accordingly. This 'palp' is the original limb basis (basipod) plus the rami, which is unclear for the 'palp' of Malacostraca. The splitting of the mandibular parts into a 'larval mandible' and 'adult mandible' (Schrehardt 1986b and subsequent papers) does not conform with the morphology of crustacean mandibles, and must be refuted. The term 'gnathobase', as a functional term, is used only for the mandibular coxal endite and is not applied to the 'proximal endite' of posterior limbs.

Dorsal shield. – In *Rehbachiella* a shield covering the anterior body region is present from the first instar. During development only the segments of the maxillae are incorporated dorsally, while the posterior edge of the shield continues its growth backwards eventually to extend freely beyond the eighth or ninth trunk segment at TS12, the latest stage known with preservation of a complete shield. The shield is termed 'cephalic' throughout, regardless of its size and segment equipment, because it refers to the simple arthropod head shield as the product of fusion of the dorsal segmental sclerites of the anterior body region. The term 'duplicature' (Lauterbach 1973 and subsequent papers) should be avoided because it is pre-occupied for the ventrally flexed rims of trilobites or ostracode shells and because nothing is duplicated. There are no morphological changes in shape that necessitate the use of different terms for the different morphogenetic stages.

As is the case with the appendages, the discussion concerning the presence of a 'carapace' and/or/versus a 'cephalic shield' has a long history in crustacean literature, and the dispute seems endless (cf. Newman & Knight 1984 for further references). Various definitions are available which are not repeated here. In my view, the major problems arose from the early misinterpretation of the structure of growth of the various shields, the focussing on 'carapaces' of Eumalacostraca, in particular the hypothesis of a 'carapace fold', and the neglect of the criteria of homology. Herein, all dorsal shields of Crustacea, whether effaced, small, large or bivalved, are considered as representing merely modifications of the ancestral euarthropod head shield by allometric growth and different incorporation of subsequent body segments (see chapter on Cephalic shields and carapaces).

Caudal end. – Up to and including the latest instar known, the posterior end of the body of *Rehbachiella* is unsegmented but buds off segments continuously. In early larvae this part is named the 'larval trunk' or 'hind body'. From the delineation of the first trunk segment, considered as the 1st thoracomere, onwards, the caudal end is named 'abdomen', because this part obviously contains at least the budding zone, internal segment anlagen, and the telson with terminal anus and furcal rami.

Although the non-somitic nature of the telson is long known (cf. Calman 1909, p. 7; Kaestner 1967, p. 885), this term is still inconsistently used in the literature. With regard to the variable appearance of caudal ends in Eumalacostraca and some confusion in the descriptions of fossil Phyllocarida, Bowman (1971) discussed the (in his view) non-homology of caudal ends and their outgrowths among Crustacea. His arguments have been invalidated in detail by Schminke (1976) and Dahl (1984), and also other authors have remarked upon the difficulties of Bowman's scheme (e.g., McLaughlin 1980; Williamson 1982, p. 61–66). Nevertheless Bowman's terminology has never been abandoned entirely in the literature (see, e.g., Schram 1986 or Martin & Belk 1988). 'Telson' and 'furcal rami' (in preference of 'caudal rami' or 'caudal furca') have been demonstrated to belong to the set of constitutive characters of Crustacea s. str. (= crown-group Crustacea, sensu Walossek & Müller 1990), validating their use in accordance with their established definition (cf. Siewing 1985, pp. 839–840, Fig. 903).

Neck organ. – The earliest stages of *Rehbachiella* possess a watch-glass shaped smooth area on the apex of their arched shield, which is surrounded by a faint ring structure and with two pairs of pits, one pair inside the area and one pair on the posterior rim. It is termed 'neck organ' because of its structural identity and corresponding position to this organ occurring in all Recent branchiopods (see subchapter on this organ in the chapter 'Significance of morphological details').

Systematic palaeontology

Taxonomic status. – Crustacea Pennant, 1777; Branchiopoda Latreille, 1817; Anostraca Sars, 1867, inc. sedis; *Rehbachiella kinnekullensis* Müller, 1983

Rehbachiella kinnekullensis Müller, 1983
Fig. 4

Synonymy. – *Rehbachiella kinnekullensis* – Müller, 1983, pp. 102–105, Figs. 7, 8. *Rehbachiella kinnekullensis* Müller, 1983 – Müller & Walossek (1985a, Fig. 6c). *Rehbachiella kinnekullensis* Müller, 1983 – Lauterbach (1988, Fig. 2d [not c]).

Type locality and stratum. – Road cut between Haggården and Marieberg at NW slope of Kinnekulle, Västergötland, Sweden; Upper Cambrian, *Olenus* zone (2), subzone with *O. gibbosus* (sample 6404); co-ordinates N583355 E132601 (according to Müller & Hinz 1991).

Material examined. – Holotype UB 644, Paratype, UB 645, and 132 more specimens of different growth stages (see Tables 1, 2); the great bulk of the material is not from the type locality and zone 2a, but from zone 1 (*Agnostus pisiformis* zone; see Table 1).

Emended description. – The diagnosis and description by Müller (1983) were based on larval specimens up to a TS12 stage. Even the largest stage now known, with 13 trunk segments, was obviously still immature, and several features recognized for the largest instar, stage TS13, may not necessarily reflect the shape of adults, which remains unknown. Because much more material and evidence is now

Table 2. Reference list of examined specimens, illustrated ones with repository numbers (UB), others with internal specimen numbers (ST); large specimens marked by an asterix.

	N	Registration numbers of specimens
L1A	1	UB 3
L2A	7	UB 4–9, ST 4520
L3A	20	UB 10–18, ST 3265, 3463, 3580, 3584, 3590, 4014, 4284, 4289, 4566, 4635, 4637
L4A	10	UB 19–24, ST 3549, 3597, 4096, 4649
TS1iA	12	UB 25–33, ST 2693, 3029, 4325
TS1A	2	UB 34, 35
TS2iA	1	UB 36
TS2A	3	UB 37, ST 3573, 4043
TS3iA	2	UB 38, 39
TS3A	1	UB 40
TS4A	1	UB 41
TS5A	4	UB 42–45
TS6iA	2	UB 46, 47
TS6A	2	UB 48, 49
TS7iA	1	UB 50
TS7A	2	UB 51, 52
TS8iA	1	UB 53
TS8A	1	UB 54
TS10A	2	UB 55, 56
TS13iA	1	UB 57
TS13A	1	UB 58
L4B	2	UB 59, 60
TS2B	2	UB 61, 62
TS3B	3	UB 63, 3017, 4020
TS4iB	2	UB 64, 65
TS4B	10	UB 66–74, ST 4092
TS5iB	3	UB 75, ST 2857, 4579
TS5B	6	UB 76–79, ST 2045, 4536
TS7iB	1	UB 80
TS8iB, TS9iB	3	UB 645 – paratype, UB 81, 82
TS10B	1	UB 771
TS11B	3	UB 82–85
TS12B	1	UB 644 – holotype
TS13B	3	UB 86, 87, ST 4647

fragmentary, not definitely assignable: ST 2048(TS10), ST 2412(?), UB 92 (TS13*), ST 2710(large), UB 88–91 (3 specimens, ca. TS4), UB 93 (ca. TS1), ST 3098(TS13?*), ST 3466(TS5–7), ST 3554(TS13?*), ST 3992 (TS9–10?), UB 95 (TS5), UB 94 (TS3), ST 4644(TS13*), ST 4886(TS7)

destroyed: UB 80 (?TS8iB), UB 52 (?TS8A)

available, the description differs in some respects from the original one (more details in the next chapter):

Body of an instar with 13 thoracomeres about 1.7 mm long including furcal rami. Cephalic shield elongate and simple, covering 8–9 thoracomeres freely. Thoracomeres without clearly developed tergitic pleurae, 12 of them carrying phyllopodous appendages (last 3–4 showing progressively less differentiation than anterior limbs). Second podomere of antennal endopod subdividing during early ontogeny. Mandibular palp (basipod and rami) largely atrophied at TS13, the two pairs of antennae at least reduced in size. First maxilla shorter than 2nd, with four specialized endites on the corm, the proximal being the largest and serving as a brush, the next one elongated medially and serving as a pusher. Second maxilla basically similar to thoracopods, but with 6 rather than 8–9 lobate

endites on its corm; proximal endite large and similar to that of 1st maxilla, slightly angled against the posteriorly flexed more distal endites; enditic setae of mature limbs arranged in three sets: double row anteriorly (closure setae), set of spines or spine-like setae on enditic crest (brush and comb function), and row of pectinate setae posteriorly (filtration).

Thoracic sternites deeply invaginated to form a triangular food channel which becomes progressively shallower posteriorly; each sternite made of a pair of rounded plates. Caudal end cylindrical, including the telson at TS13; terminal anus covered by short, faintly pointed supra-anal flap; furcal rami leaf-shaped, margin with double row of spines (specific character?) affiliated by pores ventrally; pair of large ventrocaudal outgrowths with marginal spines similar to those of furca (specific character?).

Development: strictly anamorphic, comprising a true nauplius and 29 more instars to reach the TS13 stage; appearance of limbs succeeds delineation of segments; maturation of maxillae and thoracopods requires 6–8 stages (supposedly moults); 'neck organ' and 'dorsocaudal spine' at the hind body of nauplius are transient features that are lost after few stages; other naupliar structures on the way to reduction (antennae, mandibular palp), modification (labrum), or being lost eventually (gnathobasic seta on mandibular grinding plate).

Life habits: marine, benthic or epibenthic, presumably living on a flocculent bottom layer (fluff), at least up to the largest instars known. The absence of larger stages and adults may be explained by their size and a better swimming ability, which limited their preservability. Most probably, the latest stages were filter feeders eating suspended, particulate matter once the thoracopods had achieved their definitive shape.

Remarks. – *Rehbachiella* differs in design and occurrence from other *orsten* forms, in particular the coexisting Skaracarida (Müller & Walossek 1985b), representing the copepod lineage of Maxillopoda (Müller & Walossek 1988b), and *Martinssonia elongata* (Müller & Walossek 1986a), now recognized as a representative of the stem group or lineage of the Crustacea (Walossek & Müller 1990). *Bredocaris admirabilis* (Müller 1983) is from zone 5 (Müller 1983); moreover, its recognition as a representative of the thecostracan Maxillopoda, as documented in a different tagmosis, limb morphology and ontogeny pattern (Müller & Walossek 1988b, 1991; Walossek & Müller 1992), precludes closer alliance with *Rehbachiella*. Maxillopod affinities have also been suggested for *Dala peilertae* (Müller 1983; cf. Müller & Walossek 1988b, p. 30; see also Fig. 48L herein), another fossil from zone 5.

Several species of the Phosphatocopina coexist with *Rehbachiella*, but their bivalved shield, already present in the earliest larvae known, the appendage morphology (e.g., the minute 1st antennae, non-differentiated postmandib-

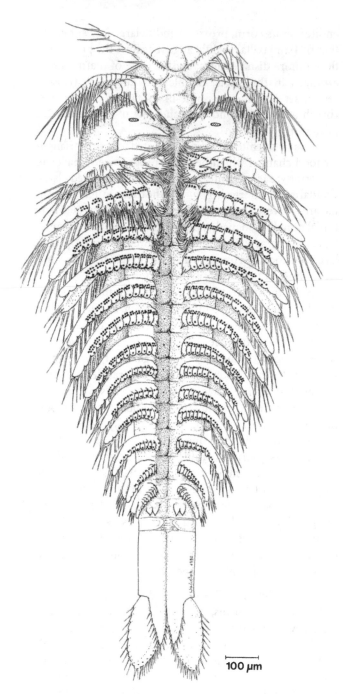

Fig. 4. Reconstructed ventral view of *Rehbachiella kinnekullensis* Müller, 1983, at largest stage known (TS13). Mandibular palp and most of setation and setules omitted for clearness; head details from earlier stages.

ular limbs), the capsule-like eye area and the feebly developed trunk region (Müller 1979, 1982, and unpublished observations) argue against any closer relationship with *Rehbachiella*.

Only *Walossekia quinquespinosa*, exclusively known from zone 1 (Müller 1983), shows more than a superficial similarity, for example in details of its appendage morphol-

ogy and the existence of a pair of ventrocaudal processes. Larger specimens, recently discovered, even indicate a similar kind of feeding ability. Again, the thoracopods have posteriorly curved endites and posteriorly curved lateral edges that suggest the presence of sucking chambers, as developed in *Rehbachiella*. The mandibular coxal body is also very large, and atrophy of its palp is apparent. In particular, *Walossekia* differs from *Rehbachiella* in the anterior head region, which comprises small egg-shaped eye lobes, and a posteriorly directed, pointed labrum. Again, the dorsoventrally slightly flattened caudal end has posterolateral spines and elongate furcal rami with setae only along the curved inner margin (in part Müller 1983, Figs. 5, 6;, and personal observations). Postnauplii of *Walossekia* can be readily distinguished from *Rehbachiella* and all other *orsten* forms by their characteristic spine-bearing shield (Müller 1983, Fig. 6). On the other hand, the existence of ventrocaudal processes is a weak taxonomic character, since they occur also in various other crustaceans. Hence, further assumptions on affinities of *Walossekia* with *Rehbachiella* and Branchiopoda must await the restudy of the entire material of this fossil.

Comparisons of the much larger Burgess Shale-type arthropods with crustacean-like appearance remain problematical. Relationships with phyllocarid Malacostraca have been proposed for a number of forms, such as *Plenocaris* (Whittington 1974), *Perspicaris* (Briggs 1977), or *Canadaspis* (Briggs 1978; also Briggs 1983), but such assignment has been questioned in particular by Dahl (1984). Again, the proposed placement of, for example, *Branchiocaris* within the Branchiopoda has been convincingly rejected by Fryer (1985). Only *Waptia*, with its broad shield, covering an unknown number of limbs, an apodous trunk and a paddle-shaped furca, appears branchiopod-like in its gross design. While a detailed description of *Waptia* is still lacking (cf. Whittington 1979; Conway Morris *et al.* 1982, p. 18), its lamellar exopodal spines of the trunk limbs, a character linking all 'trilobitomorphs' (in the sense of Bergström 1980, and personal communication, 1990), would rule out even stem-group crustacean affinities for *Waptia*.

Hence, there is virtually no Burgess Shale form with more than a superficial resemblance to *Rehbachiella*. As a whole, the differences between these two important sources of Cambrian arthropods with preserved cuticular details are substantial. This may at least be in accordance with very different environmental conditions, different ecological demands and also different preservation potential (cf., e.g., Conway Morris 1979 and Briggs & Whittington 1985 for the Burgess Shale fauna; Butterfield 1990 for preservation and taphonomy; and Conway Morris 1989a, b, for a summary of the variety of Burgess Shale type faunas now known).

Postembryonic development

General remarks

Rehbachiella shows a long series of growth stages, with very gradual increase in size and differentiation. The existence of successive sets, the anamorphic development as a whole, and the morphometric data, suggest that missing stages result from lack of preservation rather than from developmental 'jumps'. In animals in general, the number of individuals declines from young to later instars. In *Rehbachiella*, the abundance of stages is rather uneven; the majority of specimens are from stages L2 to TS4, while the nauplius and advanced stages are known from single or only a few specimens (Tables 1, 2). An explanation for the low occurrence of later stages may be that they were active swimmers well-above the bottom and, hence, were not likely to be preserved. On the contrary, from the possibly infaunal *Bredocaris*, all stages save for the third are equally represented (10–20 specimens each; Müller & Walossek 1988b, their Table 1).

While there are many details available of the early phase, not all could be monitored continuously throughout ontogeny. Since the recognized appearance of a particular structure does not always imply that it was first introduced at that stage – this may well have occurred earlier – the descriptions may not exactly follow the precise time scale of all morphogenetic changes.

The first stage is a nauplius with three pairs of functional appendages ('orthonauplius'). The largest stage known with certainty comprises 13 postmaxillary segments, regarded as thoracic. Twelve of these carry limbs, of which only 8–9 are well-developed. The last 3–4 pairs remain at a less-developed to rudimentary state. By this largest instar, the caudal end is still unsegmented but carries hinged, paddle-shaped furcal rami.

Characteristic of trunk development is the formation of its segments in two steps. In crustaceans various changes may also occur in the interphase between two successive moults, but for *Rehbachiella* the steps more likely represent moults rather than early and final stages of an intermoult stage. At first a new segment is partly delineated from the caudal end by a fissure on the dorsal surface which becomes blurred laterally ('incipient segment'). By the second step the segment is fully separated. Stages with an incipient trunk segment are intermediate in development to those with completed segments and are marked by an 'i'. With a delay of between one and several stages the limb buds appear and develop in regular anterior–posterior order.

Claus (1873), Hentschel (1967) and in particular Weisz (1947) found no sharp demarcation between instars in Recent Euanostraca, just as in the development of *Rehbachiella*. Nevertheless, the former two authors distinguished two major phases in the ontogeny: a larval phase with addition and development of segments and appendages, and a postlarval differentiation phase when besides sexual maturation other changes also occur, such as the reorganization of the naupliar head features, completion of the development of eye stalks, segmentation of the abdomen, and development of the furcal rami.

With regard to this mode, the reconstructed sequence of *Rehbachiella* is regarded to represent the complete larval phase prior to the segmentation of the limbless abdomen and further differentiation. A distinction is made here between a 'naupliar phase', including the instars L1–L4, and a 'postnaupliar phase' between stage TS1i, with incipient 1st trunk segment, and TS13. This permits the phase of delineation of postmaxillary segments to be enhanced (Fig. 5), which equals the 'thoracic phase' of Weisz (1946, 1947) for *Artemia salina*. From the 2nd instar, development of the maxillae begins on the larval trunk (= metanauplii). With the 5th instar (i.e. presumably after four moults) and appearance of the incipient 1st thoracomere, the hind body, or postthorax, is named the 'abdomen'. The abdomen includes the telson, which is not delineated externally within the whole larval phase, as in euanostracan Branchiopoda.

The two larval series (A, B) differ in length between about 10 and 25%, depending on the stages. Some individual variability seems to occur, but intermediate specimens have not been observed. The main differences are in the head development, which is also apparent in its length increase, measured as distance between 1st antenna and 2nd maxilla (hl = 'head' length). This parameter could be obtained even from fragmentary specimens (Fig. 2B).

Development of series A

Descriptions of ontogeny from the second stage onwards include major details and changes from the preceding stages; structures will be described at greater length only when especially well-preserved (principal changes are also marked by arrows in the figures of this chapter).

Naupliar phase

L1A (Pl. 1:1–4; reconstruction in Fig. 6A). – Material: One specimen, fairly complete but slightly shrunken (Table 2). Major measurements: total length (tl) 160 μm, length of shield (csl) 100 μm (further data in Table 3). Body pear-shaped. Head shield (cs) cap-like but with weakly developed margins; outline almost circular in dorsal view, reaching back to rear of mandibular segment (Pl. 1:1–3). Central area of shield (apex) smoother, gently vaulted, and bordered by ring wall, identified as 'neck organ' (no; see subchapter on this organ in the discussion, chapter 'Significance of morphological details'). Organ covering 50% of the shield (50 μm), with two pairs of pores (po): one set

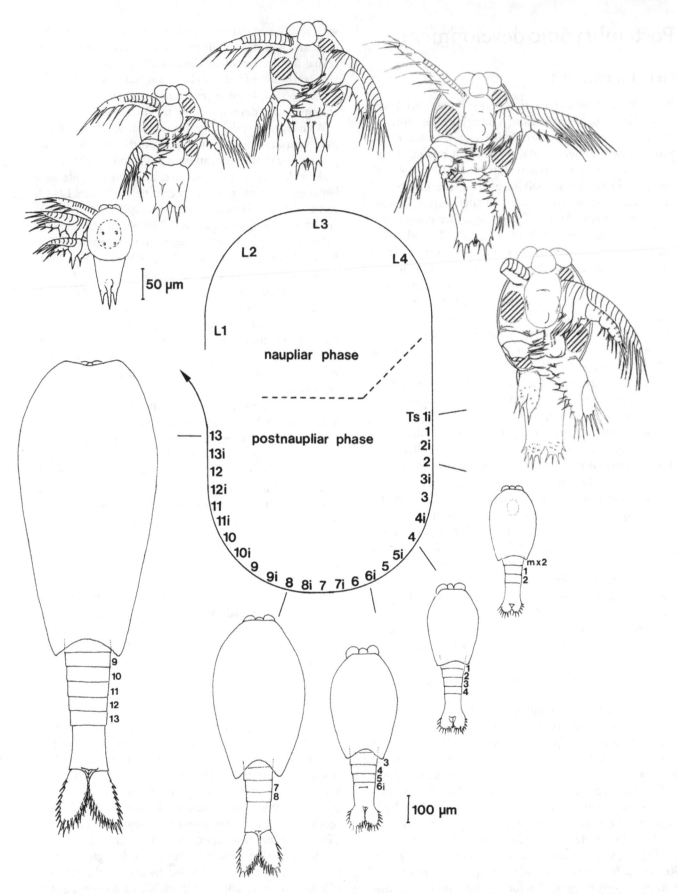

Fig. 5. Postembryonic development of *Rehbachiella kinnekullensis* up to stage TS13. Nauplius (L1) and stages of 'postnaupliar' or 'thoracic' phase from dorsal, larvae L2–TS1i from ventral; appendages omitted in part (early stages) or completely (thoracic stages). Scheme adopted from Dahms (1987a).

Table 3. Measurements of body parts and details (A and B); values approximated when necessary (data from actual specimens measured to the nearest 5 μm); data given as span, or, when obtained from single specimen, as a single value; Ø = diameter; ? = no data; – = not developed; void = stage not known; uncertain values in brackets (abbreviations see list on the last page).

Series A

	tl	csl	hl	lal/law	abd/Ø	trl	frl/w	p/sfsp
L1	160	100	–	100/40–45	50–45	?	rud	1/–
L2	190–200	110–120	–	100/40–45	45–50/55–60	?	10	2/–
L3	250–260	145–165	–	110–115/55–60	45–50/70	?	20–25/30–35	3/–
L4	310–330	195–220	180–200	130–135/60	60–80/?	?	25–30/40–50	4–5/–
TS1i	370	205–230	195–205	140/55–65	65–85/75	(30)	30–35/45–55	5/–
TS1	420	240	190–250	?/65	85–90/?	40	?/?	?/–
TS2i	440	260	230	?/65	65–70/80	70	35/60	5–7/–
TS2	440–450	260–270	220–240	130–150/60–65	90/80	75–85	40–45/60–70	7/–
TS3i	450–460	270	220–240	130–150/60–65	50–60/?	115	0/70	7/–
TS3	480	280	220–240	130–150/60–65	90–95/75	125	40–45/70	7/–
TS4i								
TS4	?	>300	230–240	?/65	?/?	>160	?/?	?/?
TS5i								
TS5	>600	330–340	220–240	140/65–75	90/?	220–240	70/60–70	12/3
TS6i	600–650	?	230–240	?	>80/?	260	?/?	?/?
TS6	650–680	375–400	220–240	140–045/80–85	90–100/70–85	250–290	75–80/65	9–11/4–5
TS7i	?	?	?	?	90/80–90	320–330	100–70	13/6
TS7	750–800	485–500	240–250	(140)/90	100/80	(350)	?/?	?/?
TS8i	>750	?	?	?	90–95/95	(400)	>100/?	?/?
TS8	900	590–600	250–300	(150)/?	105/100	(460)	?/?	?/?
TS9i								
TS9								
TS10i								
TS10	1200–1300	?	300–330	170/?	145/120	625–675	170/80	16/>6
TS11i								
TS11								
TS12i								
TS12								
TS13i	(>1600)	?	?	?	145/135	(1000)	?/?	?/?
TS13	?	?	350–380	?/(110)	?/?	?	?/?	?/?

Series B

	tl	csl	hl	lal/law	abd/Ø	trl	frl/w	p/sfsp
L1								
L2								
L3								
L4	260–270	160–165	120–130	110/50–55	60–65/60	?	30/35	7/–
TS1i								
TS1								
TS2i								
TS2	330–340	190–200	150	110–120/50	70–75/75	60–65	30–50	9/–
TS3i								
TS3	375–400	200–225	160–170	120–130/60	85–65	100	30–35/50	8–9/–
TS4i	420	230–240	180–190	110–130/50–60	50–60/75	120	35–50	8–9/1
TS4	440–450	240–250	180–200	135–140/60	80–90/70	140–150	40–45/55–60	9–11/1–2
TS5i	460–480	240–270	?	140/60	65–70/?	160	60/60	11/1
TS5	500–530	270–280	190–220	140/60–70	80–90/75–80	180–200	50–60/40–60	9–11/2
TS6i								
TS6								
TS7i	?	>300	?	?/70	80–85/?	250–260	?/?	?/?
TS7								
TS8i	?	(400)	?	?/70	95/?	?	120/75	12/5
TS8								
TS9i	?	?	250	?/70	100–105/80	?	?/?	?/?
TS9								
TS10i								
TS10	?	?	?	?	>110/?	560	?/?	?/?
TS11i								
TS11	(1300–1400)	(800)	?	?	135–150/90–100	640–650	?/?	?/?
TS12i								
TS12	(1450)	980	300–330	?	>100/95–100	750–800	?/?	?/?
TS13i								
TS13	(>1450)	>800	?	?	>130/100–115	860–880	?/?	?/?

on the surface and another at the posterior margin (Pl. 1:2, 3).

Region in front of labrum not known. Labrum (la) cylindrical and oval in cross-section, about 100 µm long (about $\frac{2}{3}$ of body length) and with rounded tip; sides with row of setules along long axis of organ. Posterior projection of labrum in UB W3 may be preservational but oriented more ventrally during life. Sternum swollen (stn), made of the sternal bars (st) of antennary and mandibular segments. Antennary bar broader than that of mandible; sternum somewhat constricted between the portions, furnished with tiny setules (Pl. 1:4).

Details of 1st antenna (a1) unknown save for its insertion at about the anterior edge of the labrum (Pl. 1:1); possibly about as long as the subsequent limbs. Second antenna only slightly longer than the mandible (100/85 µm). Insertion area of 2nd antenna extending from 1st antenna post the labrum (Pl. 1:1).

Corm (co) of 2nd antenna with distinctive coxa (cox) and basipod (bas). Both portions and 1st endopodal podomere (en1) with elongate processes (end) which terminate in a rigid spine or spine-like seta (esp; 'gnathobasic seta' in euanostracan nauplii). Spines of coxa and basipod (esp) accompanied by a more anteriorly and distally inserting seta (s) (Pl. 1:1, 4). Tip of endopod (en) not known. Exopod (ex) arising from narrow, sloping outer edge of basipod, about as long or slightly shorter than that of mandible (50 µm), made of 7–8 ring-shaped podomeres and with five rigid setae medially. Numerical difference results from the missing seta on the proximal annuli and the fact that the setal sockets are thicker than these (characteristic feature of series A larvae). Terminal segment almost spine-like.

Mandibular corm also bipartite, but coxa markedly smaller than the basipod and terminating in two short spinules (Pl. 1:1, 4; Fig. 9A). Basipod with elongate, blunt enditic process which is drawn out into a long masticatory spine accompanied by a single seta anteriorly. Proximal endopodal podomere similar to that of 2nd antenna, being slightly drawn out medially, with a stout spine and a thinner one behind; 2nd podomere as long as wide, with two setae mediodistally; 3rd rounded apically, and tip distorted except for the mediodistal seta. Exopod as in 2nd antenna, with seven annuli but only four rigid setae (Pl. 1:2). The enditic spines at least are setulate distally, indicative of the feeding state of the nauplius.

Head and trunk separated by a transverse trench behind the sternum. Hind body half as long as the shield, cylindrical to slightly conical and truncate posteriorly. Incipient furcal rami visible as pair of short ventrocaudal humps (i fr), forming the bases of a short stout spine (fsp; Pl. 1:1, 4). A single long and robust 'dorsocaudal spine' (dcsp) projects posterodistally from the dorsal end of the hind-body (Pl. 1:1, 2, 4) above the T-shaped anal slit (an). Its membranous 'anal field' (anf) is puffed up artificially in UB

W3 (Pl. 1:1, 4). Instar apparently capable of swimming and feeding, most likely using all three appendages; second antenna probably slightly dominating the mandible.

L2A (Pls. 1:5–7; 2; 30:1; Fig. 6B). – Material: Seven specimens, some fairly complete (Table 2). Major measurements: tl 190–200 µm, csl 110–120 µm (Table 3). Instar about 20% longer than 1st instar, characterized mainly by appearance of a pair of spine-like setae which arise from short protuberances on the ventral side of the hind-body and represent the buds of the 1st maxillae (mx1; developmental stage = ds1; Pls. 1:5, 7; 2:1–4; Fig. 10A). Distance between the setae 35 µm in all specimens examined. Further innovations: enlargement of appendages, better development of their armature with setae and spines in particular on mandibular coxa (Figs. 8A, 9B), a 2nd furcal spinule laterally to the 1st (Pls. 1:5; 2:1–4). Shield more oval; margins differently produced, probably due to varying preservation: margins almost absent in UB W8 and ST 4520, prominent in UB W7, shield deformed and almost circular in UB W5 (compare Pls. 1:6; 2:1 and 3); neck organ slightly anterior to the centre of the shield, size as in the nauplius (Pl. 2:7, 8).

Forehead with two large, ovate blisters in front of labrum, separated by a 3rd, axially oriented lobe (Pl. 2:3). Lateral blisters are interpreted as incipient lobes of the compound eyes (width 35–40 µm). It is unclear whether this structure is already present in the 1st instar, since this region is not preserved there. 'Midventral lobe' (mvl) possibly housing the internal naupliar eye; lobe extends from the basis of the labrum toward the anterior margin of the shield. It becomes narrower between the lateral lobes and at that point carries a small node, probably with a pit (Pl. 2:9, 10). Whole structure known from five specimens but always distorted. Comparison with later stages suggests that also at this stage the lobe protruded from the forehead originally and extended well beyond the shield. Sternum slightly longer than wide, sloping orally. Portion of 2nd antenna not positively identified. Sternal surface ornamented with setules (Pl. 2:6), posterior margin somewhat swollen.

First antenna slightly longer than labrum (115–120 µm), circular in cross-section (about 30 µm at basis), slowly tapered towards the tip and bi-composite: thicker proximal part subdivided into about 12 incomplete annuli anteriorly but pliable posteriorly, distal portion made of three cylindrical podomeres and a small distal hump which forms the socket of a thick apical seta of unknown length (Pl. 2:2).

Second antenna somewhat compressed anteroposteriorly, 10–15% longer than that of L1A and than the mandible (115 µm; Fig. 8A). Endopod four-segmented and as in L1A (50–60 µm; tiny 4th podomere in Pl. 2:5). Exopod (65–70 µm) composed of 10–11 annuli carrying 8–9 setae medially (Pl. 2:2, 5, 6). Outer distal margins of annuli

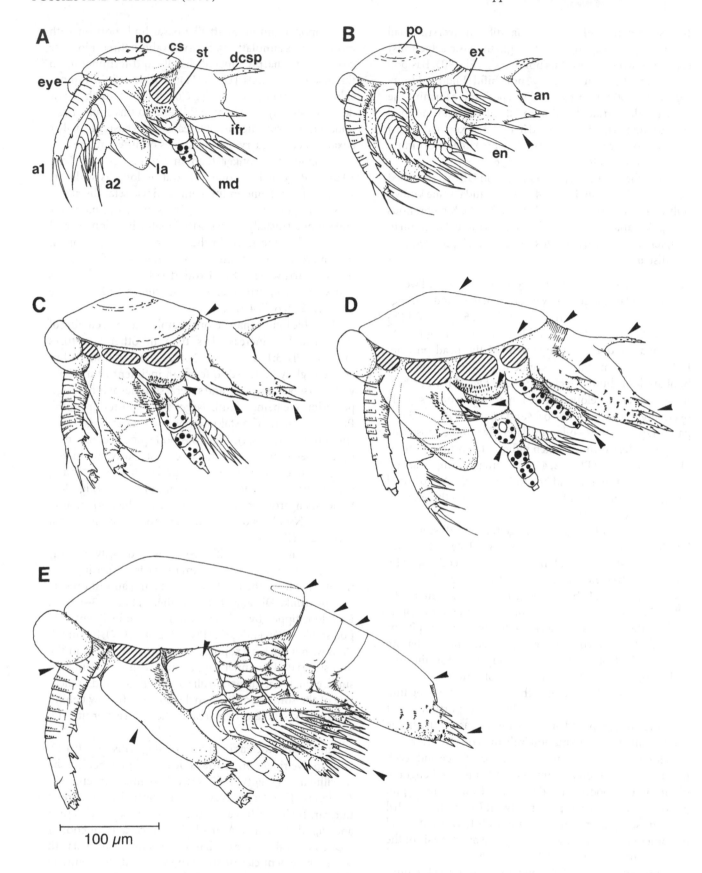

Fig. 6. Selected larvae of series A, reconstructed largely from actual specimens; lateral views; appendages partly omitted, setation in some cases drawn cut short or omitted for clarity. Short arrows in this and the following figures point to major morphological changes between the instars. □A. Stage L1A (nauplius). □B. L2A. □C. L3A. □D. L4A. □E. TS2A.

furnished with small denticles. Mandibular coxa still small compared to basipod, with two gnathobasic setae (gns) setae and two spinules at its pointed tip (Fig. 9B). Basipod prominent, tip of masticatory spine bifid. Endopod four-segmented (50 µm), its small distal podomere forming the socket of the terminal seta, as in 1st antenna and endopod of 2nd antenna (well-preserved in Pl. 2:5). Exopod (50 µm) with six to seven annuli and six setae (proximal one thinner than the others; Pls. 1:7; 2:2, 3, 4, 5, 6).

Trunk slightly thicker than in L1A (in UB W8 stretched possibly by inflation; Pl. 2:3, 4). Dorsocaudal spine known only from its wide insertion (Pls. 1:6; 2:1, 4). Membranous triangular 'anal field' sloping from dorsocaudal spine on to incipient furcal rami (not illustrated); furcal spine denticulate distally.

L3A (Pls. 3, 4; reconstruction in Fig. 6C). – Material: Twenty specimens; largest group, well-documented and homogenous (Table 2). Measurements: tl 250–260 µm, csl 145–165 µm (Table 3). Considerable increase in length (30–35%) mainly due to enlargement of the head portion. Delineation of maxillulary segment completed, now with bilobate limb buds (mx1 rud; ds3a; Fig. 10B). Furthermore: better developed shield margins, which slightly overhang the body posteriorly and laterally (Pl. 3:1, 3–5), pairs of striae at the swollen and laterally rounded posterior margin of the sternum, which indicate the future position of the paragnaths (Pls. 3:2; 4:2, 7), better developed armature of 2nd antenna and mandible (Figs. 8A, 9B), and elongation of incipient furcal rami, now with three spines (median one thickest).

Neck organ only slightly enlarged, ring wall seemingly less distinct than in L1A (Pl. 3:3, 5). Eye blisters 40–50 µm wide (triangular shape in UB W13 caused by collapsing; Pl. 3:1, 3–5; midventral lobe mostly distorted, pore identified in one specimen). Labrum about 13% longer than in L2A (Pls. 3:1, 2, 4, 6; 4:1, 2), with numerous fine setules at its posterolateral edges; posterior edge with a few pits or tubercles (illustrated in later stages). Antennary sternite not identified on sternum, probably now forming the orally sloping anterior surface of the sternum.

First antenna 20% longer than in L2A (150–160 µm; complete in three specimens). Proximal portion about 100 µm long, composed of 12–14 annuli, each with a fringe of tiny denticles; segmentation indistinct on posterior side. Two long setae arise from the posterior surface and reach into the gap between labrum and 2nd antenna. Length of distal, tubular podomeres decreasing from 20 to 15 µm, proximal with two setae posteriorly, 2nd without setae, 3rd with one seta postero- and one anterodistally. Small distal podomere (6 µm) with a small spinule at the basis of the robust terminal seta (Pls. 3:1, 5, 7; 4:1, 5).

Second antenna 20–25% longer than in L2A (150 µm). Corm with slightly more firmly sclerotized bands ('annulations') on outer surface, which are of the same length as the exopodal annuli. Shaft-like basal limb portion with more than six annulations, coxa with about three. Elongate coxal endite sharply tapered distally and terminating in three setulate spines (UB W11); additionally a thinner spine arises more anteriorly. Coxal surface with a few denticles slightly proximal and lateral to this spine. Basipodal endite truncate distally, reaching to basis of smaller coxal spine, and carrying four spines of different sizes.

Endopod 40% longer than in preceding stage (75–80 µm). Proximal podomere about as long as wide, its short endite with one enditic spine and two setae distally to it. Second podomere elongate; 3rd similar but narrower and rounded distally, carrying the small 4th podomere and a seta medially to it. As in the 1st antenna, the terminal podomere has a tiny spinule close to the basis of the apical seta now (arrow in Fig. 8B). Exopod with 13–14 podomeres and 9–10 setae; proximal 1–2 podomeres without setae (Pls. 3:1, 2, 4, 6, 7; 4:1, 2, 5).

Mandible longer than before, mainly due to stretching of the corm, but increase less than in other structures (115 µm; Pls. 3:1, 2, 6, 7; 4:1, 2–4, 7). Coxa still small, but endite slightly flattened to form an incipient gnathobase with two setae, as in L2A, and several spinules at inner edge, posterior one being slightly larger (pt = 'posterior tooth'; Pl. 3:8). Setulae of distal gnathobasic seta arranged in a spiral row (same Fig.). Rigid basipod spine accompanied by two setae anteriorly and one posteriorly. Rami not longer than in nauplius, but endopod with more setae and exopod with two additional annuli and setae (Fig. 9C). Setae arising from broad shafts predicting their orientation (Pl. 4:3). Numbers of setae and exopodal annuli are not congruent (Pl. 4:3, 4).

Maxillulary segment still present on hind-body (Pls. 3:1, 2,; 4:2, 7). Buds about 40–50 µm long (ds3a), arising ventrolaterally from the hind-body and lying almost on surface of trunk. Distance between them about 25–30 µm. Each lobe tipped by a short spine-like seta; incipient exopod smaller than endopod (Pls. 3:1, 2, 4; 4:1, 7), but in UB W16 with a thicker exopod carrying two spines as in the next stage (Pl. 4:2). Inner edge of buds with a few denticles, surface finely corrugated in all specimens at hand, indicative of the softness of the cuticle. Shallow furrow behind maxillulary segment demarcates the future sternite (Pl. 4:7).

Hind-body much as in preceding stage (Pls. 3:1–4; 4:1). Four to five ridges on ventral side of it closely behind the maxillulary segment (Pl. 4:7) resemble similar structures in *Bredocaris* (Müller & Walossek 1988b); their nature is unclear. Incipient furcal rami enlarged (l/w = 20/30 µm) and slightly flattened. A small pit of uncertain function is located ventrally to the median spine. Rami furnished with some single denticles or short rows of denticles ventrally (Pl. 4:6). Dorsocaudal spine slightly more anterior to anus (anal field in Pls. 3:1, 2, 3, 4; 4:1).

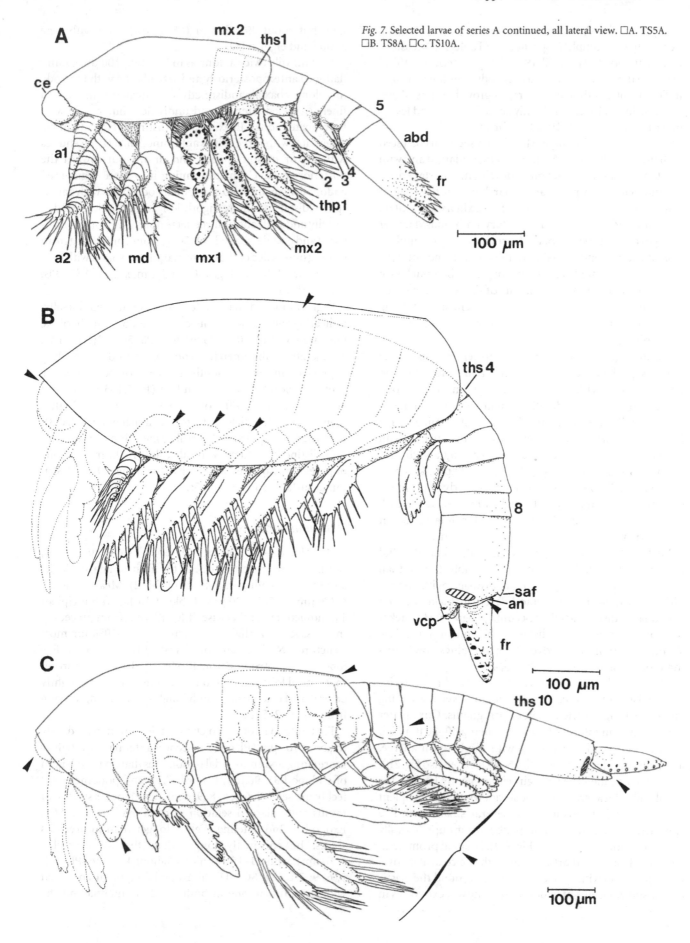

Fig. 7. Selected larvae of series A continued, all lateral view. □A. TS5A. □B. TS8A. □C. TS10A.

L4A (Pls. 5; 6:1–3; reconstruction in Fig. 6D). – Material: Ten rather incomplete specimens (Table 2). Measurements: tl 330–340 µm, csl 195–220 µm, 'head length' hl 180–200 µm (Table 3). Increase in body length about 30%, indicative of continuation of rapid growth at this phase. Increase in shield length slightly smaller (25%) and less in other features (only 5–10% for labrum).

Innovations of this stage: Maxillulary segment coalesced with the larval head (Pl. 5:1), 1st maxilla enlarged and with lobate endites and distinct rami (developmental stage ds4a; see subchapter on postmandibular limbs in the chapter 'Morphogenesis'), delineation of the maxillary segment on the hind-body carrying rudimentary limb buds. Further changes are in the shape of the labrum, beginning subdivision of the 2nd endopodal podomere of the 2nd antenna, further development of the incipient gnathobase but loss of the anterior gnathobasic seta, start of the reduction of the dorsocaudal spine, and further enlargement and flattening of the furcal rami (l/w = 25/50 µm), now with five marginal spines.

Shield slightly more elongated posteriorly, its lateral margins probably overhanging the body more than in preceding stage (Pl. 5:1, 2). Neck organ more anteriorly shifted, since correlated with the naupliar limb segments (Pl. 5:1, 6). Eye area not known in detail but seemingly large (Pl. 5:3, 5). Labrum more tapered distally than in L3A, with faint depression on anterior surface causing a slight upward orientation of the tip (Pl. 5:2, 3, 5, 8; pit-like structures in Pl. 5:7). Sternum broader posteriorly than anteriorly, and incipient paragnaths slightly more elevated, each enclosed by striations (Pl. 5:3, 4, 7, 8). Maxillulary segment with narrow sternite, that of 2nd maxilla is incipient.

First antenna inserting more medially than the 2nd antenna (Pl. 5:2). Latter now presumably longer than 150 µm. Exopod exceeding 90 µm, probably with one additional annulus and seta. Setae different in thickness: from proximal to distal ends they become progressively thicker first but decrease again distally. Position of appendage slightly more anterior, and its elongate endites now pointing more posteromedially (Pl. 5:3, 4).

Mandible (length unknown) now located at posterior edge of labrum, its coxal endite moving over the sloping anterior surface of the incipient paragnaths. Coxa better developed than in preceding instar, being 40–50 µm high at outer edge and articulating with the body in a large, abaxially oriented joint (length 75–95 µm; see also Fig. 2E). From the tip, the inner margin curves gently back to the coxal body. Between tip and 'posterior tooth', the incipient cutting edge is furnished with several small, teeth-like spinules. Surface of grinding plate with groups of orally directed setules (Pl. 5:2–4, 8; Fig. 9D). Basipod prominent, 45 µm in height, and articulating with the coxa at an oval joint; its endite with 7–8 setae arranged around the masticatory spine. Details of endopod not known; exopod as in

L3A, but about 25% longer (65–75 µm), and with nine annuli and eight setae.

First maxilla twice as long as in L2A (80–100 µm). Limb flattened anteroposteriorly and articulating with the body in a long abaxial, indistinctly demarcated joint. Corm finely wrinkled laterally and undivided, inner edge with four lobate endites. Proximal (pe) distinctly larger than the others and with three spines anteromedially. Distal endites with a pair of spines each and drawn out toward the posterior spine. Corm continuing into three-segmented endopod. Proximal two endopodal podomeres with paired spines, distal podomere slightly longer than wide, rounded distally and with about three setae apically. Exopod undivided and paddle-shaped, 40–50 µm long, projecting from sloping outer edge of the corm; margin of exopod with four rigid setae (ds4a; see Fig. 10C for 1st maxilla of TS1i; Pls. 5:1, 3; 6:1, 2).

Segment of 2nd maxilla delicately corrugated dorsally, indicating the softness of its cuticle (Pl. 5:1); its bilobate buds are about 40–50 µm long (ds3a; Pls. 5:1; 6:2; Fig. 11A). Trunk (abd = abdomen from now on) cylindrical, slightly depressed, and with smooth surface. Dorsocaudal spine probably more slender than in L3A (Pl. 5:1–3); spine and anus somewhat set off from one another (about 25 µm; membrane of anal field distorted in all specimens; Pl. 6:3). Denticles arising around furcal spines (inner and outer ones distinctly smaller). Ventrally, close to the margin, pits or pores are associated with each of the spines. Denticles on ramal surface as in L3 (Pls. 5:1–3; 6:3).

Post-naupliar phase

TS1iA (Pls. 6:4–9; 7; 30:2). – Material: Twelve specimens, status of some uncertain due to the limited data available, also some individual variability (Table 2). Measurements: tl 370 µm, csl 205–230 µm (Table 3). Instar with incipient 1st thoracomere, otherwise little different from preceding instar save for a slight size increase of 5–10% for most structures. Neck organ not identified (Pl. 6:4, 6, 7). Eye lobes large, protruding from forehead (Pl. 6:4, 5, 6; pore on midventral lobe found in one specimen). Labrum slightly enlarged. 'Head' length only slightly enlarged, varying from 195 to 215 µm.

First antenna not much changed, length increased only to about 160 µm (Pl. 6:4, 5). Second antenna with slightly more elongated endopodal podomeres than in L4A (length of endopod 85–90 µm; Pls. 6:4; 7:1); 2nd podomere still feebly incised (Pl. 6:5, 6, 8, 9; denticles on outer surface of ramus and enlarged setation of distal end see Fig. 8C). Exopod of 2nd antenna slightly longer than in previous stages, but distinctly longer than that of mandible (110 µm), with 14–15 annuli and about 12 setae (Pls. 6:4, 5, 6, 9; 7:1; Fig. 8C). Distal exopodal segment coalesced with penultimate one in both the 2nd antenna and the mandible (Pl. 7:1); from this stage the distal hump carries two setae.

Mandibular coxa enlarged (l 100 μm); gnathobase angled against the coxal body and tilted toward the labrum. Basipod and rami (forming the 'palp') also seemingly enlarged: endopod 60 μm long (details not entirely known), exopod 75–90 μm long, composed of 9–10 annuli and 9 setae (Pls. 6:4–9; 7:1, 3, 7, 8; Fig. 9E). First maxilla about 10% longer than in L4A, similar in shape (105–115 μm). Endopod about 45 μm, exopod 40–50 μm and with 5–6 marginal setae (55–54 μm; ds4a–b; Pls. 6:4, 5–8; 7:2, 3, 7; Fig. 10C). Second maxilla 20–25% longer and in preceding stage but still bilobate; incipient exopod with two spinules (ds3b; Pls. 6:6, 8; 7:2, 3, 5, 7; Fig. 11B).

Furca slightly broader but number of spines unchanged. From this stage on the rami become progressively wider attaining a more paddle-shaped outline (Pls. 6:5–7; 7:3, 4). Angle between rami about 105–120°. Anal field extending on to dorsal surface of rami. Dorsocaudal spine apparently more slender than in preceding instar (Pl. 7:7, 8; measurement of 'thorax' starts now with incipient delineation of its 1st segment). Individual variation apparent in the degree of fusion of the maxillulary segment and in the setation of the furcal rami: in UB W20 the right furcal ramus has four (2nd largest) spines, while the left has five (3rd one being the largest). Commonly the segment of the 1st maxilla is coalesced with the larval head, but in UB W32 (Pl. 7:7) this segment is free on the left side; specimen is intermediate also in other features:

- proximal endite of 1st maxilla with two frontal setae rather than three (as in stage L4A),

- exopod of 1st maxilla with five rather than six setae,

- one terminal spine on maxillary exopod rather than two (in L4A).

TS1A (Pl. 8:1–4). – Material: Two specimens, details not well-known due to distortion; UB W35 ascribed to this stage on the basis of gross size and length of shield (Table 2). Measurements: tl 420 μm; csl 240 μm (Table 3). Length increases of body and shield slowed down and about 10% only. First thoracomere delineated and with limb buds (thp1i). No significant changes in the head structures. Neck organ recognized only in one specimen, size not measurable. Eye blisters about 50 μm in width, protruding from the anterior shield margin as in preceding larvae; total width of compound eye 100 μm (Pl. 8:1–3). Margins of shield probably slightly raised in front of the mandibles. Head length between 190 and 240 μm, probably due to either wrong assignment of specimens to this stage or individual delay in development.

Anterior appendages poorly known (Pl. 8:1, 2, 4), probably similar as in preceding stage. First maxilla similar to TS1iA, but with a few more setae on the proximal endite (width of pe 25 μm; Pl. 8:2). Second maxilla not fully known, with short paddle-shaped exopod carrying four setae similar to the 1st maxilla at stage L4A (ds4a; Pl. 8:1, 2). Proximal endite with two setae anteriorly and one spine at posterior edge. Maxillary segment not coalesced with head, partly overhung by posterior shield margin (Pl. 8:1). Both maxillae with bipectinate setae on their proximal endites.

First thoracomere almost ring-shaped, except the membranous ventral side, where the limb buds insert (ds3a). Abdomen almost twice as long as in preceding stage (Pl. 8:1, 2). Furcal rami not known in detail; dorsocaudal spine not identified (Pl. 8:2). Second maxilla and 1st thoracic limb seem to arise between the segments.

TS2iA (Pl. 8:5, 6). – Material: One fragmentary specimen (Table 2). Measurements: tl 440 μm, csl 260 μm (Table 3). Second thoracomere incipient. Size increase generally less than before. Neck organ not identified. Width of eye blisters larger than 50 μm. Sternites of mandible and 1st maxilla almost fused with one another; pliable sternites present in the subsequent two segments (Pl. 8:6). 'Head' length unchanged. Appendages poorly known. First maxilla 125–130 μm long, exopod with more than seven marginal setae (at least ds4b). Most of maxillary segment appears to be covered freely by the shield; 2nd maxilla not known. Exopod of rudimentary 1st thoracopod (about 55 μm) with two terminal spines (ds3b); bud seems to stem partly from subsequent segment, as there is no clear boundary between the segments laterally (Pl. 8:5).

Furcal rami larger and further advanced, now being rounded paddles (l/w = 35/60 μm). Marginal spines showing individual variability: 7(4) on left ramus, 5(3) on right (Pl. 8:6). Dorsocaudal spine not positively identified; if still present in UB W36, the length of abdomen in Table 3 would include the distance from maxillary segment to spine with 50–60 μm and from spine to anus with another 30–35 μm (Pl. 8:5).

TS2A (Pls. 8:7, 8; 9:1–3; 30:3; Fig. 6E). – Material: Three specimens, in part exceptionally preserved (Table 3). Measurements: tl 440–450 μm, csl 260–270 μm (Table 3). Second thoracomere fully delineated. Size increase generally low, but several morphological changes are apparent, in particular in the appendages. Neck organ very feebly developed, probably 65 μm long, shifted further anteriorly due to elongation of the shield. Width of each eye lobe increased to more than 70 μm, total width more than 150 μm. Eye region protruding much from forehead and shield (almost 50% larger than in TS1; Pl. 8:7, 8). Mandibular and maxillulary sternites still not entirely fused. Paragnaths forming distinct humps and sternum depressed medially behind them (Pl. 9:1). Maxillary segment not coalesced with the head (Pl. 9:3). 'Head' length unchanged, though appendages larger (probably introduced earlier but not known due to preservation).

Appendages known in part (Pl. 8:7, 8; 9:1–3). Setation of 2nd antenna and mandible without significant progress,

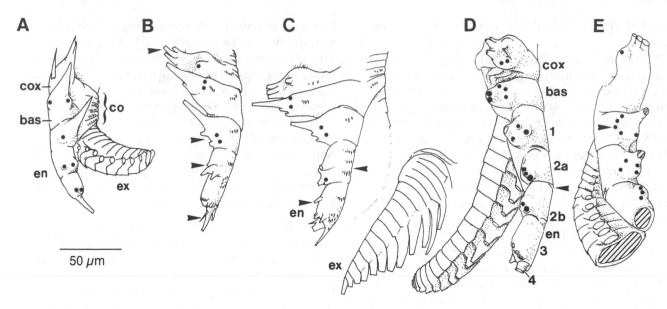

Fig. 8. Development of the second antenna of larval series A (redrawn mainly from actual specimens); completely reconstructed parts in these and subsequent illustrations stippled, or dotted, in some cases with question marks; insertion of setae, when known, indicated by hollow or black dots (diameter approximates to setal size); scale bar on lower right for all figures. □A. Stage L2A, right limb in median view. □B. L3A, left limb about from anterior. □C. TS1iA, left limb from anterior, outer view of exopod added; □D. TS5A, left limb, median view. □E. TS7A, right limb, median view.

but 2nd antenna inserting further anteriorly than in preceding stages, its length exceeding 190 μm (corm 150 μm, endopod 95 μm). Basipod drawn out mediodistally towards the endopod, its endite being similar to that of mandible but much thinner, with one spine and 4–5 setae (Pl. 9:1). Second endopodal podomere with fissure, not yet subdivided; 7–8 rows of denticles on outer side of endopod may be indicative of an original segmentation (not illustrated).

Mandible of similar size as the 2nd antenna (195 μm). Coxa not longer than in TS1A (65–75 μm), but gnathobase broader (about 35 μm). Basipod with eight setae around masticatory spine. Endopod as in TS1iA, but exopod now with 11 annuli and 10 setae (number of annuli and setae is not congruent, Pl. 9:2; Fig. 9F). First maxilla markedly enlarged, at least 180 μm long (increase from TS2i≥25%), seven-segmented along inner edge as in preceding stages comprising four endites of the corm and three endopodal podomeres. Setation further advanced (Pl. 9:1). Proximal endite only slightly larger than in TS1A but with half-crescentic row of 5–6 setae at posterior margin and two spines on median surface; still no anterior setae. Exopod 80 μm along outer margin, 55 μm along inner margin, 40 μm wide (Pl. 8:7), and with 7–8 setae (ds5).

Second maxilla 125 μm long, proximal endite 10 μm wide. Endites with paired spines. Exopod 70–75 μm long, 30–35 μm wide, and with six marginal setae (ds4b; Fig. 11C). First thoracopod 100 μm long, exopod 35 μm and with four marginal setae (ds4a; Pl. 8:7). Bud of 2nd thoracopod 30 μm long (ds3a). All developed postmandibular

limbs possess heavily corrugated shaft-like bases (Pls. 8:7, 8; 9:3). Abdomen gently convex dorsally, sloping towards the anal field. Dorsocaudal spine now absent and does not re-appear (Pl. 8:7, 8). Furcal rami longer and wider than in TS1A (see Table 3), but with the same number of spines as in left ramus (7).

TS3iA (Pl. 9:4, 5). – Material: Two fragmentary specimens (Table 2). Measurements: tl 450–460 μm, csl 270 μm (Table 3). Size seemingly unchanged: increase only about 2% for total length. Morphology incompletely known. Main progress is in development of incipient 3rd thoracic segment, which increases the length of thorax to 115 μm, while the length of abdomen is reduced to 50–60 μm. Eye lobes large, projecting markedly from forehead and shield. In UB W39 the whole eye complex seems to arise from a narrow stem-like basis (Pl. 9:5; compare with Pls. 8:3, 10:7).

First antenna about 200 μm long, its distal end made of three tubular podomeres (Pl. 9:4), suggesting that no significant changes in shape had occurred since stage TS1iA. Faint furrows on outer side may point to a former subdivision into more podomeres. Setation of postmandibular limbs seems to be arranged in more regular sets, pointing to slight progress in the development of the feeding structures of these limbs (1st maxilla in Fig. 10D, rudimentary 2nd thoracopod in Fig. 12A). No change in size of furcal rami and their armature.

TS3A (Pl. 9:6, 7). – Material: One specimen, slightly withdrawn into shield by collapse and with ventrally curved tail

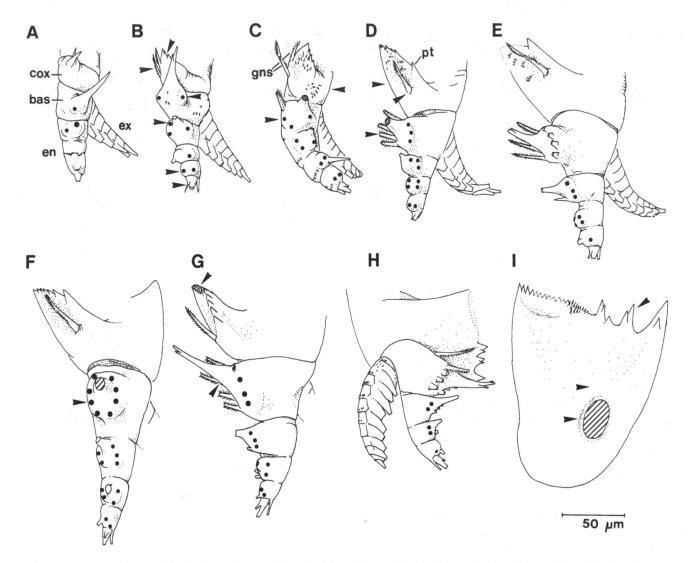

Fig. 9. Development of the mandible of series A. □A. L1A, right mandible in median view. □B. L2A, left, median view. □C. L3A, left, median view. □D. L4A, left, seen from posterior. □E. TS1iA, left, from posterior. □F. TS2A, left, from posterior. □G. TS5A, left, from posterior. □H. ?TS7, left, from anterior. □I. TS13A, right, palp (basipod and rami) unknown save for its insertion = 'palp foramen'.

(Table 2). Measurements: tl 480 μm, csl 280 μm (Table 3). Generally not much advance from preceding stages in major details, which again points to some stagnation in this period of development (increase in total length 5%). Neck organ still faintly recognizable on anterior third of shield. Labrum similar as in earlier stages, also with small tubercles at posterior surface (Pl. 9:7). Details of sternal region not known.

Appendages seemingly not changed significantly (Pl. 9:6). Antennary endopod slightly longer than in TS2A (100 μm). Mandibular basipod with nine setae around the masticatory spine; exopod 80 μm long, number of annuli and setae unchanged. Width of proximal endite of 1st maxilla unchanged, exopod slightly enlarged (90 μm along outer edge and 60 μm along inner edge), setation unchanged. Second maxilla longer than 100 μm, nine-seg-

mented along inner margin (division unclear), still with paired spines on the enditic processes, as in TS2A (ds4a, Fig. 11D).

Second thoracomere slightly produced laterally to form a feeble pleura-like structure. Thoracopods not known in detail, 2nd one with one terminal spine on its exopod (ds3a), seemingly inserting somewhat between the 2nd and 3rd segments. Third thoracomere segment supposedly apodous. Abdomen increased in length to 90–95 μm. Furcal rami similar to that of TS3iA.

TS4iA. – Unknown.

TS4A (Pl. 10:1, 2). – Material: One specimen, UB W41, assigned to this stage on basis of thoracic segments, measurements, and shape of mandibular coxa (Table 2). Measurements: tl unknown, csl probably ≥300 μm; trl probably

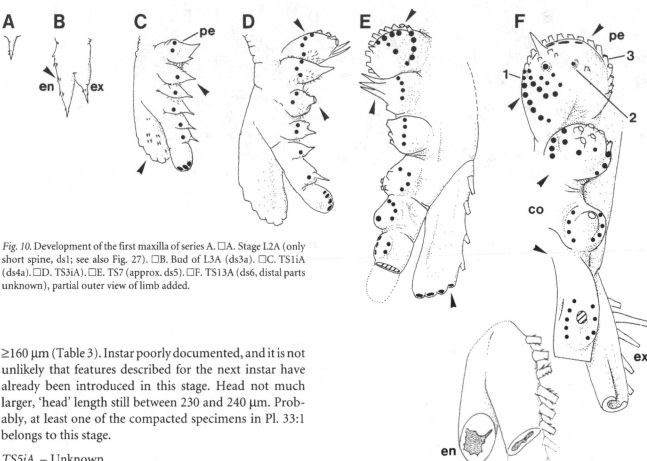

Fig. 10. Development of the first maxilla of series A. □A. Stage L2A (only short spine, ds1; see also Fig. 27). □B. Bud of L3A (ds3a). □C. TS1iA (ds4a). □D. TS3iA). □E. TS7 (approx. ds5). □F. TS13A (ds6, distal parts unknown), partial outer view of limb added.

≥160 μm (Table 3). Instar poorly documented, and it is not unlikely that features described for the next instar have already been introduced in this stage. Head not much larger, 'head' length still between 230 and 240 μm. Probably, at least one of the compacted specimens in Pl. 33:1 belongs to this stage.

TS5iA. – Unknown.

TS5A (Pls. 10:3–8; 11:1–8; 30:4; Fig. 7A). – Material: Three specimens, details fairly well-documented (Table 2). Measurements: tl about 600 μm, csl 340–350 μm (Table 3). Length increase of 20% from TS3A, i.e. not more than 5% for each of the intermediate stages. Shield moderately arched, more extended anteriorly and laterally than in earlier stages (Pls. 10:3, 4, 7; 11:6). Appendages only slightly covered by the shield (not more than 50% of the corms). Posterolateral corners of shield slightly wing-like and extended rearwards (Pl. 10:3). Neck organ not identified. Small hump with six pores located in front of the point of flexure of the excavated posterior shield margin (Pl. 11:6, 7); its nature is unknown.

Eye complex most likely extending well beyond the shield and somewhat separated off from the head by a constriction enclosed by the insertions of the 1st antennae (Pl. 10:7). Labrum prominent (Pl. 10:3, 7, 8) and modified in shape: anterior surface gently increasing first but behind a shallow depression much more steeply ascending towards the tapered tip; surface steeply descending behind tip, and deflecting inwards into the mouth tube close to grinding plates. Posterolaterally the edges of the labrum are excavated to provide space for the movements of the mandibular gnathobases. Sides slightly depressed in long axis of the labrum, which causes the middle posterior edge to be somewhat protruded as a ridge (Fig. 25C). In general

this new labral shape remains largely unaltered in the subsequent stages.

Paragnaths much elevated, with deep channel between them, reaching posteriorly to maxillary sternite which is now coalesced with the sternum (Pls. 10:3, 5, 8; 11:4). Sternum covered with numerous setules arranged in short half-crescentic rows. Sternite of 1st thoracomere also with setules but only lateral to median food groove (Pls. 10:3, 5, 8; 11:4).

First antenna not known in detail (Pl. 10:7). Second antenna stretched far anteriorly in UB W42 (Pl. 10:3), about 205–220 μm long; endopod 110 μm, exopod 150–170 μm, with about 17–19 annuli and 17 setae. Armature enhanced (Fig. 8D). Second endopodal podomere distinctly divided into two (en2a, 2b; tiny 4th podomere still recognizable; Pls. 10:3, 4, 5, 7, 8; 11:1).

Mandible with larger coxal body; surface of gnathobase slightly concave, cutting edge with several spinules and broader than before. Posterior rim of gnathobase somewhat swollen, continuing into posterior tooth. Insertion of gnathobasic seta shifted to distal edge of gnathobase, fairly

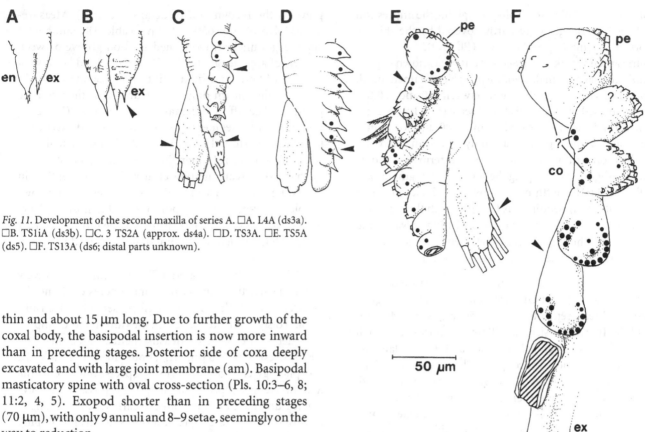

Fig. 11. Development of the second maxilla of series A. □A. L4A (ds3a). □B. TS1iA (ds3b). □C. 3 TS2A (approx. ds4a). □D. TS3A. □E. TS5A (ds5). □F. TS13A (ds6; distal parts unknown).

thin and about 15 μm long. Due to further growth of the coxal body, the basipodal insertion is now more inward than in preceding stages. Posterior side of coxa deeply excavated and with large joint membrane (am). Basipodal masticatory spine with oval cross-section (Pls. 10:3–6, 8; 11:2, 4, 5). Exopod shorter than in preceding stages (70 μm), with only 9 annuli and 8–9 setae, seemingly on the way to reduction.

Corm of 1st maxilla shorter than that of posterior limbs and with four developed endites. Proximal endite 40–50 μm wide, now with about 18 pectinate setae along inner margin and an additional short row of spines distal to it (Pl. 11:4). Next endite drawn out medially and terminating in three spines or setae. Basis of elongation crowned by more than seven setae. Third endite with three setae on either side of a central spine (Pls. 10:3–6, 8; 11:2–5).

Second maxilla about 170 μm long, its corm with six endites. Proximal one similar to that of 1st maxilla but slightly smaller (30–40 μm) and with fewer setae (13–14); armature more similar to distal endites than to corresponding endite of 1st maxilla. Distal endites all similarly equipped with setae (Pl. 11:5), with much advanced pattern, consisting of a frontal row, a setation of the enditic surface, and a U-curved posterior row (ds5; Pls. 10:4–6; 11:3–5; Fig. 11E). Exopod with 8–9 marginal setae.

First thoracopod similar to 2nd maxilla but smaller (about 125 μm); proximal endite 30–35 μm and with 8–10 setae in posterior row (3; about ds5; Fig. 12C). Outer edges of the corms of 2nd maxilla and 1st thoracopod more firmly sclerotized than the flattened side and subdivided into 2–3 portions. Second limb not fully known, seemingly smaller than the 1st one and with bifid endites (ds4). Third limb rudimentary, about 60 μm long, probably with a few incipient endites medially and with two spines on exopod (ds3b; Fig. 12B). Fourth limb rudimentary, 25–30 μm long, tipped by two spines (ds2). All developed postman-

dibular limbs are concave posteriorly, indicative of their phyllopodous shape (Pls. 10:5, 6; 11:3).

Posterior thoracic segments almost ring-shaped and lacking tergitic structures. Furcal rami much more enlarged and as long as or slightly longer than wide, with 12 spines in the primary row and three in a secondary row (sfsp; the secondary row may have been introduced in earlier but undocumented stages; Pl. 10:3). Incipient 'ventrocaudal processes' appear as a pair of small protuberances terminating in a stout, acute spine at the posteroventral margin of the abdomen (also probably developed already one or two stages earlier; Pl. 11:8).

Development apparently advanced, and anterior trunk limbs may already have functioned for some primordial kind of filtration. This change is accompanied by modification of the labrum and formation of a recessed food path between the appendages. The 'head' length, however, is still no larger than in earlier stages.

TS6iA (Pl. 11:9, 10). – Material: Two partly preserved specimens (Table 2). Measurements: tl 600–650 μm, csl 350 μm (Table 3). Morphology known only in part, but instar apparently larger than TS5A, particularly in the length of thorax. Lateral margins of shield further extended from the body (UB W46 is collapsed, giving the impression as if the appendages were partly covered by the shield). Labrum 160 µm long and 65 wide (Pl. 11:9).

First antenna unknown. Second antenna known in part, seemingly stagnant in development: endopod unchanged, exopod varying greatly in length between 140 and 170 µm and composed of 18 annuli and 17 setae (Pl. 11:9). Height of mandibular coxa increased to almost 80 µm along outer edge, but width of gnathobase only 45 µm in UB W47, which is comparatively small. Sides of anterior 3–4 thoracomeres slightly overlapping the subsequent segments laterally (Pl. 11:10). Fifth one almost ring-shaped, unclear whether it was appendiferous (incipient 6th thoracomere in Pl. 11:10). Abdomen slightly widening towards the insertions of the furcal rami (angle between them 150°). Ventrocaudal processes as in TS5A (Pl. 11:9).

TS6A (Pl. 12:1–5; 31:1). – Material: Two distorted specimens; trunk fragment UB W49 only tentatively assigned (Table 2). Measurements: tl about 650 µm, csl 375–400 µm (Table 3). Increase in size and differentiation low, although the shield has expanded laterally and posterolaterally (wing-like extensions) through the last stages (Pl. 12:1), probably also in anterior direction. Eye area not preserved. Labrum as long as in earlier stages but much broader (Pl. 12:1, 2). Sternite of 1st thoracic segment made of two rounded plates with a pore in the middle (Pl. 12:2). Progress of head development low (stagnant 'head' length).

Both antennae distorted, mandible seemingly unchanged. First maxilla known only from its large bulging proximal endite and the conspicuous elongate 2nd endite (Pl. 12:2). Second maxilla and anterior two thoracopods preserved with their proximal 4–8 endites (Pl. 12:2). Length of 2nd maxilla and 1st thoracopod exceeding 140 µm, the former probably with six endites on its corm. Distal endites successively changing their shape into that of the median surfaces of the endopodal podomeres.

Limbs anteroposteriorly much flattened, their bulging endites being posteriorly oriented. Frontal set of setae well-developed. Proximal endites decrease in width from 40 µm in the 2nd maxilla to 35 µm in the 1st thoracopod and 25 µm in the 2nd (all of ds5). Third limb markedly smaller (ds4), 4th one being a bilobate bud (ds3a; Pl. 12:1), 5th also rudimentary, uniramous (ds2; Pl. 12:4). Furcal rami slightly dorsally oriented, inarticulate, but future joint faintly recognizable (Pl. 12:4, 5). Rami slightly larger than in TS6iA, varying in the two specimens: spines ranging from 9 to 11 in the primary row and 4 to 5 in the secondary row. Pits ventrally to all marginal spines (Pl. 12:1, 3–5). Ventrocaudal processes enlarged (10 µm) and with 2–3 terminal spines (Pl. 12:3, 5). Abdomen starts to form a trench between the processes.

TS7iA (Pl. 12:6, 7). – Material: One specimen, tentatively assigned, since distortion of head and lack of appendages

prevents the recognition of details (Table 2). Measurements: tl probably 650–700 µm (Table 3). Number of appendages unclear. Deep median food groove between the sets of appendages (in UB W50, enhanced because of collapse of body and sinking into body cavity; Pl. 12:6). More details known only from trunk. Last thoracomere partly marked off by indentation on abdomen (Pl. 12:6). Posterior trunk segments almost ring-shaped except the membranous sternal area (Pl. 12:7). Ventrocaudal processes slightly longer than in TS6A, number of spines unchanged. Furcal rami now clearly hinged, larger than in preceding stage and with 1–2 more spines in both rows (Table 3). Rami upwardly pointing and almost in plane. Anal field almost vertical (feature may be introduced earlier), with faint incipient supra-anal flap (saf) dorsal to it.

TS7A (Pls. 13; 31:2). – Material: Two specimens tentatively assigned since the number of thoracomeres could not be ascertained; UB W51 is rather well-preserved, but shrinkage and distortion did not permit measurements of important details (Table 2). Measurements: tl about 750–800 µm, csl 450–500 µm; trl about 400 µm (Table 3). Morphology changed in various aspects, but due to the incompleteness of preceding stages, it is possible that several features have been introduced earlier. Shield changed in proportions and shape: anterior side steeply sloping from the apex above the mandibles at first fifth of shield length, thus, anterior margin only a little raised towards the middle; lateral margins curve gently backwards, run parallel along the body, and rise slightly towards the posterolateral rounded corners; shield somewhat narrowing posteriorly in dorsal aspect; posterior margin deeply excavated. In cross-section the shield extends widely ventrolaterally, probably covering the proximal parts of the limbs (Pl. 13:1–3). 'Head' length increased again (Table 3).

Eye blisters no larger than in TS2A, which means that their size has decreased relative to the body. Eye area probably no longer projects from the anterior margin of shield. Labrum similar to that of TS6A, probably slightly better developed, sides of labrum with setules (Pl. 13:1–4). Depression between maxillulary and maxillary segments. Pore on sternal portion belonging to 2nd maxilla (Pl. 13:3, 4, 7).

First antenna not known in detail. Fragment of 2nd antenna of UB W52 exceeding 165 µm; all elements seem to have elongated in long axis of the appendage. Antennal corm of UB W51 somewhat deformed by stretching, and due to peculiar growth of the lateral side, the proximal endopodal podomere gives rise to the outer ramus rather than the basipod (Pl. 13:5). In consequence, the proximal 3–4 exopodal annuli lack setae (Fig. 8E).

Mandible seemingly reduced in total length (140 µm long), while the coxal length has enlarged to about 115 µm. Gnathobasic seta positioned at basis of ridge-like posterior margin of grinding plate closely anterior to insertion of basipod, 25–30 μm long. Cutting edge slightly thickened where the grinding plates are facing each other (6–7 μm), covered with numerous small spinules or setules (Pl. 13:4). Posterior part of cutting edge almost unchanged. Basipodal endite shorter than in preceding stages, almost directly tapering into masticatory spine. Tip of spine split into at least three spinules. Eight setae arranged in an oval rather than in a circle, all pectinate. Exopod 85–90 μm long, carrying nine setae (Pl. 13:2–4, 6; Fig. 9H).

First maxilla exceeding 190 μm in length. Its proximal endite slightly larger than in preceding stages, almost 50 μm wide. Subsequent two endites unchanged; 4th endite elongate in long axis of limb, with typical pattern of enditic setation of posterior limbs (frontal group 1, median set of spines 2, posterior row 3). Proximal two endopodal podomeres preserved in UB W52, with fewer setae. Exopod slender and paddle-shaped, about 120–130 μm in length (Fig. 10E).

Subsequent limbs known only in part. Second maxilla longer than 155 μm, 1st thoracopod ≥125 μm, 2nd ≥140 μm, with proximal endite of 25 μm, and more than eight divisions along inner edge. Third limb known from its proximal part only. Next two broken off distally in UB W51 (Pl. 13:3), the latter most likely smaller than anterior limb (ds4?). Sixth limb probably only a uniform bud (ds2; Pl. 13:1, 3, 6–8). Trunk almost circular in cross-section; tergitic pleurae indistinct, 6th and 7th segments almost ring-shaped, pliable ventrally. Abdomen slightly widening towards the outer edges of the furcal rami, which are not known in detail. Ventrocaudal processes longer than in preceding stages, but with the same number of spines. Median trench along ½ of the abdomen (Pl. 13:1, 3, 8).

TS8iA (Pl. 14:1, 2). – Material: Possibly two specimens, with details of head and appendages not well-documented (Table 2), assigned to this stage since their thoracic length is about 400 μm. Measurements: see Table 3. Few details are known of this stage. Distance between outer edges of mandibular coxae 230 μm (each coxa about 110–115 μm long). Fourth thoracopod seemingly small, subdivision of outer surface also appearing in the 3rd thoracopod. Sixth limb probably still a bud. Thorax and abdomen thicker than in TS7A, but abdomen only little shorter despite the appearance of the incipient 8th thoracic segment (95 μm; Pl. 14:1). Ventrocaudal processes further elongated (25–30 μm), with five spines (Pl. 14:2). Furcal rami well-articulated, seemingly thicker than in preceding stages, other details not known (Pl. 14:2).

TS8A (Pl. 14:3–6; Fig. 7B). – Material: One specimen, beautifully preserved originally, but destroyed almost completely (the only advanced specimen with entire endopods of postmandibular appendages; Table 2). Measurements: tl about 900 μm, csl 600 μm (Table 3). Shield similar to that of earlier stages, but apex seemingly less developed (preservation?) and lateral margins more gently convex. Details of anterior head structures not known. Labrum similar to that of preceding stages (Pl. 14:3). Originally, much of anterior appendages were preserved in UB W54, but only the proximal parts of these are left now, still exhibiting many setae and setules on these and the enditic surfaces, which gives an impression of the former completeness of preservation of this specimen (Pl. 14:5, 6).

Frontal area with eye and 1st antenna disguised by large foreign particle. Second antenna known only from its coxal endite. Distal parts of mandible obviously shortened, exopod small and thin, probably composed of many fewer annuli than in preceding stages (Pl. 14:3), other details not known. First maxilla known from its corm, but pictures taken prior to distortion suggest that it was considerably shorter than the more posterior limbs. Setal armature well-differentiated and pectinate, demonstrating their high level of development for filtratory function. Setae of posterior row arise from thicker sockets (Pl. 14:6).

Length of 2nd maxilla increased remarkably to 300 μm. Corm about 150 μm long, with pliable between endites and more firmly sclerotized outer edge subdivided into three parts; details of median surface not known. Endopod about 150 μm long, forming the continuation of the basipod, probably four-segmented. Exopod at steeply sloping outer surface of limb corm only a little shorter, slender and paddle-shaped, with concave posterior surface (length 145 μm). More than 10 setae along outer margin up to ramal tip. Shape somewhat sigmoidal, curving outwardly first and then distally again. Joint with basipod weakly developed anteriorly but much better developed posteriorly (see also Figs. 21, 39). In Pl. 14:3 it appears as if the exopod slims distally, but this may be due to the perspective of the micrograph and the spoon-shaped curvature of the ramus.

First thoracopod of the same design, probably with slightly shorter rami. Second limb again similar, corm also tripartite on outer surface. Posterior limbs not known in detail, but appear to decrease in size progressively. Abdomen 10% larger than in preceding stage (l/h 105/100 μm). Fragments of furcal rami are 100 μm long in UB W54, with at least six secondary spines. Ventrocaudal processes longer than 25 μm and with 3–5 spines. Supra-anal flap better developed than in preceding instars and slightly pointed (Pl. 14:3, 4). Postmandibular limbs demonstrate a considerable advance towards the completion of the filter apparatus, also reflected in a slight enlargement of the head portion, as recognized in a greater 'head' length of 250–300 μm.

TS9iA–TS10iA. – Unknown.

TS10A (Pls. 15; 16; 31:3, 4; Fig. 7C). – Material: Two specimens: UB W55 is outstretched but compressed and slightly twisted, lacking shield and furca but with many limbs and setae still preserved; UB W56 is a trunk fragment with complete furcal rami, included because of its size (Table 2).

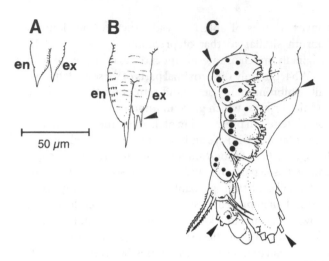

Fig. 12. Developmental stages of thoracopods of series A. □A. Second thoracopod of TS3iA (ds3a). □B. Rudimentary 3rd thoracopod of TS5A (ds3b). □C. First thoracopod of TS5A, between ds4 and ds5.

Measurements: tl approximately 1200–1300 μm; trl roughly 650 μm (Table 3). Head morphology known only in part, since it is much deformed in UB W55, affecting particularly the anterior region back to the mandibles (Pl. 15:1, 2). Outline of shield unclear due to wrinkling. Labrum about 170 μm long and 115 μm wide, similar to preceding stages (Pl. 15:4). Sternal region deformed, but one of the prominent paragnaths recognizable (Pl. 15:4). Thoracic segments clearly free from head and also shield (maxillary segment incorporated; Pl. 15:1–3). 'Head' length increased to 300–330 μm, indicating continuation of enlargement of the head region.

Mandible with coxa 125–130 μm in length. Proximal endite of left 1st maxilla bulging, about 60 μm wide and armed with numerous setae and spines of different size and equipment with setules; between the spines, a flap-like structure of unknown function is developed (final state = ds6; Pl. 15:4, 5). Second maxillae of both sides known from their proximal parts; design as in earlier stages, with anterior setae, surface armament, and crescentic posterior row of pectinate setae (Pls. 15:4; 16:2). Width of proximal endite not changed from TS8i, surface covered with numerous setules (also ds6; Pl. 16:1).

Filter apparatus much more advanced than in TS8A: seven thoracopods have developed endites, endopods, and paddle-shaped exopods, 8th limb present but seemingly shorter (ds4?); ninth rudimentary (ds2 or 3; Pls. 15:1–3, 16:3). First thoracopod longer than 310 μm, exopod and distal endopodal podomeres broken off in UB W55. Corm drawn out medially towards the endopod; because of this its distal endites are distal to the insertion area of the exopod. Length of second one unknown since limb twisted backwards. Third and fourth limbs longer than 260 μm (distal endopodal podomeres missing), 5th limb 250 μm

and 6th one 200 μm. Corms of anterior six thoracopods with 7–8 endites medially, outer edges subdivided at least in the anterior four (Pls. 15:1–3).

Endites with different types of setae and spines (Pl. 16). Setae of proximal endites (belonging to posterior row) reaching into deeply recessed sternal food groove (Pl. 16:4). Major spine of more distal endites developed as comb or scraper spine (belonging to median set; Pl. 16:3, 5; Fig. 35A). Pectinate setae of the posterior rows reach far posteriorly between the endites of at least the subsequent limbs (Pl. 16:3, 7). Setulae of filter setae closely spaced (2 μm on an average; Pl. 16:6), arranged within the set of posterior setae to point to the centre of the endite (16:2). Anterior setae possibly articulate.

Abdomen considerably enlarged compared to TS8A, now 145 μm long and 120 μm high (slightly deformed in UB W55). Ventrocaudal processes longer than 75 μm, conical, with more than 10 marginal spines and pits ventrally to them in UB W55. Ventral trench between processes reaching anteriorly to about $1/2$ of abdomen, in UB W56 anteriorly to $2/3$ of it. Furcal rami of the latter specimen are 170 μm long and 80 μm wide, carrying 16 primary and more than six secondary spines (Pl. 16:8, 9). Since the ventrocaudal processes are only 60 μm long and have eight spines, it is not quite clear whether this indicates individual variability or whether UB W56 is of stage TS9A rather than of TS10A.

TS11iA–TS12A. – Unknown.

TS13iA (Pl. 17:1). – Material: One specimen, representing a large trunk fragment with parts of the furcal rami (Table 2). Measurements: tl presumably more than 1.6 mm; trl about 1 mm (Table 3). Stage with incipient 13th thoracomere. Few details known of this stage only. Size of all thoracic segments apparently larger than in TS10A. Anterior segments lacking tergitic structures, but showing two humps laterally distal to the insertions of the limbs, which are separated by a depression; another swelling is located dorsal to the groove (Pl. 17:1; see also Pl. 15:1 for TS10A). Segment boundaries clearly developed by deep fissures, suggesting well-developed arthrodial membranes.

Posterior segments with faint pleural structures, curving inward to the sternal region. Eleventh segment almost ring-shaped save for its poorly sclerotized ventral part where the supposedly rudimentary limbs inserted originally. Possibly also 12th thoracomere with small limb buds. Abdomen about as long as in preceding stage but 10% thicker (length 135 μm; increase 2.5% for each intermediate stage only). Supra-anal flap feebly developed, probably slightly pointed.

TS13A (Pls. 17:2–4; 31:4). – Material: One fragmentary specimen, consisting of a slightly deformed and distorted head and anterior part of the thorax, all appendages being widely stretched laterally (Table 2; measurements Table 3).

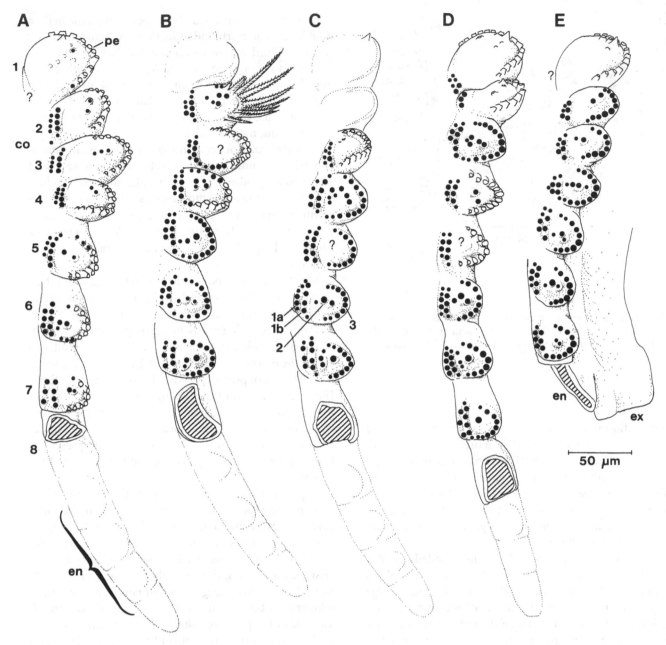

Fig. 13. First to fifth thoracopods (A–E) of TS13A (ds6; distal parts reconstructed).

UB W58 is the largest specimen at hand recognized herein as *Rehbachiella kinnekullensis.* The measurements indicate its position within this stage, but some uncertainty remains, because only data of the head region and the anterior part of thorax are available. Again, due to collapse the specimen is not only dorsoventrally flattened, but also the posterior structures have been pushed into the head, causing deformation of the sternal region in particular (Pl. 17:2).

Frontal head region with eye and 1st antennae not fully known (Pl. 17:2). Shield present but incompletely pre-

served (Pl. 17:3). Labrum broken distally, but obviously enlarged (width 120 µm) and further modified in shape: basis now gently merging with the body wall rather than inclining steeply. Labrum seemingly better sclerotized save for the cuticle of the anterior hump in front of the constriction (see also Fig. 25D).

A broken surface discloses the labrum of UB W58, where coarse phosphatic fillings form two strings which reach towards depressions on the sides of the labrum (preservation of musculature?). Deep lateral excavations for mandibular grinding plates reaching anteriorly towards the

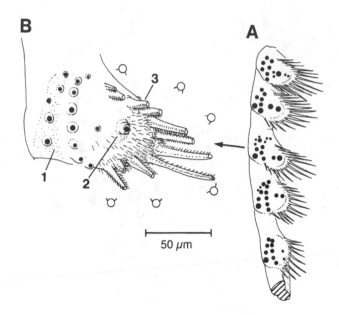

Fig. 14. Large limb fragment redrawn from Pl. 33:3, 4 (UB 92). □A. Close-up of one of the endites to show the arrangement of setae and setules (numbers in brackets refer to groups of armature). □B. Overview of endites.

constriction and indicating a further anteriorward shifting of the mandible. Ceiling of 'atrium oris' partly preserved, probably rather pliable (Pl. 17:2, 4). Sternum deeply excavated and with prominent paragnaths (Pl. 17:4). Postcephalic sternal region with separate sternites made of two rounded plates which form the slopes of the deep, V-shaped food path, as known from younger stages. 'Head' length now 350–380 μm.

Mandibular coxae huge and broadly rounded laterally, about 150–160 μm long and 100 μm wide. Grinding plates with recessed surface, widening towards the cutting edge (right one artificially inflated in UB W58), which is about 125 μm long. Teeth much larger than in earlier stages and of varying size, some being flattened and accompanied by smaller spinules. Gnathobasic seta not identified. Palp not known, but the small size of its insertion area ('palp foramen') on the distal surface of coxa (30×20 μm), which is about that of earliest instars, suggests that it has undergone considerable reduction (Pl. 17:2, 4; Fig. 9I).

Median surfaces of eight postmandibular limbs fairly well-known, regarded as fully functional (ds6; Pl. 17:2, 3; Figs. 11F, 12F, 14). First maxilla larger than 250 μm but apparently smaller than posterior limbs. Proximal endite 70 μm wide and inflated (increase 14% within six moults). Surface slightly recessed distally to provide space for the 2nd endite. More than 20 pectinate setae standing around posterior and inner margin, proximal ones arising from slightly below the endite and joining the anterior set, which is split into two rows (arrow in Fig. 10F). Enditic surface originally covered with numerous setae and spines.

Elongate 2nd endite with 7–8 setae or spines medially, a double row anteriorly and more than six setae posteriorly (arrow). Third endite much smaller than second one, drawn out into a spine and with four setae on either side. Fourth endite elongate in long axis of the limb and with oval enditic surface. Endopod not known. Exopod broken off distally in UB W58, seemingly similar to that of posterior limbs but more slender.

Second maxilla longer than 300 μm, its corm most probably with no more than six endites. Proximal endite of similar shape and size as in 1st maxilla, but distal endites designed as those of the thoracopods, with two rows of 2–4 anterior setae, a median set of a few thin setae, and a posterior row of 9–12 pectinate setae. Endites progressively becoming longer than wide and more distally projecting (Fig. 11F).

Next five thoracopods similar to 2nd maxilla, but their corms with at least 7–8 endites. First three limbs longer than 325 μm. Proximal endites smaller than in 2nd maxilla, width decreasing gently from 50 μm in the 1st limb to 40 μm in the 5th. Setal armature of all endites similar to that of the more distal endites of the 2nd maxilla, changing somewhat from proximal to distal and between the limbs (Fig. 13A–E). Towards the endopods the endites become progressively more elongate in longitudinal direction, while their median surfaces become more triangular, decreasing in length and size. Sixth and last limb preserved, with a distinctly smaller proximal endite than that of the preceding limb.

All postmandibular limbs insert abaxially. Corms phyllopodous, except for the median endites and the outer edges (see preceding stages; lower right of Pl. 17:2, 3). In cross section, the limb bodies are concave posteriorly, with the endites and the outer edges pointing backwards. At least these limbs together with the 2nd maxilla probably were already functioning as filter organs. While the mandibular coxal body has much enlarged, the palp is reduced. The size of the preserved shafts of the 2nd antennae (Pl. 17:4) points to reduction also of these appendages and their diminishing relevance in locomotion and feeding now taken over by the postmaxillulary limbs.

Development of series B

This smaller-sized series is documented by fewer stages than series A but has also sets of successive instars. No earliest larvae have been found of series B, but the size spans of specimens from L1A–3A and particularly their morphogenesis gives little evidence to assume that individuals of these instars have been mistakenly placed into series A.

Naupliar phase

L1B–L3B. – Unknown.

L4B (Pl. 18:1–6; reconstruction in Fig. 15A). – Material: Two slightly distorted specimens (Table 2). Major measurements: tl 260–270 μm, csl 160–165 μm (Table 3). Smallest stage recognized of series B. Gross shape similar to L4A, but most structures only about as large as those of L3A.

Shield with feebly developed margins which do not extend much laterally (Pl. 18:2, 3). Neck organ present. Eye area prominent and protruded; size of lateral blisters about as large as in L2A (Pl. 18:1–3). Labrum similar to that of L4A but size as in L3A (Pl. 18:1, 2, 4, 5). Anterior labral margin slightly excavated at connection with 'midventral' lobe. Sternum more similar to L3A than to L4A, still with distinctive remains of antennary sternite, recognizable as a triangular plate at the posterolateral basis of the labrum (Pl. 18:2). Mandibular portion of sternum large, bulging posteriorly and gently sloping orally.

First antenna known only from its proximal part which is similar to that of L4A (Pl. 18:3, 4). Second antenna about 25–30% smaller than in L4A (110 μm along endopod). Length of endopod 55–60 μm, exopod longer than 75 μm. Median armature as in L4A, but endites and setae shorter (Pl. 18:1–5). Coxal endite with three spines and a thinner one more anteriorly. Basipod with one strong spine and three thinner ones (Fig. 16A).

Endopod four-segmented, proximal podomere continuing into one stout spine medioproximally and with two more setae anteriorly at the basis of the spine. Subdivision of 2nd endopodal podomere indicated. Enditic surfaces covered with fine denticles or setules; more occur as short rows or groups on the flattened sides. Each of the three tubular proximal podomeres with rows of denticles on their outer edges, probably indicative of a lost subdivision. Distal podomere as in series A.

Mandible 95–110 μm along the endopod; shape much as in L4A. Coxa 75–80 μm long and 30 μm wide, carrying two gnathobasic setae (Pl. 18:1, 4, 5; compare with L4A). Basipod with six pectinate setae arising in a circle around the median masticatory spine rather than eight in L4A. Endopod as in L4A (40–45 μm; Pl. 18:5), exopod not known (Fig. 17A).

First maxilla similar to that of L4A but smaller (80 μm; ds4a), seven-segmented. Proximal four endites probably belonging to the corm. Proximal endite as in L4A but with two anterior spines or setae only; 2nd endite less elongate, with one seta; subsequent endites progressively less drawn out and more symmetrically. Again, posterior spine progressively more similar to anterior seta, while both approach each other. Distal endopodal podomere rounded apically, bearing a set of setae. Exopod not known in detail (reconstructed in Fig. 18A).

Deep transverse furrow delineating posterior end of larval head. Maxillary segment still at anterior part of trunk, pliable dorsally, as in L4A. Second maxilla rudimentary, of similar shape and size as in L4A (ds3a), but standing very close together (Pl. 18:1; Fig. 19A). Clearly different

from L4A are the 'head' length, which is only about 65% of that of the latter, and the caudal end:

- a dorsocaudal spine is missing in L4B, and the dorsal surface of the trunk curves gently toward the terminal anus (Pl. 18:2, 4, 6),

- the furcal rami of L4B are shorter but broadly rounded (l/w = 20/35 μm) and carry already about seven marginal spines, and

- the angle between the rami is about 90°, while they are almost in plane in L4A.

Post-naupliar phase

TS1iB–TS2iB. – Unknown.

TS2B (Pls. 18:7, 8; 19:1–3). – Material: Two slightly distorted and partly preserved specimens (Table 2). Measurements: tl 330–340 μm, csl 190–200 μm (Table 3). Increase in total length about 20–25% from stage L4B (7–8% for each intermediate stage; size 25% less than TS2A). Two trunk segments delineated (Pl. 18:7), but only 1st trunk limb present (in TS2A also the 2nd limb). Eye region not known in detail, shield distorted in both specimens at hand, presumably similar to TS2A. Labrum unchanged. Sternal region not known. 'Head' length 150 μm (increase of about 16–17% from L4B only), which is more than 30% less than in TS2A, mainly resulting from the smaller size of all head appendages (in length as well as width; at least as is known from their insertions).

First antenna not known. Second antenna slightly increased in length (less than 10%; 115–120 μm), much smaller than in TS2A but similar in shape, also 2nd endopodal podomere almost subdivided into two portions (Pl. 19:2, 3). In contrast to series A, the exopodal segmentation is always strictly correlated with the setation in series B (Fig. 16B).

Mandible known in part, about as long as the 2nd antenna. Shape similar to that of TS2A save for its smaller size. Coxa forming the major portion of the appendage, smaller than in TS2A, but larger relative to the whole limb (80–85 μm long); width nearly the same in both series (33 and 36 μm, respectively). Grinding plate large, sharply deflected and twisted towards the labrum as in series A (Pl. 18:7, also showing distinct basipodal joint with large membrane). Basipod with seven setae around the median masticatory spine (7–8 in TS2A; Pls. 18:8; 19:2). Mandibular exopod about 60 μm long, with 8 annuli and 6–7 setae (Pl. 18:7; Fig. 17B).

First maxilla 100 μm long, which is slightly more than half the size of that of TS2A. Corm probably with one more endite than in series A (ds4b; Fig. 18B). Proximal endite 23 μm wide, setation not yet clear. Maxillary and 1st thoracic segments with faintly developed pleural extensions. Appendages of these segments known more or less only

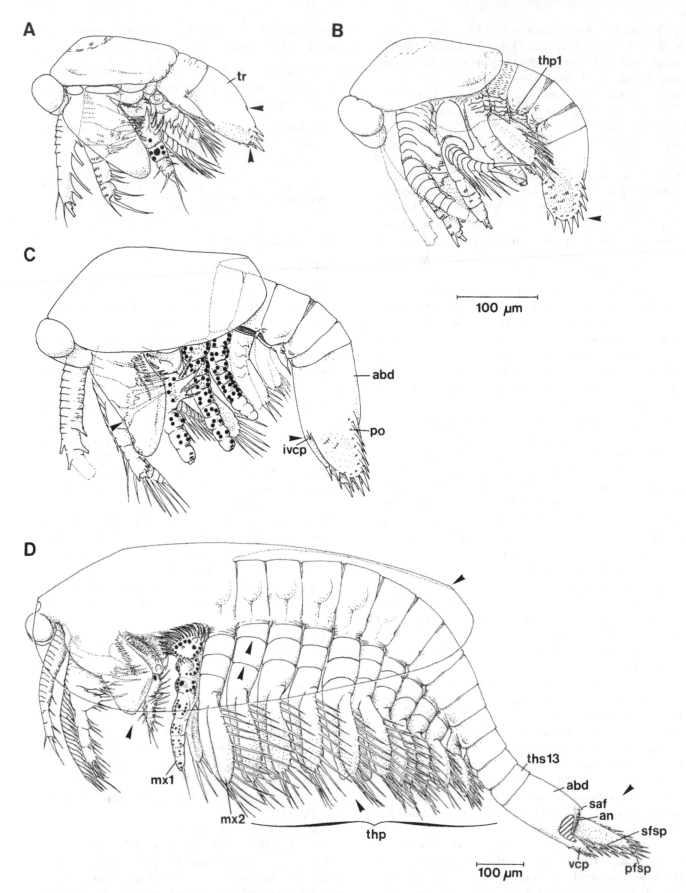

Fig. 15. Selected larvae of series B; lateral views. □A. Stage L4B, dorsocaudal spine, present in L4A, missing here, shape of the furca also different (compare with Fig. 6D). □B. TS3B. □C. TS4B. □D. TS13B (largest stage known at present).

from their insertions; 1st thoracopod being a bilobed bud (Fig. 20A). Second thoracomere almost ring-shaped and without appendages. Abdomen unchanged and similar to that of TS2A, but furca different to the latter: rami much shorter (l/w 30/50 µm) but much more paddle-shaped and with nine marginal spines rather than seven in TS2A (Pls. 18:7; 19:1). Accordingly, the increase is only two setae from L4B (4 stages).

TS3iB. – Unknown.

TS3B (Pls. 19:4–7; 32:1; Fig. 15B). – Material: Three specimens, UB W63 almost complete (Table 2). Measurements: tl 375–400 µm, csl 200–225 µm (Table 3). Instar similar to preceding larva, but about 10–15% larger, mainly resulting from addition of a further thoracomere. Shield not much enlarged. Margin distinctly extended ventrolaterally but not much over the limb bases. Posterior end of shield covering the maxillary segment in part, its margin slightly concave. Neck organ present. Eye region poorly known (Pl. 19:5). Posterolateral sides of labrum covered with thin setules. 'Head' length slightly increased to 160–170 µm, which is still 28% smaller than in TS3A.

First antenna known only from its proximal part (Pl. 19:4, 5). Second antenna and mandible larger than in TS2B, both still of similar length (150 versus 135 µm; Pl. 19:4, 5). Surface of 2nd antenna with many denticles. Mandibular coxa 85 µm long, 50 µm high and 35 µm wide, width of gnathobase 30 µm (Pl. 19:4, 5). Basipod with 7–8 setae around the median spine. Endopod 45–50 µm, exopod not known in detail, seemingly thin (Pl. 19:4; Fig. 17C).

Total length of 1st maxilla unknown (fragment >110 µm; Pl. 19:4, 7). Proximal endite 25 µm wide, setation better developed than in preceding larva (about ds5; Pl. 19:6; Fig. 18C). Length of 2nd maxilla increased to 110 µm along endopod. Endopod 65 µm, paddle-shaped exopod 60 µm along outer margin, not reaching to tip of endopod and with five marginal setae (ds4a; Pl. 19:4; Fig. 19B). Corms of maxillae with fine corrugations laterally as in TS3A; also surface of segments adjacent to the limbs finely wrinkled. Maxillary segment not coalesced with the head and finely wrinkled dorsally (Pl. 19:7).

In contrast to TS3A, the 1st thoracopod is still a bifid bud, but 50–60 µm long (ds3a–b; Fig. 20B). Incipient exopod with one terminal spine (Pl. 19:7). Second limb supposedly rudimentary, but details not known. Anterior two thoracomeres with feebly developed posterolateral margins, fixation points (pivot joints) between them clearly recognizable in UB W63 (Pl. 19:7). Last thoracomere almost ring-shaped lacking limbs. Abdomen unchanged. Furcal rami with groups of denticles ventrally as in series A; size and armature unchanged (Pl. 19:5) but angle between rami still much steeper than in TS3A (90°).

TS4iB (Pl. 20:1–4). – Material: Two specimens, only tentatively assigned due to limited data available (Table 2).

Measurements: tl 420 µm, csl 230–240 µm; trl 110–120 µm (Table 3). Incipient 4th trunk segment on abdomen (Pl. 20:1–3). Size increase from instar TS3B about 10%, but body considerably smaller than TS3A. Eye region not known in detail (Pl. 20:4). Neck organ still present (not figured, but see series A). In further contrast to series A, the 'head' length has enlarged slightly (Table 3).

Few details are known of the anterior appendages. Mandibular gnathobase about as wide as in TS3A (Pl. 20:4). First maxilla known in part, its corm probably carrying four endites. Proximal endite 27 µm wide, with at least 10 marginal pectinate setae, several spines forming a secondary row distal to the former row, and a prominent spine medially. Second endite somewhat drawn out medially, tipped by a spine and with three setae on either side. Third endite with two setae on either side of a central spine. Shape of these two endites already resembling those of the late stages.

Maxillary segment free from larval head (Pl. 20:3). Second maxilla with at least five endites on its corm, armature of these not much advanced (ds4b?). First thoracopod similar to 2nd maxilla (ds4b?), 2nd one rudimentary (ds3). Abdomen shorter than in preceding stage due to incipient delineation of the new segment. No further progress in the development of furcal rami; angle between them still about 90° (Pl. 20:1).

TS4B (Pls. 20:5–8; 21, 22; 32:2; Fig. 15C). – Material: Nine specimens, some well-preserved; group somewhat inhomogeneous, probably not all specimens satisfactorily assigned (Table 2). Measurements: tl 440–450 µm, csl 240–250 µm (Table 3). Instar characterized by appearance of incipient ventrocaudal processes (Pl. 22:1, 3). Shield roof-shaped in anterior view, lateral margins seemingly more extended than in preceding stages (Pl. 20:6–8). Apex in the first third or quarter. Eye region protruding from shield and arising from a stem-like basis (Pls. 20:6–8; 21:1), as in TS4 and TS5 of series A (see Pls. 9:5, 10:7). Anterior shield margin seems to be even slightly recessed to provide space for the eye lobes (Pl. 20:7).

Labrum slightly enlarged. Shape modified as described for TS5A (Pls. 20:5; 21:5; 22:4; see also Fig. 25C). Posterior edge also with pores and tubercles. Sternum deeply incised medially to form a 'paragnath channel', surface covered with numerous thin setules, most of them arranged in short, slightly curved rows. Groove continues backwards to the still separate maxillary sternite (Pls. 20:8; 21:5, 7).

First antenna known only from its insertion (Pl. 21:1, 2), but next two limbs completely preserved in UB W66. Second antenna much smaller than in TS5A, but larger than in earlier stages of series B (180 µm along exopod; 150 µm along endopod). Coxal and basipodal endites unchanged (Pls. 20:4; 21:3, 5; 22:6). Endopod 90 µm, somewhat nesting in distal margin of basipod. Endite of proximal podomere elongate, reaching to basipodal endite (Pl.

21:3); 2nd endopodal podomere now definitely subdivided (2a, b), both about equal in size and with a few setae mediodistally on a shallow enditic elevations (setal pattern of endites see Fig. 16C). Exopod 135–140 µm long, with 16–17 annuli and 15 setae (only 13 and 11, respectively, in UB W69; Pl. 21:5). Exopodal setae different: proximal three thin, next eight thicker, subsequent ones progressively more slender again (Pls. 20:5, 7, 8; 21:3, 5, 6; 22:5, 6).

Mandible still about as large as the 2nd antenna. Coxal length increased to 100–105 µm (width 45–50 µm, height 55 µm, gnathobase 35–40 µm wide). Double row of setules or spinules at cutting edge (feature introduced as early as in TS2B with appearance of two spinules in front of the posterior tooth). Number of pectinate basipodal setae unchanged, but major spine more oval in cross-section and split or bifid distally (compare with TS5A). Enditic surface of basipod also more oval than circular, as in earlier stages (Pls. 20:4, 5, 7, 8; 21:4–7; Fig. 17D). Endopod 60–70 µm long, exopod 65 µm, with eight annuli and same number of setae. Exopodal setae of 2nd antenna and mandible fringed with opposing rows of setules (Pl. 22:5; the same figure also shows the small 4th endopodal podomere of these two appendages, carrying the robust apical seta which, in all cases, is broken off, as in the 1st antenna).

First maxilla 130–140 µm long. Limb markedly more advanced than in preceding larva (ds5) and with eight divisions along inner edge, as in TS2B. Proximal endite much larger than all others, 35 µm wide, and with more than 12 pectinate setae around proximal margin (Pls. 21:5, 6; 22:8). Subsequent endites also with more setae: second one with three setae medially and 3–4 anteriorly and posteriorly on inner edge; Fig. 18D). Second maxilla 120–130 µm long. Unclear whether five or six of the nine median divisions belong to the corm, and four respectively three are endopodal. Proximal endite 25–30 µm; exopod 70 µm along outer edge and with 7–8 setae (about ds5; Fig. 19C).

Thoracopods poorly documented. First limb larger than 90 µm, but with small endites similar to the maxillae in their early stages of development (ds4). Second limb partly known, most likely with more than three endites carrying paired spines (also ds4). Third limb rudimentary (ds3; not illustrated in detail). Trunk segments connected by pliable arthrodial membranes (Pl. 21:8); dorsal side slightly more sclerotized than the ventral part (Pl. 22:1).

Shape of abdomen unchanged, but again slightly longer. Membranous field around T-shaped anus (Pl. 22:2) extending on to the dorsal surface of the furcal rami, as in series A. Size and shape of rami varies individually: length ranging from 40 to 55 µm, width from 55 to 65 µm; number of spines in primary row ranging from 9 to 11. First occurrence a secondary row by 1–2 spines dorsal to primary row. Angle between rami wider than in preceding stage (Pls. 20:5; 22:1–3).

Shape of postmandibular limbs indicates the initiated change in life style. Enditic armature now clearly subdivided into a front row of setae, a median set of spines or setae, and a posterior, half-crescentic row of setae (Pl. 21:5; ds5). Some of the pectinate setae are very long and tapered distally (Pl. 22:7). Endites are also covered with numerous fine setules, in particular where they contact each other. The two maxillae at least are apparently capable of a primordial type of filter function. This may also explain the higher degree of development in the cephalon as compared to series A.

TS5iB (Pl. 23:1). – Material: Three incomplete specimens, tentatively assigned to this stage (Table 2). Measurements: tl 460–480 µm, csl 240–270 µm (Table 3). Limited data available. Eye region as in preceding instar, large and separated from head by constriction. Labrum similar to that of TS4B. Groove located in the middle of the maxillary sternite (see TS5B). Appendages incompletely known. Length of 2nd antenna about 210 µm, which is an increase of more than 20% from the preceding stage and about as in stage TS5A. Exopod slightly longer and with one more annulus and seta than in TS4B. Mandible seemingly like that of TS4B. Little is known of the posterior appendages. Probably a rudimentary 4th thoracopod is developed. Abdomen shorter than in preceding stage due to release of the incipient 5th thoracomere. Furcal rami almost as wide as long (l/w = 50–60/60 µm), still unhinged; margins with 11 primary spines and one secondary spine. Ventrocaudal processes unchanged.

TS5B (Pls. 23:2–7; 24:1–3). – Material: Six fairly well preserved specimens (Table 2). Measurements: tl 500–530 µm, csl 270–280 µm (Table 3). Instar similar to preceding stage in gross shape, and few changes particularly in the head region. Shield extending freely beyond the 1st trunk segment, with wing-like extended posterolateral margins (Pl. 23:4, 5). Small humped area with 4–6 pores medially in front of posterior margin (not figured).

Eye region projecting from forehead, seemingly large but size not measurable. Labrum similar to that of TS5iB and TS5A, with tapered and rounded distal end (Pl. 23:7). Deeply recessed sternum with prominent paragnaths (Pls. 23:7; 24:3). Sternite of 2nd maxilla still not fused with the sternum but made of two rounded plates, as in the subsequent segments. Groove with two slits on maxillary sternite, and two pores on sternitic surface (Pl. 24:3). 'Head' length slightly larger than in TS4B (190–220 µm), thus difference between the two series reduced to less than 15%.

First antenna not known in detail. Second antenna complete in UB W77 (Pl. 23:3). Shape similar to that of TS4B and TS5A, about 170 µm along endopod and 200 µm long. Endopod 100–110 µm long, setation unchanged (Fig. 16D). Exopod slightly longer than in TS5iB (150 µm), but with one less annulus and seta.

Size of mandibular coxa unchanged (Pl. 23:6). Gnathobases approaching the labral sides with the whole distal surface when anteriorly turned (Pl. 23:7). Distal surface of grinding plate slightly concave (gnw 35–50 µm). Gnathobasic seta present, reaching almost to cutting edge (approx. 20 µm; Pl. 23:6; tooth-like spinules of cutting edge in Pl. 23:6, 7). Basipod smaller than in earlier stages relative to coxal body (Pl. 23:3, 6). Endopod 65–70 µm, its podomeres decreasing rapidly in size distally; exopod 75 µm, with 7–8 setae; Fig. 17E).

Both maxillae similar to those of series A but considerably smaller: length of 1st maxilla more than 135 µm, 2nd maxilla more than 105 µm long. Outer edges of both limbs subdivided (Pls. 23:3, 4; 24:2; Figs. 18E, 19D). Rows of pectinate enditic setae well-differentiated. Proximal endites of both limbs bulging and about as large as in TS5A (40 µm in mx1, 30–35 µm in mx2). Subsequent endites progressively more distally oriented. Exopod of 1st maxilla seems to be thinner and shorter than that of the subsequent limbs (Pls. 23:3, 4; 24:2); endopod unknown.

Number of trunk limbs as in series A, but also these limbs are shorter. First thoracopod longer than 100 µm (pe 20 µm; ds4–5; Fig. 20C); 2nd limb with bifid endites, more than nine divisions along inner edge (ds4a); 3rd one rudimentary, about 40 µm long (ds3a); 4th only a small, probably uniform lobe (ds2, but ds3a in series A). Furcal rami changed in proportions, now being slightly longer than wide. Hence the rami approach those of TS5A in shape but are still 20% smaller. Armature made of 9–11 primary furcal spines and two secondary ones. Ventrocaudal processes not changed (Pl. 23:3).

TS6iB and TS6B. – Unknown.

TS7iB (Pl. 24:4, 5). – Material: One distorted specimen with dorsoventrally depressed trunk, providing only limited number of details (Table 2, 3). Instar with incipient 7th thoracomere (Pl. 24:4). Shield similar to TS5B (24:4), hump with pores anterior to the point of flexure of the excavated posterior margin as in TS5A (not figured). Maxillary segment still not coalesced with the head dorsally. Probably five pairs of thoracopods present, 2nd one now also subdivided on outer surface. Ventrocaudal processes not further advanced, but their spine is accompanied by a thinner spinule on inner edge and a pit ventrally (Pl. 24:5; compare with the incipient furcal rami of L2A).

TS7B and TS8B. – Unknown.

Specimens from stages TS8iB to TS9iB (Pls. 24:6–8; 25; 26:1–4). Material: Three incomplete specimens (Table 2). Measurements: tl ranging from 600 to 800 µm, csl larger than 400 µm (Table 3). All three have an incipient thoracomere, and most data indicate a position between >TS7iB and <TS10iB. However, the thoracic region is either distorted or disguised, which prevents a clear recognition of segmentation and assignment (Pls. 24:8; 26:3). UB W645 may be

of stage TS8iB, status of UB W81 is unclear, and UB W82 may be of TS9iB (lost prior to detailed examination). Hence, the description covers a range of three stages.

Anterior head region distorted in all three specimens, including eye region, labrum (width 70 µm), and both antennae (Pl. 26:4). Shield large but preserved only in part (Pl. 24:6, 8). Sternal region not known. Mandibular coxa prominent, 105–110 µm long, 50–60 µm wide and 65 µm high. Gnathobase widening medially (w 55 µm; Pl. 25:1). Teeth of cutting edge bifid or with secondary spinules (also Fig. 17F). Basipod as in preceding stages, but rami seemingly thinner and smaller, 'palp foramen' 30×50 µm wide.

Lengths of maxillae unknown, 2nd one possibly about 200–220 µm. Proximal endites of both maxillae 45–50 µm. Endites carrying many, mostly pectinate or setulate setae, indicative of their advanced state (about ds6; Pl. 25:3, 4; Fig. 19E for 2nd maxilla). Five developed thoracopods are preserved in UB W81, the last with short paddle-shaped exopod carrying five marginal setae (about ds4; Pl. 24:7, 8). Sixth limb a bilobate bud (ds3), 7th one present as a small uniform lobe (ds2). Postmandibular limbs seem to have elongated in long axis compared to earlier stages.

Exopods of anterior thoracopods slender and paddle-shaped (Pls. 24:8; 25:2) with very robust marginal setae, furnished with opposing rows of setules (Pl. 25:5). Setal sockets with coronary row of acute denticles (Pl. 25:6). Width of exopods decreasing from 40 µm in the 1st to 20 µm in the 5th limb, length 70 µm in 3rd limb, 7–8 marginal setae in 3rd limb, 4th with 5–6 setae. Exopodal surfaces curved posteriorly, possibly also during life.

In UB W82 the posterior thoracomeres are slightly pleura-like produced laterally, but this may be caused simply by distortion of the ventral surface (Pl. 26:4). Abdomen ranging from 95 to 105 µm, with median trench between the ventrocaudal processes, as in series A. Furcal rami known only from the smallest of the specimens, UB 645, where they articulate in well-developed joints. Rami considerably larger than in preceding stages (l/w = 120/75 µm), now similar to series A in size as well as in the number of primary and secondary spines (12 and 5).

Ventrocaudal processes also enlarged and similar to those of series A (l 25–35 µm). Process of UB 645 with five (3rd thickest) spines, that of UB W82 with three spines on right process and four on left, and UB W81 with six (3rd largest) on right and five on left process. Variability might also indicate their belonging to different instars. Number of pits ventrally to spines increased to three, as in series A (Pl. 26:1, 2). Spines and adjacent surface of processes denticulate (Pl. 25:7, 8). Anal field with feebly-developed supra-anal flap (Pl. 26:4).

TS9B and TS10iB. – Unknown.

TS10B (Pls. 26:5–7; 32:3). – Material: One fragmentary specimen missing head and trunc end, including furca and ventrocaudal processes (Table 2). Measurements: trl

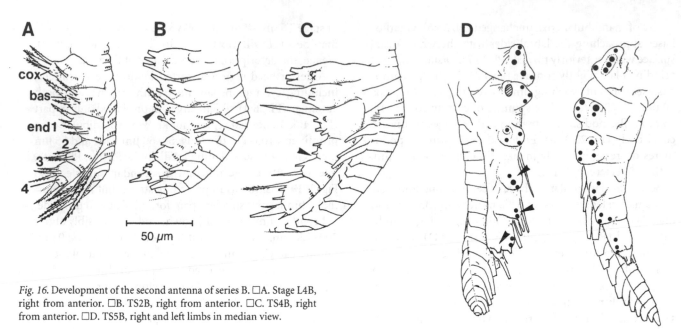

Fig. 16. Development of the second antenna of series B. □A. Stage L4B, right from anterior. □B. TS2B, right from anterior. □C. TS4B, right from anterior. □D. TS5B, right and left limbs in median view.

560 µm (Table 3). Nine of the 10 thoracomeres with appendages. Details of head not known, but from the preserved trunk details it is obvious that the development of the postnaupliar feeding apparatus has further advanced, though all appendages are considerably smaller than in TS10A (probably not more than ⅔ of the lengths of series A). Thoracic food groove deeply recessed. Each sternite composed of two rounded plates, progressively decreasing in size rearwards and set off from one another by membranes.

Proximal endite of 1st maxilla bulging and ball-shaped, almost 60 µm wide, as in series A. Row of pectinate setae forming almost a circle (Pl. 26:5, 6; Fig. 18F). Secondary row of spines slightly distal to medial setae (introduced earlier). Second endite with 4–5 spines medially, all endites apparently more developed as compared to earlier stages (length approximately 240 µm; ds6).

Proximal maxillary endite similar in size, but more like the distal endites in shape and armature, and with less pronounced median surface than in the 1st maxilla. Anterior setae further anterolaterally positioned, similar to those of the thoracopodal endites. Subsequent three endites with distinct median protuberance, originally with one or a few spines. Anterior group of setae consisting of two rows, as in series A. Fifth endite less developed; more distal parts of corm and rami not known in detail. Limb at an advanced state of development (ds6; length 280 µm; Fig. 19F).

Enditic surfaces of thoracopods similar to those of 2nd maxilla, progressively decreasing in size and armature (e.g., fewer pectinate setae in the posterior row (anterior ones of ds6; Fig. 20D for 3rd thoracopod). Lengths about 280 µm for 1st limb, 250 µm for 3rd one; 4th limb exceed-

ing 165 µm, 5th still larger than 120 µm. Sixth limb less developed than in TS10A and with much shorter and bifid endites (ds4). Next two limbs known only from their insertions. Ninth limb rudimentary, inner ramus slightly thicker than the outer one (35–40 µm; ds3; Pl. 26:5). Tenth thoracomere pliable ventrally, lacking limbs. Outer surfaces of anterior thoracopods better sclerotized and split into three portions (Pl. 26:7). Abdomen broken off posteriorly, presumably slightly longer than 110 µm originally. Ventral trench reaching anteriorly to last thoracomere. Furcal rami and ventrocaudal processes unknown.

TS11iB. – Unknown.

TS11B (Pl. 27). – Material: Three rather distorted specimens (Table 2). Measurements: tl probably 1.4–1.45 mm; csl probably 800 µm; trl 640–650 µm (Table 3). Instar with 11 thoracomeres (Pl. 27:1, 3, 4). Shield roof-like, most likely reaching back to 6th thoracomere (Pl. 27:3). Details of cephalon poorly known. Trunk further elongated, now carrying ten appendages. Anterior six well-developed (Pl. 27:5), 7th and 8th limbs progressively decreasing in size and shape. Ninth limb small but still with endites on median edge (ds4b); 10th rudimentary (?ds3; Pl. 27:1).

Developed thoracopods (ds6) with at least eight endites on their corms. Rami known only from their most proximal parts. First limb longer than 300 µm, 250 µm to upper edge of insertion of exopod. Corm 35 µm thick and 130 µm in abaxial extension; decrease in this dimension to 70 µm in the 6th limb. Fifth limb still longer than 180 µm, and 6th longer than 160 µm. Tenth and 11th thoracomeres almost ring-shaped save for their most ventral part. Development of the serial filter apparatus has obviously progressed, as can be seen in the advanced setation and enlargement of the

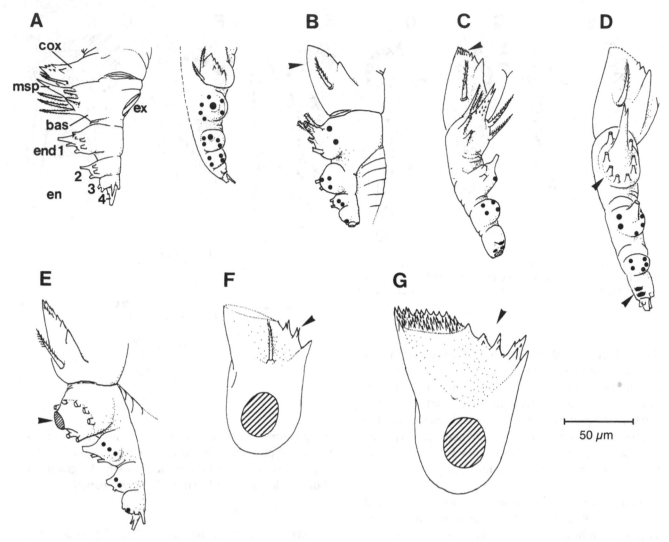

Fig. 17. Development of the mandible of series B. □A. L4B, left from posterior and median view. □B. TS2B, left from posterior. □C. TS3B, left in median view. □D. TS4B, left in median view. □E. TS5B, left from posterior. □F. TS9iB. □G. TS12B (distal parts of last two stages unknown save for the 'palp foramen').

limbs. Degree of development now comparable to that of stage TS10A.

Abdomen cylindrical but slightly depressed dorsoventrally. Length similar to series A, varying from 135 to 150 µm, thickness 100 µm. Ventrocaudal processes much more elongate than in preceding stages and oar-shaped, 60–80 µm long, and with nine marginal spines (Pl. 27:2). Due to strong inward folding of the ventral cuticle of the abdomen along the trench in UB W83, the ventrocaudal processes are slightly inversely angled against one another, with the inner margins more dorsally oriented. Short supra-anal flap seemingly pointed, as in series A.

TS12iB. – Unknown.

TS12B (Pls. 28; 32;4). – Material: One slightly distorted specimen, the largest one with complete shield and designated as holotype (Müller 1983) (Table 2). Measurements:

tl presumably slightly longer than 1400 µm, trl 780–800 µm (Table 3). Twelve thoracomeres. Due to collapse, the body lies deeply recessed within the shield in UB 644; other details also are deformed by shrinkage effects.

Shield enlarged to almost 1000 µm in length and 300 µm in height, reaching back to about the 8th thoracomere, but sides do not extend beyond the limb corms (Pl. 28:1, 2). Shape roof-like in anterior aspect (Pl. 28:3), greatest height and width at about between $^1/_5$ and $^1/_4$ of the shield length above the mandibles, from there gently narrowing and decreasing in height rearward. Dorsal line almost straight. Anterior margin bluntly rounded in lateral view, but almost straight medially in dorsal aspect and with short, but distinct indentation at midlevel (Pl. 28:4). Lateral margins gently convex, posterior corners slightly drawn out, and posterior margin excavated. Due to deformation, the lateral margins are slightly rolled inward in UB 644.

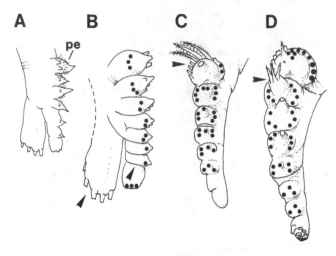

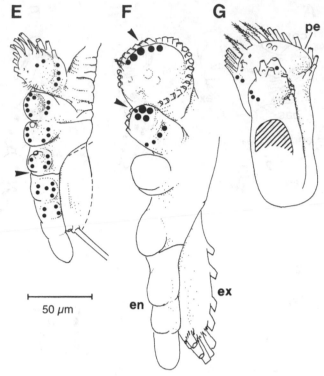

Fig. 18. Development of the first maxilla of series B. □A. L4B (ds4a). □B. TS2B. □C. TS3B (about ds5). □D. TS4B. □E. TS5B. □F. TS8i–9iB (ds6), distal part obtained from different specimens. □G. TS12B (ds6).

Details in front of labrum poorly known (only fragments of eye and the two antennae are preserved in UB 644). Labrum arising from broad basis, about 95–100 μm in width. Anterior part slightly humped similar to labrum of TS10A. Distal part behind constriction broken off in the specimen at hand. Posterolateral margins deeply excavated proximally (Pl. 28:3–5). Sternum not known in detail, since it is hidden by appendages and alien particles in UB 644.

Mandibles with huge, laterally rounded coxal body (l 135, w 65 μm; Pl. 28:9; Fig. 17G). Anterior and posterior margins of grinding plate somewhat ridge-like (Pl. 28:5, 6). Cutting edge distinctly divided in two, as in TS13A. Gnathobasic seta also not positively identified. It is not clear whether there are differences between the margins of right and left mandibles. Insertion area of basipod smaller than in ?TS9i (30×35 μm), indicative of the diminution of the 'palp'.

Only the proximal endites are known of the two maxillae, pointing to further enlargement of these and differentiation of their setation (Pl. 28:7; Fig. 18G for 1st maxilla). Setae at inner margin of the proximal maxillulary endite with numerous setules, seemingly more brush-like than pectinate (mechanical food transport activity?; Pl. 28:8). Deformation of the limbs in UB 644 indicates their phyllopodous nature. Thoracopods known from their proximal parts only, all being similar as in TS10B but with more setae and more triangular enditic surfaces (Pl. 28:8; Fig. 20E for 3rd thoracopod). Exact number of limbs unclear, but at least the anterior 6–7 are at stage ds6.

Posterior two trunk segments almost ring-shaped, 9th and 10th segments with faintly developed pleural structures laterally, slightly overhanging the subsequent segments. This may, however, at least in part be caused by the

sharp ventral flexure of the trunk. Cylindrical abdomen broken off posteriorly in UB 644. Furcal rami and ventrocaudal processes not known, accordingly. Fragment of abdomen longer than 100 μm, 95–100 in diameter.

TS13iB. – Unknown.

TS13B (Pl. 29; reconstruction in Fig. 15D). – Material: Three differently preserved specimens: UB W86 with distorted head and trunk pushed anteriorly, but still with many thoracopods, ST 4647 with distorted head and complete trunk lacking appendages, and UB W87 with distorted head but entire trunk save for the furca and with some of the posterior thoracopods well-preserved (Table 2). Measurements: tl probably 1.45–1.5 mm; trl 860–880 μm (Table 3). Head not known in detail. Size of shield unknown, crushed in UB W87 and almost rubbed off, while the posterior part is still present in UB W86 but also deformed and partly broken off. Length possibly >800 μm, height about 300 μm (Pl. 29:1). Sternum poorly known, obviously deeply incised, as in earlier stages (same figure).

Size of thoracopods much smaller than in TS13A, but at least the anterior three also exceeding 300 μm by far. Gross shape similar to series A, but comparatively more compressed (Pl. 29:1, 3–5; Fig. 20F for 3rd thoracopod). At least eight limbs are well-developed, indicating that the development of the filter apparatus has progressed again. This is recognizable in particular in the shape of the thoracopodal endites and their armature: anterior set now with two distinct rows made of 2–3 and four setae (Pl. 29:2, 3; less

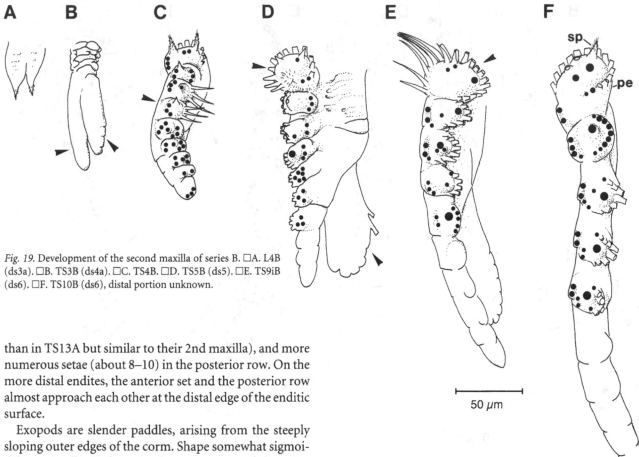

Fig. 19. Development of the second maxilla of series B. □A. L4B (ds3a). □B. TS3B (ds4a). □C. TS4B. □D. TS5B (ds5). □E. TS9iB (ds6). □F. TS10B (ds6), distal portion unknown.

than in TS13A but similar to their 2nd maxilla), and more numerous setae (about 8–10) in the posterior row. On the more distal endites, the anterior set and the posterior row almost approach each other at the distal edge of the enditic surface.

Exopods are slender paddles, arising from the steeply sloping outer edges of the corm. Shape somewhat sigmoidal proximally preforming the orientation of the rami. Joint almost effaced anteriorly (outer side), but well-developed posteriorly (inner side; Pl. 29:4, 5). Marginal setation starting slightly distal to the joint and reaching around the tip of the rami. Surface of rami slightly curved inward (see also Pl. 25:5 for UB W81 of possibly TS9iB). Shape much as in TS8A, except that in the posterior limbs of TS13B, the exopods appear slightly longer relative to the whole limbs. Setae projecting from paddles, each with opposing rows of setules originally (Pl. 29:5; see also Figs. 21, 39).

Width of corm at least 135 µm in the anterior limbs. Of the 4th and 5th limbs only the corms are known in UB W87, being 150 and 130 µm long. Sixth limb 295 µm long, its exopod 175 µm along outer edge and 110 along inner edge; 7th limb 270 µm long, with exopod of 165 µm. 8th limb 210 µm long, its exopod 140 µm along outer edge and 100 µm along inner edge. Ninth limb still 140–150 µm long, and its exopod 100 µm (UB W87; about ds5; see also Fig. 21).

Tenth limb smaller, with nine divisions medially, similar to early larval 1st maxillae (e.g., bifid endites); endopod indistinctly four-segmented. Exopod 80 µm along outer edge and 55 µm along inner edge, with about 11 setae (UB W86; about ds4b; Pl. 29:6; Fig. 20G). Eleventh limb about 70 µm long, little sclerotized, its feebly developed endites carrying most likely only one spine each (ds3b-4a; Fig. 20H). Twelfth limb seemingly uniramous, about 30 µm

long (ds2; Fig. 20I). Last trunk segment lacking limbs, almost annular save for the membranous ventral part (Pl. 29:3). Abdomen similar to TS13A in length (>135 µm) but slightly thinner, median trench as in preceding stages; furca and ventrocaudal processes not known.

Unassigned specimens and further instars

A number of fragmentary specimens could not be placed into a particular stage, though assignable to *Rehbachiella* (Table 2, 3). Nevertheless, some of these individuals provide interesting details, illustrated in Pls. 33 and 34:

UB W88–W91 are four aggregated specimens supposedly of about TS4 of either series (csl about 280–290 µm; Pl. 33:1, 2). Though much wrinkled and not clearly assignable, details are exceptional, for example the distal ends of two antenna and mandible (Pl. 34:1; 2nd endopodal podomere subdivided: en2a, b), denticles or setules on the furcal rami (Pl. 34:2), the extruded hindgut (Pl. 34:3), or the probable sensory organs at the posterior surface of the labrum (Pl. 34:4).

UB W93 is badly distorted, but assignment to an early stage is indicated by the small mandibular coxa and rudimentary 1st maxilla (ds4b). On the

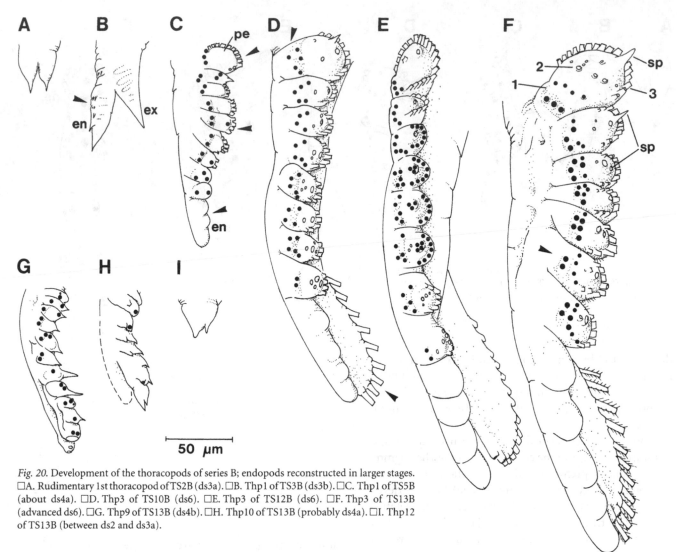

Fig. 20. Development of the thoracopods of series B; endopods reconstructed in larger stages. □A. Rudimentary 1st thoracopod of TS2B (ds3a). □B. Thp1 of TS3B (ds3b). □C. Thp1 of TS5B (about ds4a). □D. Thp3 of TS10B (ds6). □E. Thp3 of TS12B (ds6). □F. Thp3 of TS13B (advanced ds6). □G. Thp9 of TS13B (ds4b). □H. Thp10 of TS13B (probably ds4a). □I. Thp12 of TS13B (between ds2 and ds3a).

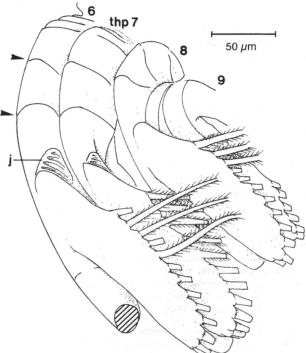

distal part of its left 1st antenna, the setation of the tubular podomeres is partly preserved (Pl. 34:3), a feature rarely present in the material and adopted in the reconstructions.

UB W94 is a fragment of the anterior body region, most likely larger than stage TS3. Appendages are partly preserved, showing the different degree of rigidity of the limb bases from 2nd antenna to 1st thoracopod (Pl. 34:6). The 1st maxilla is slightly more firmly sclerotized indicating the initiated process of stiffening of the outer edges of the limb bases.

UB W95 is badly distorted, yet it exhibits a peculiarity: the labrum is preserved as a mass of phosphate representing an internal filling; thereby the ceiling of the atrium oris (underneath the labrum) can be traced anteriorly up to the entrance of the esophagus (Pl. 34:7). Due to collapse, the sternal region is sunken in and flat. This allows observation of the grooves between 1st and 2nd maxillae, the pits on the sternites of the maxillary, and of the 1st thoracomere (Pl. 34:8).

Fig. 21. Sixth to ninth left thoracopods of TS13B (redrawn from UB 87) as viewed from outer side; arrow points to joint at posterior side of exopod basis; endopods reconstructed.

A number of fragments, most likely belonging to *Rehbachiella*, are from specimens that were larger than the largest definitely assignable ones (marked with an asterix in Table 2). If truly belonging to this species, they give evidence for a continuation of growth beyond stage TS13. UB W92, for example (Pl. 33:3, 4), is a fragment about 570 μm long, comprising two appendages with remains of the rami. The enditic armature shows much resemblance in arrangement but is apparently further advanced (Fig. 14). Assuming a similar shape to that of a trunk limb of known *Rehbachiella* specimens, the whole limb would have exceeded 1 mm. Extrapolating a similar increase in length of the whole body, the animal could easily have exceeded 4–5 mm in total length.

Life history

Ontogenetic stages

Development of *Rehbachiella* is strictly anamorphic, and increase in size between the stages is slow (life cycle in Fig. 5). During the 'naupliar phase' (L1–L4), the maxillae appear on the larval hind body. The segment of the 1st maxilla becomes fused dorsally with the 'head' after, presumably, about two more moults (some individual variability is possible). Ventrally this fusion occurs later, supposedly shortly after TS2, recognizable by the incorporation of the maxillulary sternite into the sternum. The maxillary segment remains free from the 'head' for a longer period. Its fusion occurs at different stages on dorsal and ventral sides, as in the 1st maxilla, but in a reverse manner: ventrally around the TS5 stage (slightly later in series B), dorsally between stages TS7 and TS10.

With the 5th instar, TS1i, the 1st thoracomere appears by partial separation from the trunk. Accordingly, the characteristic development of thoracomeres within two steps may be also roughly applicable to the two maxillae (L1–L4). The 'thoracic phase', from TS1i to TS13, embodies 26 stages, which makes a total of 30 instars (29 moults). The anlagen of thoracopods appear generally 1–2 stages later than the segments. This points to a different ontogenetic process for the development of limbs and to the existence of two independent mechanisms. All postmandibular limbs, including the two maxillae, develop in regular anterior-posterior order. It generally takes eight stages (six in the 1st maxilla) until a limb becomes functional (ds5), but many more stages to reach a mature shape (advanced ds6).

Two larval series

The incompleteness of the two series is regarded as preservational: of 30 possible stages up to TS13, 21 are found for series A (nine missed) and 13–14 for series B (at most 16 missed, because specimens from TS8i to TS9i may belong to different instars). Alternatively, the missing stages in the two series may be regarded as 'developmental jumps', as found in extant crustaceans, probably even at different stages. This cannot be excluded beyond all doubt, but there are strong arguments in favour of the working hypothesis:

- two-step formation of postmandibular segments,

- very gradual morphogenesis of limbs and other structures,

- occurrence of successive sets of larvae in both series which even supplement each other (of the first 20 possible stages only two are missing in series A, of series B all stages save for one are known between TS2 and TS5),

- taking both sets together, only four stages are missing, all from the later phase of development (TS9, 10i, 11i, 12i),

- the growth curves (see below) are continuous; they would be quite uneven when developmental 'steps' had occurred, and

- the details of the growth data, in particular the lag-phase, which is in line with changes mainly seen in the head.

In general, the later stages are less well represented or even missing. Of the last ten stages of series A only three are known. In series B, on the other hand, 50% of the larger instars are represented (about every 2nd instar). Since a TS13i stage is found for series A, it is concluded that the two-step development is typical for the whole series.

Several observations on this fossil material cannot be explained in full, such as the lack of early stages of series B. On the other hand, also from series A only a single specimen of the 1st instar has been found. The growth curves would leave at least the possibility open that specimens of L2 gave rise to either series, but the enormous increase in size and progress in development hereafter leaves little doubt that the specimens of L3 belong exclusively to the larger series A.

The size differences between series A and B range from about 10% to 25% depending on the larval stages. Intermediate specimens have not been observed. Though some individual variability has been recognized (see below), most of the younger individuals could be grouped within the series with a sufficient degree of confidence save for very fragmented specimens. It was more difficult to group later specimens, since also measurable data of these are scarce. The major differences between the series concern the early growth phase, recognizable in:

- the development of the head,

- the dorsocaudal spine, which is retained in series A maximally until TS2i, but already lost in L4 of series B,

- the morphogenesis of the furca, and

- the non-correlation between exopodal podomeres and setae in the 2nd antenna and mandible of series A (not identifiable later).

Various differences are merely related to the time of occurrence of features. The first three at least become more or less balanced eventually, most clearly in the furca (Table 3; see below). Discrepancies, which last into later stages, are in the coalescence of the maxillae with the head, in the morphogenesis of postmandibular limbs, and in the degree of setation of the filter limbs. Series B in some ways seems to be delayed relative to series A, while in others it precedes the latter, particularly in the loss of the caudal spine and in the development of the furca. The considerable size differences in the early larval phase may be the reason for differences in the head development, balanced roughly at TS6. On the other hand, numerous details are shared between the two series, such as the:

- progressively changing shape of the labrum,

- separation of the trilobed anterior head region,

- subdivision of the 2nd podomere of the antennal endopod,

- two gnathobasic setae on the mandibular gnathobase of early larvae, of which the anterior one becomes reduced (but at different stages),

- number of setae around the masticatory spine of the mandibular basipod, and the cross-section of the spine,

- gross design of the four endites of the 1st maxilla and the double row of setae/spines of its proximal endite,

- sclerotization and lateral subdivision of the postmandibular limbs,

- gross setal pattern on the endites of postmaxillulary limbs,

- shape, pore pattern and setation of the furca in later stages, also with double row of furcal spines, and

- the slightly pointed supra-anal flap.

These shared details suggest that the two series belong to one species. Additional support comes from the co-occurrence of the two series (Table 1). Such intraspecific variability may have different reasons, such as sexual, seasonal or environmental. All three modes of variation are typical of Recent Branchiopoda, to which *Rehbachiella* is considered to be affiliated. Conchostraca, for example, show sexual differentiation in their appendages (e.g., claspers in males, cf. Botnariuc 1947; Battish 1981; Martin & Belk 1988). Sexual differences and seasonal changes in morphology are also described from Cladocera (e.g., Kaestner 1967, pp. 955–970; Siewing 1985, pp. 886–890). These may

also adjust their number and shape of filtratory setae in accordance with seasonal availability and size of food (e.g., Koza & Korinek 1985; Korinek *et al.* 1986; Fryer 1987b).

Notostraca are very variable in their number of limbless abdominal rings (Linder 1952; Longhurst 1955; Bushnell & Byron 1979), which may even grow as spirals and carry up to six pairs of legs. They also differ in the sexes. Anostraca may show strong sexual dimorphism in the head (e.g., 2nd antenna in Euanostraca, 1st maxilla in Lipostraca), in the trunk (brood pouch of females, furca), and in size. Growth is in general much affected in branchiopods by environmental factors (salinity, temperature) and seasons (e.g., Bushnell & Byron 1979), which may also modify the ontogenetic pattern (e.g., Hentschel 1967, 1968).

Besides morphological and physiological specializations (Potts & Durning 1980), this high plasticity of response to environmental changes recognizable in the Recent Branchiopoda may have been laid down very early in their evolution. It may have greatly facilitated their radiation and ability to survive even in the extreme environmental systems now inhabited by the various extant members of this group. The variability of *Rehbachiella*, proposed here as a marine ancestral branchiopod, might thus be understood as a step toward such strategies, but the reasons for the variability in the fossil remain uncertain.

For both series of *Rehbachiella*, growth cannot be followed into those instars, where Recent Euanostraca show the various well-known modifications, in particular of the larval head, and the sexual differentiation. Also, in the other Branchiopoda the sexual characteristics appear very late during ontogeny (according to the segmentation pattern, not to the moulting sequence). Since features referable to reproduction could not be recognized, it remains unclear whether or not the two series indicate sexual differences.

Growth

In addition to morphological parameters, measurements of various body portions were used for grouping specimens into particular stages (Table 3). Data of body length (tl, including telson and furca), cephalic shield (csl), 'head' length (hl = distance between insertions of 1st antennae and 2nd maxillae) and thorax (thl) are given as hand-fitted curves in Figs. 22–23.

Body length of series A (Figs. 22A, 23A, line 1) shows a sigmoidal growth curve. This effect is less obvious in series B (line 2) and recognizable only in relative growth (Fig. 23A). The same trend is present in the shield length (Figs. 22C, 23D) and 'head' length (Figs. 22D, 23C). Its absence in thorax length (Figs. 22B, 23B) points to an exclusive feature of head development. Lag of growth in series A approximately between stages TS2i and TS6 is closely associated with various morphogenetic changes in the head region, in particular in the oral area and the append-

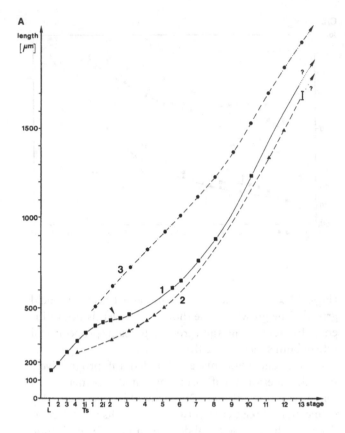

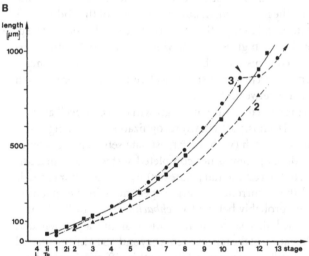

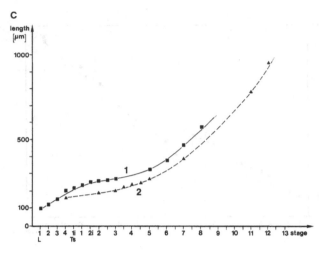

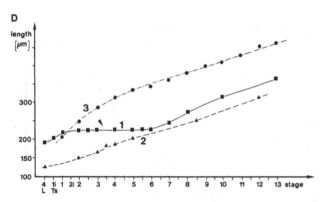

Fig. 22. Absolute growth of total length (A), thorax (B), cephalic shield (C), and length of head (D; for *Rehbachiella* measured as distance between 1st antenna and 2nd maxilla); 1, larval series A of *Rehbachiella*; 2, series B; 3, *Artemia salina* (data from Weisz 1946); short arrows point to characteristic events during growth.

ages. By the end of this 'lag phase' – roughly at TS5–6 – the labrum has approached its 'final' shape (with regard to the sequence known). In the sternal region, the paragnaths develop as paired humps on the mandibular sternite, the sternites of both maxillae are progressively added to the sternum, and a deep V-shaped median food path extends

from the atrium oris ('paragnath channel') backwards to the anterior of trunk by median invagination of the sternites.

The same reorganization takes place in series B but is more continuous. This can also be seen in the differences in growth of the appendages, at least as far as could be obtained from the limited data available (in particular of 2nd antenna and mandible; Fig. 24). In series A, rapid growth of the 2nd antenna slows down after L3A and seemingly continues at low rates of increase, while the mandible grows more continuously until TS2A and decreases in size afterwards. This is, however, mainly due to progressive reduction of the palp. The progressive growth of the coxal body compensates this reduction slightly, and it may be possible that after degeneration of the palp the growth curve of the mandible increases again to some extent.

Growth of the 1st maxilla is rapid until about TS2A and slower beyond this stage. From the preserved proximal parts of this limb it is supposed that growth continues but to a lesser degree than in the 2nd maxilla. After a more rapid increase in the first stages, growth of the 2nd maxilla seems to slow down during the lag phase. Hereafter growth is continued, and size at TS13 is assumed to have exceeded

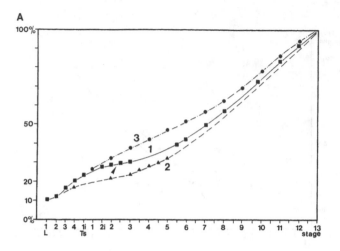

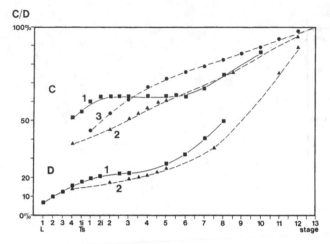

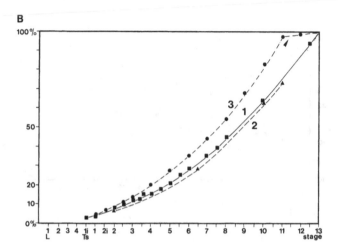

Fig. 23. Relative growth (cumulative percent values) of total length (A), thorax (B), head (C) and shield (D; only for *Rehbachiella*); 1, series A of *Rehbachiella*; 2, series B; 3, for *Artemia salina* (data from Weisz 1946; arrows as in Fig. 22).

400 µm. The limited data of thoracopods suggest a similar growth.

These fluctuations in growth are not apparent in series B, where the size increase of all four appendages is slower than in series A but continuous. A striking difference in size is recognizable at TS2, when the 1st maxilla is about 100 µm in series B but almost twice as large in the other series. Later stages of growth are largely unknown. Extrapolation suggests a size of about 350 µm for the 2nd maxilla at TS13B.

A possible explanation for this difference in growth strategy may be the smaller size of series B at its start and consequent slightly different food preferences: series A has an enormous growth increase in the earliest stages, while series B grows continuously also during the 'lag-phase' between TS2 and TS6, the phase of the reorganization of the feeding apparatus (slight curvature in total length only, no lag in 'head' length). Only in the development of the shield is the cessation in growth similar to that of series A

(Figs. 22C, 23D), indicating some effect also on shield growth. The growth of the thorax, however, is not influenced by these changes and grows at the same proportional rate in both series (Fig. 23B).

Postcephalic segments are budded off progressively from the anterior end of the unsegmented abdomen which remains as such until the 13th segment is developed. This delineation of segments in two steps results in 'staircase-like' growth increase for abdomen and thorax: each time when the new segment appears, the size of the abdomen is reduced and the length of thorax increases, by the next step, the abdomen grows again, while the thorax length is unchanged. This effect becomes indistinct when the increment differences are small relative to the general length increase between the stages.

This special feature of the growth curves as well as their general trends, particularly recognizable in total length and thoracic length (with the largest data sets), gives evidence that development is not completed at stage TS13, in addition to the rudimentary state of the posterior thoracopods and the occurrence of larger specimens in the material, which probably belong to *Rehbachiella*. Accordingly, it is assumed that segment formation of the abdomen and delineation of the telson occurs beyond TS13 in a postlarval phase still undiscovered. Since in this differentiation phase important changes occur in Recent Euanostraca (modification of head – e.g., eye lobes become pedunculate eventually, naupliar appendages atrophy or change considerably – segmentation of abdomen, completion and maturation of thoracopods, development and articulation of furca, sexual differentiation), similar changes may be expected also for later stages of *Rehbachiella*.

Morphogenesis

Body. – The shape of the body develops progressively. The nauplius is pear-shaped, with the hind body slightly set off from the anterior portion (Pl. 1:1). Its prominent struc-

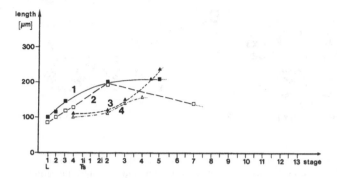

Fig. 24. Growth of second antenna and mandible; 1, 2nd antenna of series A; 2, 2nd antenna of series B; 3, mandible of series A, 4 mandible of series B; no data of larger stages available.

tures are the circular, slightly arched shield, the large labrum, and three pairs of appendages of about the same size (Fig. 6A). Subsequently, the body elongates progressively with sequential addition of body segments and limbs (see Figs. 6B–E, 7, 15).

Cephalic shield. – The shield, which covers only the naupliar head portion at first, elongates very gradually. Its extension also in anterior and lateral directions leads to a roof-like form. In early instars the shield is truncated anteriorly to give space for the protruding forehead (Pls. 3:3; 6:4, 7; 8:3, 7, 8; 20:6, 7). Eventually, this margin elongates somewhat and probably extends beyond the forehead (Pls. 13:1; 14:3; 28:1, 3, 4, 8) but is still almost truncate in dorsal aspect. A rearward tapering leads to an elongated drop shape of the shield (e.g., Pls. 13:1, 28:8), very similar to that of certain, but much larger Notostraca (Linder 1952, Pl. 3:1, 2, for *Lepidurus packardi*).

Incorporation of body segments terminates behind the maxillary segment, but already and much earlier the shield has continued its rearward elongation. This is recognizable first by a wing-like extension of the rounded posterolateral corners (Pls. 10:1, 3; 11:6; 23:4; 24:4). By TS12, the shield freely covers about 8–9 thoracomeres (Pl. 28:8). The excavation of the posterior margin is retained to give space for the trunk. A similar shape is very common in crustacean shields, recognizable for example in Notostraca (e.g., Claus 1873, Pl. 6:2c; Linder 1952, Pls. 1–5; Longhurst 1955, Fig. 13C) as well as in fossil and Recent members of the thecostracan lineage of Maxillopoda (cf. Müller & Walossek 1988b, Figs. 4, 10, Pl. 3:1). At no stage do the lateral margins extend much beyond the limb corms.

The early larval neck organ forms the apex of the naupliar shield (Pls. 1:3, 6; 2:1, 4). During growth, it shifts successively anteriorly relative to the shield length and in accordance with the shifting apex, supposedly due to its correlation with internal structures of the anterior head region (compare Pls. 3:3; 5:1; 8:8; Fig. 6A–E with, e.g., Fig. 45D–F for notostracan larvae). Eventually, after stage TS4,

this organ becomes invisible externally and does not reappear again.

Head. – The beginning of the postnaupliar phase is largely characterized by transformation of various ventral structures, causing a lag of head growth in series A. Principal changes, also illustrated in Fig. 25, affect the labrum, the sternum, the position of naupliar appendages relative to the posterior edge of the labrum (entrance of atrium oris), the proximal parts of mandible and maxillae (marked by arrows), and the initiated reduction of the naupliar appendages in the latest stages.

The ventrally projecting naupliar labrum is large, conical and with a rounded tip (Fig. 25A). Both the 2nd antenna and the mandible are postoral. By this stage the 2nd antenna seems to be slightly dominant. In the mandible, the major portion of the corm is the basipod with its developed armature, while the coxa carries only two short spine-like setae (Pl. 1:4). Remarkably, the antennal segment has a distinct sternite (Pl. 1:4; Fig. 25A), supposedly in accordance with the feeding function of the appendage.

After a few stages (about L3; Fig. 25B) the 2nd antenna has shifted anteriorly, with its long endites pointing posteromedially around the corners of the labrum (Pls. 3:2, 7, 8; 4:1, 2). The mandible has also shifted anteriorly, and its coxal endite has become enlarged and flattened. Two gnathobasic setae support the prominent basipod (Pls. 3:2, 7–9; 4:1, 2, 7; Fig. 9C). On the sternum, the antennal portion is no longer recognizable. It is possibly coalesced with the anterior part of the sternum which slopes steeply into the 'atrium oris'. At its rear (now mainly the mandibular sternite) short furrows represent the first signs of the developing paragnaths (Pl. 4:2, 7).

Up to about TS5 (Fig. 25C) the anterior surface of the labrum processes a distinctive bend on its anterior surface, separating the raised posterior part from the shallower anterior one (compare Pls. 8:4; 9:6; 10:1, 8). The posterolateral sides, adorned with setules from the earliest stages on, are slightly deepened, while the posterior edge is slightly ridge-like enhanced medially (Pl. 10:3). This edge bears characteristic papilliform tubercles, often associated with tiny setules and pores (Pls. 5:7; 9:7; 23:7; 34:4). Similar structures are also known from other *orsten* forms, such as the phosphatocopines (Müller 1979, Figs. 21B, 35; Müller & Walossek 1985a, Fig. 2f). Possibly the pores relate to openings of labral glands, while the tubercles were some kind of chemoreceptors.

Both the 2nd antenna and mandible have shifted farther anteriorly. Proximally, the posterolateral edges of the labrum are excavated to provide space for the enlarged and sharply angled mandibular gnathobases. The maxillulary proximal endite moved around the raised paragnaths, transporting food particles towards the gnathobases with its anteriorly curved setae (compare Pls. 9:1, 6; 10:1–3, 5, 8; 11:4; 21:5, 6; 23:5, 6). The maxillulary sternite is now fused

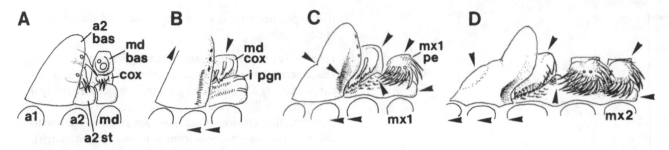

Fig. 25. Morphogenetic changes in anterior body region; short arrows point to major events as in preceding figures (not to scale). □A. Status at naupliar stage (L1) with postoral 2nd antenna and sternite separate from its segment. □B. Stage L3, with anteriorly shifted 2nd antenna and mandible, elongation of labrum, and disappearance of antennal sternite. □C. Development up to TS5, with changes in the shape of labrum, (e.g., lateral excavation for mandibular gnathobases), fusion of maxillulary sternite with sternum, and enlarged paragnaths. □D. Development up to latest instar (TS13), with further modified labrum, widely anteriorly shifted insertions of all naupliar appendages, deep excavations at labrum for mandibular gnathobases, fusion of maxillary sternite to form a single cephalic sternum, and highly elevated paragnaths.

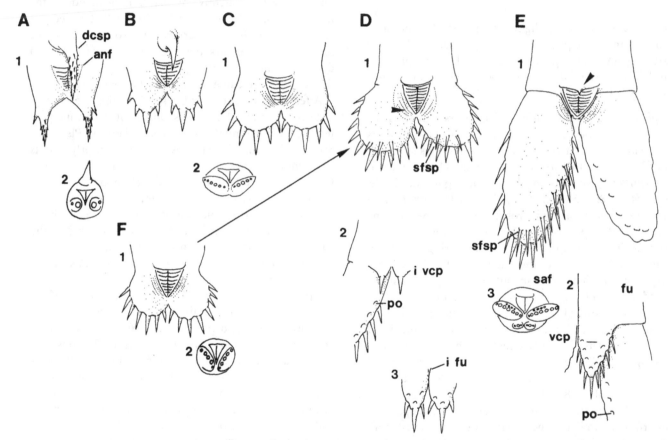

Fig. 26. Morphogenesis of caudal end (not to scale). □A. Shape at stage L2A, dorsal (1) and posterior (2) view. □B. At L4A from dorsal. □C. Between TS2A and TS3A, from dorsal (1) and posterior (3). □D. Between TS5 and TS6; 1, dorsal view; 2, ventral view; 3, ventrocaudal processes of TS8. □E. At TS10A, dorsal view (1; short arrow points to short supra-anal flap with tiny spinule), ventral view (2), and posterior view (3). □F. Caudal end of L4B, from dorsal (1) and posterior (2). Long arrow indicates transgression of series B into shape of series A. Beyond about TS4 the furcal rami and ventrocaudal processes of both series are almost identical.

to the sternum, which becomes deeply recessed medially and covered with many setules arranged in short crescentic rows (Pl. 11:4; 21:5; 23:1, 3; 24:3).

The anterior shifting of all naupliar appendages progressed up to the late stages (Fig. 25D). Eventually, the 1st antenna inserts almost in front of the labrum, while the 2nd antenna inserts at about its anterior edge. The mandibles

have also shifted, and their huge grinding plates reach into deep excavations at the posterior edges of the labrum (Pl. 17:2, 4). The gnathobasic seta is lost and the small size of the 'palp foramen' indicates considerable atrophy of the distal parts of the limb (Pls. 17:4; 28:8).

The labrum no longer projects straight from the ventral surface but merges gently with the body. Its anterior part is

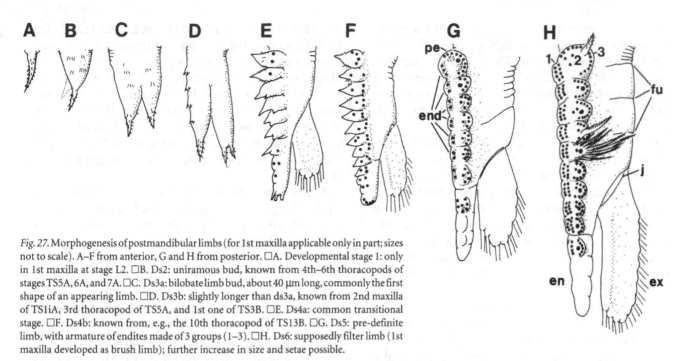

Fig. 27. Morphogenesis of postmandibular limbs (for 1st maxilla applicable only in part; sizes not to scale). A–F from anterior, G and H from posterior. □A. Developmental stage 1: only in 1st maxilla at stage L2. □B. Ds2: uniramous bud, known from 4th–6th thoracopods of stages TS5A, 6A, and 7A. □C. Ds3a: bilobate limb bud, about 40 μm long, commonly the first shape of an appearing limb. □D. Ds3b: slightly longer than ds3a, known from 2nd maxilla of TS1iA, 3rd thoracopod of TS5A, and 1st one of TS3B. □E. Ds4a: common transitional stage. □F. Ds4b: known from, e.g., the 10th thoracopod of TS13B. □G. Ds5: pre-definite limb, with armature of endites made of 3 groups (1–3). □H. Ds6: supposedly filter limb (1st maxilla developed as brush limb); further increase in size and setae possible.

humped, presumably anterior to the bend. A similar transverse bend is common to the labrum of various crustaceans, characterizing the insertions of its musculature (e.g., Hessler 1964, Fig. 2 for Cephalocarida, Fig. 29 for Notostraca; Boxshall 1985, Fig. 73 for Copepoda; see also Fig. 44C herein for the euanostracan *Branchipus stagnalis*). The sternum is now fused with all cephalic sternites, and the paragnaths have developed into prominent bulging lobes immediately behind the gnathobases (e.g., Pls. 13:4; 15:4; 17:4).

Eyes. – From the 2nd instar a set of three lobes projects from the head, which become larger and more bulging progressively (Pls. 2:3; 3:1; 5:4; 6;4, 5, 7; 8:3, 7, 8; 18:2, 3, 19:5; 20:6, 7; 21:1). From about stages TS2–3 they undergo no further increase in size, which leads leads to a reduction of this set of structures relative to the whole body. In parallel, from about TS2–3i the whole region becomes set off from the head (Pl. 8:7, 8; 20:7) and raised on a narrow, socket-like basis (Pl. 9:6), most clearly seen when the frontal part is torn off (Pl. 9:7). This structure is, however, not well-enough preserved in later stages for recognition of its further fate (Pls. 13:2; 14:3; 28:1, 2, 4, 8).

The pair of blisters separated by the 'midventral lobe' are identified as the compound eyes, not only being in the same position as those of euanostracan Branchiopoda, but also having a similar mode of development. In these crustaceans, the forehead including the incipient compound eyes also extends beyond the anterior margin of the neck organ, here having taken over the place of the shield almost completely (e.g., Claus 1873, Pls. 1:4', 5", 2:5, 7; 3:8; Fig. 44B, C herein; see also Fig. 53B). The anlagen of the

compound eyes are already present at hatching, while ommatidia are not formed before the 4th moult (Weisz 1947, p. 52). The constriction appears in advanced euanostracan larvae (e.g., Claus 1873, Pl. 4:11, 13; Jurasz *et al.* 1983, Fig. 5b–d), while development of the peduncles does not begin before the postlarval phase. Since in Euanostraca the internal naupliar eye is located medially between the lobes of the compound eye (above figures), it is not unlikely that the bulging midventral lobe of *Rehbachiella* had encased the internal naupliar eye (Pls. 18:3; 19:5; 20:6, 8; 21:1).

The nature of the pit at the anterior end of this structure remains unclear (Pls. 2:9, 10; 3:6). A similar pore is known from Notostraca and Conchostraca (Eberhard 1981, Figs. 9, 13, 79; Martin & Belk 1988, Fig. 2d, e; pp. 478, 479). In these it relates to the eye chamber enclosing the compound eye. Rhizocephalan cirriped larvae (personal observations) and ascothoracid larvae, on the other hand, possess a similar pit or node, which seems to demarcate the position of the internal naupliar eye (e.g., Grygier 1985, Figs. 3, 5, and personal communication, 1988). If the median lobe and/or the pore could be correlated with the naupliar eye, the recognition of either of the structures would indeed help to recognize the presence and approximate position of the internal naupliar eye not only in *Rehbachiella* but also in other forms, such as *Bredocaris* which has a similar lobe and pit (Müller & Walossek 1988b, Pls. 8:2; 10:4, 5; 14:8) or the Skaracarida which have a pit below the frontal 'rostrum'; Müller & Walossek 1985b, Pl. 4:5, 6).

Trunk. – All thoracic segments are poorly sclerotized and separated by pliable arthrodial membranes (Pls. 14:1; 17:1;

21:8; 29:4). In some specimens, the posterior margins of
the posterior segment may be slightly raised and overlapping the subsequent segment (Pls. 14:3; 15:1; 23:2; 24:8;
29:4), but, as in Euanostraca (e.g., Brendonck 1989, Fig. 1
for *Streptocephalus proboscideus*), they never form clear
pleural extensions, such as, e.g., in the Cephalocarida (e.g.,
Sanders 1963b, Figs. 14, 18–24).

Development of the furca and associated structures, as
shown in Fig. 26, is different in the two larval series, but
only in the early stages. Starting with a single spine with a
slightly thickened socket in the nauplius of series A (Pl. 1:1,
2, 4; Fig. 26A$_1$, A$_2$; see also Fig. 44E, G, for larvae of
spinicaudate Conchostraca), the furcal rami grow out progressively to attain an oval, paddle-shaped design (Pls. 3:1–
4; 4:1:6, 6; 5:1–3; 6:3–5; 7:6, 6; 8:5–8; 9:6. At about TS4A
these are held almost in plane (Fig. 26B–C$_1$, C$_2$; Pl. 10:3;
11:8).

The number of spines also increases continuously (see
Table 3). From an early stage, pits appear ventral to each
spine. Their nature is unclear; in some cases it seems as if
thin setae had arisen from them originally (Pls. 4:6; 7:6; also
Pls. 12:5; 25:7; 34:2), in others they appear as pores (e.g.,
Pls. 16:8; 26:1).

While in L4A the incipient furcal rami are elongate and
have four spines, those of L4B are already short paddles,
being as long as they are wide, rounded, sharply angled
against one another (90°) and carrying seven spines (Fig.
26F$_1$, F$_2$; Pl. 18:2, 4, 6; see also Pls. 18:7; 19:1; 20:1). These
differences becomes equalized during further development, and roughly by TS4 the rami have reached about the
same shape and degree of setation in both series (arrow
pointing from Fig. 26F to D: Pls. 20:5; 22:1; 23:1–3).

With further enlargement of the rami, more spines are
added to the marginal row, and a second row appears
between TS4 and TS5 (Fig. 26D). Each furcal spine is
furnished with denticles at its base, some more are positioned close to the spines (Pl. 25:7, 8). At about TS5–TS6
the furcal rami show faint incisions at their bases (Pl. 12:1,
3, 4, 5), and by the next one or two stages they are hinged
(Pl. 12:6, 7). Further development includes stretching,
thickening, better definition of the joints, and increase in
the number of marginal spines and pits (e.g., Pls. 14:3;
26:2). Eventually the rami are subtriangular in cross-section (Pls. 14:1, 2; 24:8), their margins being fringed with
more than 16 spines in the primary row and more than six
in the secondary row (Fig. 26E; Pl. 16:8, 9).

First recognizable at about TS5, the furcal rami become
more dorsally oriented (e.g., Pls. 16:9; 23:1, 2; 26:1). This
habit may be influenced by the developing 'ventrocaudal
processes' and is maintained to the latest stages. These
processes appear first in TS4 of series B (Pl. 22:1–3),
probably about at the same level also in A. Strikingly similar
to the furcal rami, the processes become progressively
elongated while receiving more marginal spines, denticles,
and pits (Fig. 26D$_2$, F$_3$, E$_2$, E$_3$; compare Pls. 11:8; 12:1, 4,

5, 7; 13:8; 14:2, 3; 15:2; 16:8, 9; 24:5, 8; 25:7; 26:1; 27:2). On
both structures, the spines never transform into setae.
Similar outgrowths occur in various Crustacea (see subchapter on ventrocaudal processes in the discussion, chapter 'Significance of morphological details'), but have never
been described with regard to a particular function. Since
the transverse musculature of the hind gut is located in this
region, such outgrowths, in accordance with the median
trench, could have participated in the opening mechanism
of the anus.

A typical transient larval feature of *Rehbachiella* is the
rigid and slightly curved spine covered with denticles distally and arising dorsally to the anus. In series A it is
retained until TS2i (Pls. 1:1, 2; 2:1, 4; 3:1–4; 4:1; 5:1, 2; 6:7,
8; 7:7, 8), while it is already absent in L4 of series B (Pl. 18:6;
also Fig. 26A, B, F). Such a spine is also known from
Bredocaris where it is reduced in size progressively but is
retained as a small pimple in the adult (Müller & Walossek
1988b, Pls. 6:3–5; 7). Again, dorsocaudal spines occur in
early larvae of various Recent crustaceans, particularly in
Cirripedia (e.g., Bassindale 1936; Dalley 1984; Moyse 1987;
Anderson *et al.* 1988; Egan & Anderson 1988, 1989), Copepoda (e.g., Onbé 1984), Mystacocarida (Hessler & Sanders
1966; Lombardi & Ruppert 1982), and penaeid decapod
Eumalacostraca (e.g., Cockcroft 1985). This suggests that
such spines are an ancient larval structure at least of Crustacea s. str.

The anus is a T-shaped slit enclosed within a triangular
membranous field ('anal field'; Fig. 26; Pls. 10:3; 11:9; 12:3;
14:2; 18:7; 19:1; 22:2). Its position at the rear of the hind
body in the triangle between dorsocaudal spine and furcal
rami is retained throughout development, and there are
only minor changes recognizable, such as the progressively
more vertical orientation of the anal region (compare Pls.
9:6; 10:3; 19:1 with Pls. 14:3, 4; 24:8), a slight extension of
the anal field onto the dorsal surface of the rami, and the
development of a short, faintly pointed supra-anal flap
(arrow in Fig. 26E$_1$; Pls. 14:3; 15:1; 24:8). In some specimens the anus or its membranous cover is artificially
protruded, possibly due to decay and gas production at the
time of burial of the animal (Pls. 6:3; 7:8; 8:2; 9:6; 20:1;
34:3).

Naupliar appendages. – These are more or less completely
developed and functional by the first stage. Few data are
available for the development of the 1st antenna; they
indicate a slow but continuous increase in size within the
early stages. During this period only an addition of ringlets
on the proximal 'shaft' could be observed with certainty,
while setation and distal portion showed little progress
(Pls. 2:1; 3:7, 7; 4:5; 6:4–6; 18:2, 4; 19:4; 34:5). Specimens of
later stages never have preserved these appendages. The
reason for this is unclear.

The 2nd antenna shifts progressively more anteriorly
(see above) and increases in size considerably. Accord-

ingly, its armature is enhanced (Figs. 8, 16). Significant morphological differences between the two series have not been observed (apart from the correlation of exopodal setation and segmentation: Pls. 1:2; 4:3, 4; 9:2 for series A; Pls. 19:3; 21:3, 4 for series B). Already in the early stages the endopod becomes elongated by an increase in length of the podomeres and by a subdivision of the 2nd podomere, which is completed approximately between TS3 and TS4 (Pls. 10:8; 19:3; 21:3; 23:3; 34:1). This process may be in accordance with progressive growth of the labrum during this phase and the necessity to elongate the endopod to reach toward the mouth.

The exopod, being only slightly longer than that of the mandible at first (Pl. 1:2, 4), is much enlarged during further growth. At about TS5 of both series are than 150 μm, comprising 17 ringlets and 15–17 setae (e.g., Pls. 6:4, 5; 7:1; 10:3, 8; 11:1; 21:3; 23:3; 34:1), while that of the mandible does not exceed 100 μm, having nine ringlets with 8–9 setae. This discrepancy probably relates to the major locomotory function of the 2nd antenna, while the mandible progressively transforms into a masticatory organ. In the later phase of development, the 2nd antenna, however, seems to undergo reduction. This is deduced from the size decrease of its insertion area and the fragments preserved (e.g., Pl. 17:2, 4). It suggests a progressive loss of importance of this appendage as the postmandibular limbs become increasingly functional.

The mandibles start with a similar design and size to the 2nd antennae (Figs. 9, 17), but the coxa is poorly developed, carrying only two setae medially. In the nauplius the basipod is the principal structure, having a huge body which is drawn out medially into a rigid masticatory spine (Pl. 1:4, 5, 7). Within the next few stages the coxal endite grows considerably and develops a triturating surface (cutting edge) with acute denticles (Pls. 2:2, 5, 6; 3:2, 3:7, 9; 4; 1, 2, 7; 18:1, 4, 5).

From the two gnathobasic setae of the second instar only the distal one remains. With progressive growth, the coxal body enlarges significantly, and the grinding plate becomes angled against the coxal body and turned obliquely towards the labrum (Pls. 5:2, 3, 6, 8; 10:2; 20:4, 8; 21:5, 6, 8). The posterior spine of the inner edge is slightly set off and is referred to as the 'posterior tooth' (e.g., Pls. 5:6; 11:10; 17:4; 23:5; 25:1; 28:2). From about TS5 the cutting edge differentiates further into a broader anterior part with many small spinules and setules, the 'pars incisivus', and a posterior part with rigid spinules or teeth, the 'pars molaris' (Pls. 5:6; 11:2, 4, 10; 17:4; 23:5; 25:1; 28:2).

The basipod is halted in growth after having increased in size for a few stages, while the number of setae surrounding the basipodal masticatory spine increases continuously from 1–2 to nine. Moreover, with progressive growth of the coxal body and its gnathobase, the basipod – and most likely the rami too – undergo reduction (e.g., Pl. 14:3), as is recognizable by the reduction in size of the 'palp fora-

men' even when the basipod is missing (Pl. 25:1; 28:5); by the largest stage the foramen is only as large (or small) as in the nauplius (20×30 μm; Pl. 17:2, 4). This process of atrophy can also be derived from the limited growth data of the rami, which are shorter from the beginning than those of the 2nd antenna (e.g., Pls. 6:4; 7:1; 8:7; compare Pl. 21:3 and 4). The gnathobase, on the other hand, developes into an enormous blade-like structure with a concave surface (Pls. 13:4; 17:2; 24:8; 25:1; 28:5; Fig. 9I, 17G), while the gnathobasic seta has not been recognized in the latest stages of both larval series.

Postmandibular limbs. – Development of these limbs occurs as a more or less slowly increasing process to a primordial functional state at first, and to a definite one later. A simplified and idealized scheme of their development, grouped into six major categories (ds1–6) is given in Fig. 27. Particular limbs as well as the developing limbs of late instars may deviate from this. Again, sequential decrease in definition within the series and from stage to stage result in a variety of minute differences at each stage and for each limb.

It proved difficult to adopt Benesch's (1969) categories for limb development of *Artemia salina* due to his consideration of anatomical evidence and the fact that the limbs of the Recent form show a mixture of delay and advance relative to those of *Rehbachiella*. Thus, the division used here corresponds only approximately to Benesch's stages. It ends at stage 6 since it is, of course, not known when the limbs of *Rehbachiella* are of truly mature shape (which would be approximately Benesch's stage 7, external appearance of a limb is approximately at his stage 3 in *Artemia*).

Principally, all postmandibular limbs pass through these stages. The 1st maxilla, however, deviates considerably from the very beginning. It is the only limb which appears first as a single spine (ds1; Fig. 27A; Pls. 1:5, 7; 2:2–4). The subsequent limbs start with the bifid lobe of developmental stage 3, the second stage of maxillulary development at instar L3 (Fig. 27C, D; Pls. 3:1, 2, 4; 4:1, 2, 7 for mx1; Pls. 8:5, 6; 18:8; 19:7 for thp1, Pl. 9:4, 6 for thp2). The fourth to sixth limbs of TS5A, 6A, and 7A at least start with a short, supposedly uniramous bud (ds2; Fig. 27B; Pl. 12:4). The 2nd maxilla shows another speciality, passing through the two rudimentary stages ds3a and ds3b (Fig. 27C, D; Pls. 5:1; 6:7; 7:2, 3, 5; 18:1, 4).

A transitional stage toward functionality is stage ds4a, with bifid endites on the corm and developed rami (Fig. 27E). This limb may already be functional in a primitive state, i.e. supporting locomotion. At this stage the 1st maxilla has reached its final level of subdivision with four endites on the corm and three endopodal podomeres (Pls. 5:3; 6:1, 2, 4–7; 7:2, 3, 7; 18:8). During further growth each endite becomes quite individual and different from those of the subsequent appendages (Figs. 10, 18; Pls. 9:1; 10:3, 5,

6; 11:3; 13:6; 14:5; 17:2; 21:5, 6; 23:3; 24:1, 2, 8; 26;3, 4; 28:5). In stages TS2 and TS4 of series B eight subdivisions were recognized medially, but it remains unclear whether this results from individual variability or indicates a further distinction between the two larval series.

The 2nd maxilla develops six endites maximally within a span of eight stages (Fig. 11, 19; Pls. 11:3; 12:2; 13:6; 14:6; 17:2; 23:3; 24:1, 2, 8; 26:3, 4). The thoracopods attain a final number of eight to nine endites, but the mode of definition varies within the set: at TS13 the 10th thoracopod is in the transitional stage 4b (compare Figs. 13C, 21C and 21G) but already having all endites and at least eight exopodal setae (Pl. 29:3; Fig. 27F is a mean).

By stage TS5–6 (at the end of the 'lag phase') the armature of the 'oldest' endites definitely consists of three sets of setae and/or spines (ds5; Fig. 27G; e.g., Pls. 25:3; 26:4). Various setae are already pectinate, but a definite filter function may not yet be achieved. During further development the outer edges of the postmandibular limbs become progressively more firmly sclerotized and subdivided (Pls. 34:6 compared to 11:5; 14:3; 15:3; 26:5; 27:6, 7; 29:4, 5).

The last category, ds6, is characterized by the subdivision of the sclerotized outer edge of the corm into three portions, the anterior group of enditic setae (set 1 in Fig. 27H) consisting of a double row of setae (Pl. 17:3; 29:2, 3; see also 33:3, 4), more than six setae in the posterior row (set 3; proximal endites develop differently; secondary row of brush spines in the 1st maxilla; Pl. 13:6; 15:4, 5; 26:4; 28:7; Fig. 33), and a very slender exopod with more than 10 marginal setulate setae (Pls. 14:3; 24:6; 25:2, 5; 29:4, 5). By this stage these limbs have a length at least of 300 μm. Further increase in size and definition occurs, but on accord of the limited data no further category has been erected.

This last level is reached at about TS7–8 in the anterior three thoracopods (Fig. 7B), at TS10 for the anterior 5–6 limbs (Fig. 7C), and the anterior 7–8 at the last stage TS13 (Figs. 13, 15D). The more posterior thoracopods are still at a lower level of differentiation: thp9 at ds5, thp10 at ds4b, thp11 at ds4a, and thp12 at ds2–3a (Figs. 20G–I, 21 for TS13B). The discovery of isolated limbs, tentatively assigned to *Rehbachiella*, shows a further development of the shape of the endites and enhancement of their armature (Fig. 14 and Pl. 33:3, 4). This indicates that the largest stage of the sequence known at present still represents an immature state of development.

In summary, the anterior four pairs of head appendages back to the 1st maxilla seemingly become reduced in size during late ontogeny, at least proportionally, and/or modify their shape. The 2nd maxilla is principally of the shape of the following limbs, save for its proximal endite, which is more like that of the 1st maxilla, and its lower number of endites on the corm, being six rather than 8–9.

Intraspecific variation

Besides the occurrence of two separate larval series, discussed above, the size and structures in general vary only little. However, in such rare fossil material it is rather difficult to identify differences as the result of intraspecific variability. Clearly individual habits can be seen in the asymmetrical arrangement of furcal setae in specimens of several stages, and in both series (e.g., UB W20, W36, W81, W82). Another example is the strange shape of the basipod of UB W52 (Pl. 13:5), where the proximal endopodal segment seems to be so much enlarged that it carries the exopod laterally rather than the basipod. It is not unlikely that this shape is an individual artifact.

Functional morphology and life habits

Early larvae

Remarks

The physical world of organisms in the millimetre range is, according to Koehl & Strickler (1981) 'dominated by viscous forces rather than the inertial forces that large organisms like humans encounter when moving through fluids'. This is particularly true for early crustacean larvae. In such a regime at low Reynolds numbers (a measure of the ratio of the forces against a solid object) the body is enclosed by water that reacts as a viscous mass (as if it were moving in liquid honey). Any disturbance, by movements of limbs, for example, will be damped out rapidly.

Assuming that the characteristics of water were not different at any time and that the physical demands at least should have been comparable, there should be similarities between recent and fossil crustacean larvae in their locomotory and feeding habits, particularly in the creation of flow fields. Generalizing Strickler (1985), the perception of food of swimming larvae occurs in the sensory core (signals to locate food are still unknown), while a re-routing within a reactive field brings food particles into the capture area for selection (proximo-reception), seizing, ingestion or rejection.

The overall resemblance of *Rehbachiella* larvae to those of Recent Crustacea is used to reconstruct the habits of the fossil from strategies of Recent forms, at least in a generalized way. Comparisons are based on comparative morphological studies on shape, motion and feeding habits of recent crustacean larvae, as presented by Gauld (1959), Sanders (1963b), and especially Fryer (various papers). Important new information about particular groups or species and life habits at low Reynolds numbers has been added by the application of high-speed cinematography (e.g., Barlow & Sleigh 1980: movements and food intake of *Artemia* at different stages; Koehl & Strickler 1981 and Strickler 1985: Copepoda; Moyse 1987: larval lepadomorph barnacles; Fryer [various papers]: different Branchiopoda, including ontogenetic changes and functional morphology).

Gauld (1959) and Sanders (1963b) distinguished two nauplius types, the branchiopod type and that of all other Crustacea. The former author, however, considered only the slender 2nd stage of *Artemia*; in fact, the hatching nauplius is much less elongate and non-feeding (Barlow & Sleigh 1980; Rafiee *et al.* 1986; Schrehardt 1986b, 1987a; Go *et al.* 1990; Fig. 53A–C herein). Accordingly, various of its specialities must be seen in the light of this habit (e.g., huge labrum for yolk storage, large antennal corm, closed hind gut). Other euanostracan nauplii may well have longer 1st antennae and better developed mandibles and be distinctly waisted between head and trunk (e.g., Heath 1924, Pl. 3:18 for *Branchinecta occidentalis*), or pear-shaped (Fig. 44A). As to the shape of larvae of other branchiopods, and the variety of other crustacean nauplii, there is no 'gulf' between naupliar types.

Not only in *Artemia* but also in various maxillopod nauplii and in cephalocarid metanauplii the 2nd antenna is at least the principal locomotory organ. There is little information available about malacostracan nauplii, but the size of their appendages indicates a similar habit. Hence, the size and prominence of the 2nd antenna of an *Artemia* larva may be largely influenced by functional needs rather than be of great phyletic importance. 'Exclusive' use of the 2nd antenna in Euanostraca is merely the extreme of 'dominant' function of it, as in Conchostraca or Cladocera (Fryer 1983) and other Crustacea.

Superficially the 'hydrodynamically more disadvantageous' pear-shaped nauplii, such as those of lepadomorph cirripeds (Moyse 1987) or ovate types, such as those of many Copepoda, euphausiids (Mauchline 1971) and penaeids (e.g., Cockcroft 1985, Fig. 1) use all appendages in a metachronal rhythm. Nevertheless, the ovoid copepod nauplii may be very mobile (Dahms, personal communication, 1989). Euanostracan nauplii, on the other hand, are reported to make rather slow movements ('inefficient in terms of propulsion' according to Barlow & Sleigh 1980; see also Fryer 1983, p. 256).

Another argument for the isolated status of the branchiopod type of nauplius concerns the 1st antenna and its reduced appearance. Admittedly, in Cladocera it is small and unsegmented. It is also unsegmented in various Euanostraca, but in other species it can be at least as long as the 2nd antenna and much longer than the mandible. Again, remnants of segmentation can be recognized in various Recent and fossil Branchiopoda, particularly in their larvae. This suggests strongly that the effacement of segmentation and size reduction are most likely rather 'modern' inventions and evolved parallel within the branchiopod

taxa (see subchapter on appendages in the discussion, chapter 'Significance of morphological details'). Barlow & Sleigh (1980) report that in *Artemia* the 1st antenna at least beats in rhythm with the 2nd antenna (see below), which seems to reflect earlier stages of its evolution when it was fully functional. Important feeding aids are the 1st antennae of planktotrophic lepadomorph cirriped nauplii (Moyse 1987), while the 1st antenna is not greatly involved in locomotion of Copepoda, if at all, neither in the larvae nor the adult, according to Perryman (1961).

While it is the non-feeding habit that seems mainly responsible for a small size of the mandible (also lacking a developed coxal portion and endite), it is known that, as in the nauplii of other crustacean groups, the mandibular palp also of euanostracan larvae could curve and rotate inward farther than the antennal one and thus is a sweeping device in this group (e.g., Barlow & Sleigh 1980, Fig. 2, for the feeding stage 2C of *Artemia salina* and Fryer 1983, Figs. 1, 2, for *Branchinecta ferox*). All naupliar limbs of Euanostraca are moved in a metachronal rhythm, as in other crustacean larvae.

Locomotion and feeding of the nauplii

In general the outline of early larvae ranges from an egg to a pear and comma shape and, apart from specialized larval types, all three appendages are of the same size order. According to Gauld (1959), the two processes of swimming and feeding are intimately connected with one another in crustacean larvae. While feeding habits are very diverse, swimming is fairly uniform among crustacean nauplii. Due to the physical constraints of the environment, mobility of a nauplius seems to be affected largely by size and number of locomotory organs rather than by its shape. Again, a slimmer body would first be of advantage when size exceeds about 0.5 mm. Below this, even long, slender appendages would not greatly enhance efficiency, as seen in the *Artemia* nauplius.

Mobility can be deduced also from the development of a natatory setation on antennal and mandibular exopods (slender, setulate setae; Fig. 35E). However, this is not necessarily coupled with feeding, as can be seen in non-feeding eumalacostracan nauplii for example. On the other hand, feeding ability cannot be deduced just from the presence of enditic spines on both the larval 2nd antenna and mandible or the presence of mouth and anus: a nauplius may still feed on yolk (lecithotrophy), while the development of these structures has preceded functionality.

Feeding ability of a nauplius can be recognized more readily when special aids are developed, such as in particular delicate setules on all setae concerned with feeding (basically on all naupliar appendages), brush-like sides of the prominent labrum, and a well-developed and 'hairy' sternum. The non-feeding 1st *Artemia* nauplius (Schrehardt 1986b, 1987a) has naked setae which become

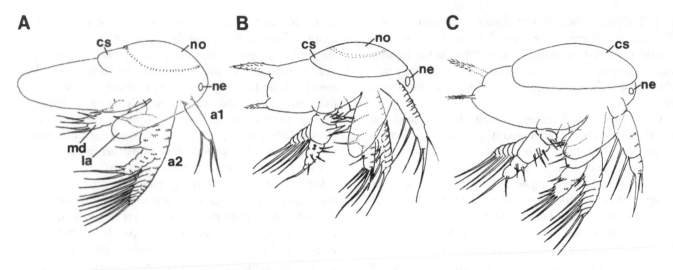

Fig. 28. Types of swimming nauplii in profile (sizes not to scale). □A. Nauplius of *Artemia*, as an example of the elongate, non-feeding type (modified from Barlow & Sleigh 1980; Rafiee *et al.* 1986, Schrehardt 1987a; Go *et al.* 1990; arrow points to larval shield at rear of neck organ). □B. Nauplius of *Rehbachiella*, as an example of the pear-shaped type. □C. Copepod nauplius as an example of the ovoid type (generalized from different authors; dorsocaudal spine present in some species drawn stippled).

equipped with setules at the moult to the feeding 2nd stage. Poorly developed armature also characterizes the lecithotrophic larvae (e.g., Moyse 1987 for lepadomorph cirripeds; Dahms 1989b for a harpacticoid copepod).

Few Recent crustacean larvae start with feeding immediately after hatching but do so after the next one or two moults. Such an instar may still look like a 'nauplius', but in many cases it is already an advanced larva with several or many trunk segments and limbs buds. It is, thus, necessary to restrict comparisons of larval stages to the same level of development, since their design and habits are highly dependent on it.

With regard to the structural demands of locomotory and feeding mechanisms of Recent crustacean larvae, the nauplius of *Rehbachiella* possesses essentially the same features. Again, this instar and the first larva of *Bredocaris* (actually a metanauplius; Müller & Walossek 1988b, Fig. 4A) share the pear shape of the body and the similar size of all three naupliar appendages. Use of all naupliar appendages as well as feeding from the beginning is indicated for both Upper Cambrian larvae by the sizes, full development (e.g., of corms and rami), and differentiated armature of their appendages equipped with swimming setation and feeding devices, the setulate sides of the labrum (and size) and the setulate sternum. In both the 1st antennae are well equipped with long setae on their posteromedian edges besides the distal group of setae. This points to their collaboration in the feeding process, probably in a way described by Moyse (1987) for nauplii of lepadomorph barnacles, which wipe their 1st antennae against the brush-like sides of the labrum (cf. Müller & Walossek 1988b, Pls. 3:4; 7:8; 8:5 and Figs. 14, 17 for *Bredocaris*; Pls. 2:6; 3:8; 5:5; 6:4, 6; 18:2–4 herein for *Rehbachiella*).

In both larvae the labrum forms the anterior wall of a short feeding chamber. The posterior wall is made by the somewhat ventrally flexed hind body (Fig. 30 for *Rehbachiella*; Müller & Walossek 1988b, p. 22 and Pl. 7:3 for *Bredocaris*). Another typical structure of early larvae is a gnathobasic seta on the distal surface of the mandibular grinding plate, used as sweeping device (for function see Fryer 1983, also his Figs. 8–10). This seta is present in both fossils. In *Rehbachiella*, the earliest stages have two such setae (e.g., Pl. 2:6), but only one is retained until approximately TS12. Its peculiar spiral row of setules is shown in Pl. 9:3.

Similar armature to that of *Rehbachiella* occurs among the Maxillopoda in most cirriped nauplii (cf. e.g., Costlow & Bookhout 1957, 1958; Crisp 1962; Dalley 1984; Anderson *et al.* 1988; Egan & Anderson 1988, 1989; Fig. 45H), and in particular among nauplii of planktotrophic lepadomorph barnacles, such as *Capitulum mitella* or *Pollicipes polymerus* (Moyse 1987), to a varying degree also in copepod nauplii (e.g., *Calanus armatus* in Gauld 1959, Fig. 3b, c; *Bryocamptus pygmaeus* in Dahms 1987b, Fig. 1; *Drescheriella glacialis* in Dahms 1987a, Fig. 2). According to Dahms (1989b) all harpacticoid nauplii, save for one from the Antarctic area, are well equipped right after hatching and feed at least from the 2nd instar onwards.

The major difference between the two Upper Cambrian nauplii is the less developed mandibular coxa in *Rehbachiella* (compare Pl. 1:1–4 and Müller & Walossek 1988b, Pl. 8:7). Since the 1st larva of *Bredocaris* corresponds, however, already to the L3 stage of the former (see subchapter on Maxillopoda in the chapter 'Comparative ontogeny'), the differences accord well with its advanced larval state (see Pls. 3, 4 for L3 of *Rehbachiella*).

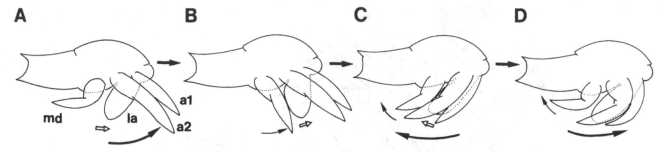

Fig. 29. Suggested movement phase of a _Rehbachiella_ nauplius (A–D); long arrows indicate movements of the antennae, short hollow arrows movements of mandibles, short black arrow points to movements of labrum.

Both larvae belong to the pear-shaped type (Fig. 28B), as present among cirriped nauplii. This type is intermediate between the 'slender nauplius' type, as is represented by the 2nd instar of _Artemia_ (Fig. 28A; 1st one is still pear-shaped) and the 'egg-shaped nauplius' type of various copepods or eumalacostracan nauplii (Fig. 28C; the shield may even be much better developed and extending posteriorly above the hind body, such as in _Bryocamptus pygmaeus_, Dahms 1987b, Fig. 1). In the intermediate type (B), the prominence of all naupliar appendages is as in type C, while the outline of the body is more like that of type A.

Possible phases of a _Rehbachiella_ nauplius in a moving cycle are reconstructed (Fig. 29) after illustrations of Barlow & Sleigh (1980, Fig. 1) for _Artemia_ and by partial inclusion of Fryer's (1983, Fig. 20) motion cycle of a _Branchinecta ferox_ nauplius. Similar beat sequences, also with ranges of each limb, have been illustrated by Moyse (1987, Fig. 13) for the lepadomorph cirriped _Lepas pectinata_. It is likely that the _Rehbachiella_ nauplius was oriented ventral side up while swimming, as described for euanostracan nauplii (Fryer 1983, p. 256). Furthermore, three major characteristics are also adopted for the fossil:

- the collaboration of the 1st and 2nd antennae (long arrow),
- the phase difference of the mandible (short arrow), and
- the large labrum moving in accordance with the swing of the limbs (hollow arrow), as extrapolated from different modes of orientation in actual specimens (see, e.g., Pls. 1:1; 2:1–4; 3:4; 4:2; 6:4; 18:1–4).

When the two antennae swung anteriorly (A), the mandible reached its posterior maximum and then started with its 'recovery stroke' (B). At this phase the labrum was raised passively to enhance the opening of the atrium oris. During the 'power stroke' (C) the antennae met the mandible which then also swung backwards. This caused the labrum to be lowered again to cover the atrium oris. Lastly, the two antennae moved anteriorly again (D), being flexed far backwards (facilitated by their external annulation), while

the mandible still continued its backward–inward movement.

In the light of high-speed cinematographic studies, various traditional interpretations on motion and food intake of animals in the viscous regime (incl. terminology) may no longer be unequivocal (cf. Strickler 1985, and for further references). Due to viscosity and laminar flow of the surrounding water body, a back swing generates a steady flow alongside the larva. Food particles embedded in the water medium follow passively. Hence a back swing does not result in the catching of food – water and food would simply pass the body (antennal exopod setae are purely natatory, according to Fryer 1983). Again, their movement comes to a halt immediately when the larva stops beating (Koehl & Strickler 1981). According to these authors the bristled limbs act as solid paddles in Copepoda, which would negate a flow around the minute setules on the setae. This may be valid for this group, while Fryer (1987b, p. 428, also for further references) provides convincing arguments that sieving (filtration) is still likely in branchiopods (see also Barlow & Sleigh 1980; Korinek _et al._ 1986).

On the other hand, the flexure of the naupliar appendages during the anterior swing ('recovery stroke') is not so much to reduce drag, but produces a lower pressure behind the limb, which sucks water and nutrient particles towards the body ('re-routing' of Strickler 1985). The further inward swing of the mandible observed in the _Artemia_ nauplius may enhance this effect. Once the particles are close enough, they are trapped in the capture area, here in the postlabral feeding chamber (see also Fig. 30). It is not surprising that the same 'trick' of re-routing can be observed in the ciliary movements of bivalve larvae, which operate at similar Reynolds numbers (Gallager 1988, in particular his Fig. 3).

In summary, the two mechanisms of locomotion and feeding obviously operate hand in hand in crustacean nauplii, but with some competition with regard to functional needs of the different structures. The resulting compromise in construction explains the large variety of larval types reflecting specific adaptations. Since this apparently refers to modifications at lower taxonomic levels, gross

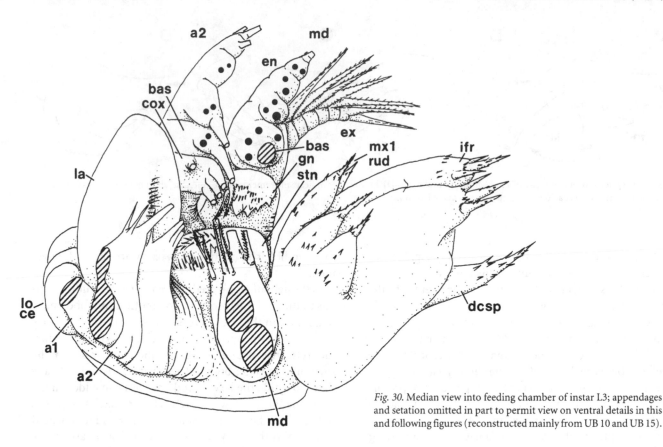

Fig. 30. Median view into feeding chamber of instar L3; appendages and setation omitted in part to permit view on ventral details in this and following figures (reconstructed mainly from UB 10 and UB 15).

body morphology is of rather limited value for recognizing 'gulfs' between larval types of higher taxa. This is even more apparent when taking the nauplii of the Upper Cambrian crustaceans *Rehbachiella* and *Bredocaris* into account. These not only are clearly intermediate in their shape between the ovoid and the slender types, but also demonstrate the primordial feeding state from the beginning, while this is lost in many Recent types.

It is apparent that these ancient nauplii were mobile swimming and feeding larvae. The eumalacostracan nauplii are exceptional in the way that, while retaining the shape and use of their appendages, they have no feeding structures, such as endites, setae, setules, large labrum and sternum. It is not unlikely that this is due to very early loss during the evolution of this particular group.

Advanced stages

Functional ontogenetic changes

During postnaupliar growth of *Rehbachiella* the larval hind body elongates posteriorly. With this, more stability in terms of locomotion is progressively achieved. Whereas the naupliar appendages grow rapidly at first, they apparently cease to grow thereafter save for the mandibular coxa with its grinding plate. The postmandibular limbs are added sequentially and progressively come into action. Eventually they transform into locomotory and feeding organs, while a primordial type of functionality was most likely achieved as early as at developmental stage 4a (Fig. 27E).

The gradual process of attaining functionality from the front to the rear, eventually diminished the prominence of the transient naupliar apparatus (at least relative to the body) and modified its components. The morphogenetic changes in the locomotory and feeding apparatus are reconstructed in Figs. 30–33, drawn as if viewed from the ventrolateral side towards the sternal region, with most of the right series of appendages omitted. They exhibit the continuous addition of postmandibular limbs to the adult locomotory and feeding mechanism once they are functional. Even at the last instar recognized as yet, the naupliar limbs are still supporting the now well-developed posterior feeding apparatus, as known from euanostracan Branchiopoda (cf. Fryer 1983, p. 231). Parallel changes occur in the head region.

From about TS7 the furcal rami are hinged (Fig. 34), most likely acting as stabilizers or rudders of the trunk by flapping up and downward. Their ventral flexure is progressively limited by the enlarging ventrocaudal processes. Outward flexure might have been possible by the latest stages when the basal joints become slightly narrower (for

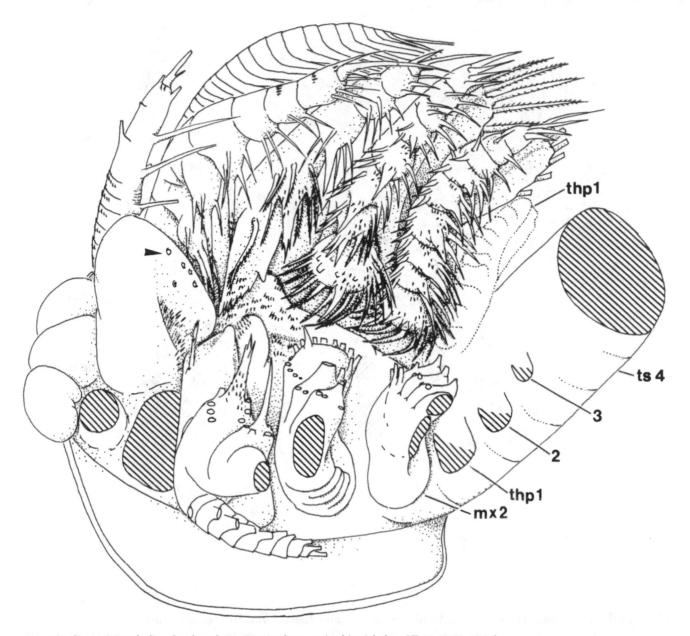

Fig. 31. Median view into feeding chamber of stage TS4; trunk rear omitted (mainly from UB 66, 69, 70, 71, 73).

Euanostraca see Fryer 1983, pp. 278–279 and Figs. 38–41; Jurasz *et al.* 1983, Fig. 8; Schrehardt 1986a, Fig. 16).

With addition of functional thoracopods, the 1st maxilla successively transforms into a 'pusher limb', transferring food from the sternal food channel toward the mandibles. Its function in mechanical transport of food particles around the paragnaths into the 'paragnath channel' and toward the cutting edges of the mandibular grinding plates can be deduced from their well-developed armature, especially by the presence of spines, irregularly furnished with setules ('brush spines', see, e.g., Pls. 15:5; 28:7; Figs. 33, 35B–D), and the design of the proximal setae, which are more setulate than those of the posterior limbs (in part Fig. 35F, G). The 2nd maxilla retains the shape of a trunk limb

save for its proximal endite which is more like that of the 1st maxilla, while its proximal endite is also equipped with several brush-like setae (Pl. 16:1).

Setae and spines of different types are added progressively to the limbs. They start as pappose or setulate spines (Fig. 35F, G) and eventually transform into pectinate setae (Fig. 35H–J). It remains speculative at what stage filtration started, and up to which stage the pectinate setae were still used for mechanical particle transport. At about stage TS4 (Fig. 31) the setules on the enditic setae are still rather widely spaced which makes definite filtration rather unlikely (Pl. 21:5; Fig. 35H).

At about stage TS8 3–4 four postmaxillary limbs are in an advanced stage of development (ds6; Fig. 32), 7–8 at TS10,

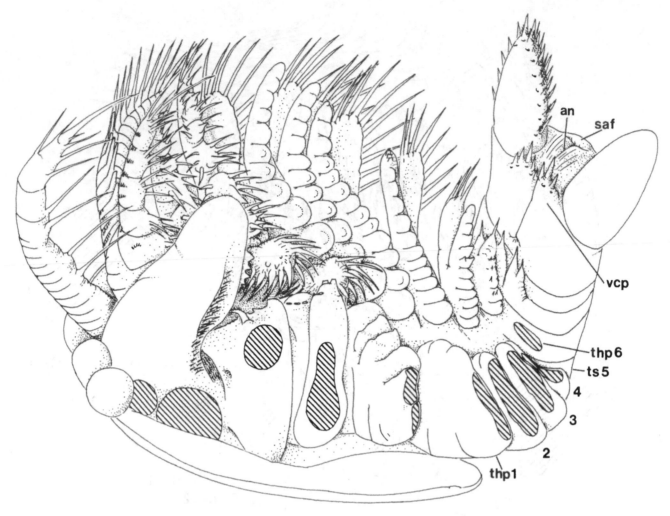

Fig. 32. Median view into feeding chamber of about stage TS8 (mainly from UB 82).

and 8–9 by the largest stage TS13 (Fig. 33). At this largest stage recognized, the 2nd maxilla and the thoracopods have a large corm with the maximum number of endites, a slender, four-segmented endopod and an elongatedly leaf-shaped exopod (Figs. 4, 36). The limbs insert almost ventrally at the border of the deep food channel provided by the deeply recessed sternitic plates. The corm, with an oval cross-section originally, has become elongated and flattened in an abaxial direction, attaining a fleshy to phyllopodous habit with pliable sides (the posterior one is concave). The inner edge carries the endites, while the outer edge is slightly better sclerotized and posteriorly bent. Incisions appear progressively on the outer side (e.g., Pl. 27:7; Figs. 27, 36). Such interrupted sclerotization might represent a functional compromise between the enhancement of rigidity and the retention of flexibility. Similar structures can be found, for example, on the thoracopods of cypris larvae of facetotectan Maxillopoda (Ito 1989b, e.g., Figs. 3, 7).

Function of the advanced apparatus

The oval to sub-triangular surface of a typical endite is furnished with many setules (Fig. 37A; see also Pl. 14:5, 6 for the maxillae). Its armature is made of three distinct sets of differentiated setae or spines in an advanced state of development. The anterior set (set 1) is composed primarily of one and later two rows of closely spaced setae, oriented in the long axis of the corm. It is not unlikely that these pectinate setae were articulate, as indicated by ring-like sockets (e.g., Pl. 29:2). The two opposing setule rows on these setae point posteriorly (Fig. 37B). The distal setae or spines of this set may be more like brushes having a distal tuft of setules and a coarse grid of setules proximally (Fig. 35D).

The median armature (set 2) varies along the corm from proximal to distal and from limb to limb. Typically, the enditic surface is slightly humped and bears one or a few larger spine-like setae and some smaller ones around it (Fig. 37A). One of the larger ones, at least on the more distal

Fig. 33. Median view into anterior part of feeding chamber at stage TS13 (in part from UB 87).

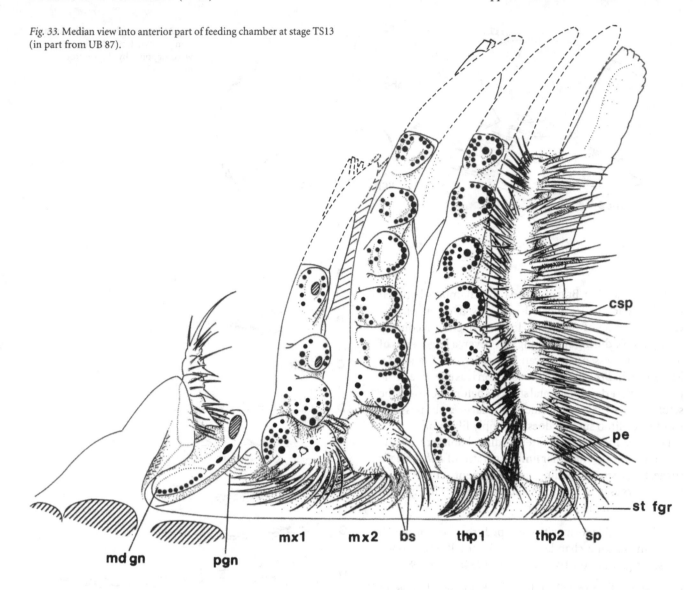

endites, is a 'comb spine' with a setulate proximal part and serrate distal part (Figs. 33, 35A; Pl. 16:3, 5). Such a spine might have been used to collect or groom particles from the setae of the posterior limbs.

The posterior set (set 3) is a row of bipectinate setae arranged in a semi-circle. Towards both ends of the row the setae decrease in size. The median ones are the longest and gently tapered, reaching at least between the setation of the subsequent limb (Fig. 35H–J; e.g., Pl. 16:7). All setae of this set arise from broad, slightly curved and firm sockets (Pls. 14:6; 16:2). The opposing rows of setules are arranged in such a way that the angle between them opens always to the centre of the endite (Fig. 37B). The setules are evenly spaced, the distance between them being about 2 μm. From proximal to distal all the posterior setae more or less form a close grid (particularly Pl. 16:6).

The setal pattern of living Branchiopoda shares all major details with *Rehbachiella*. Taking the Euanostraca, the anterior set (set 1) of the latter corresponds to the set de-

scribed as 'Medialborsten' by Eriksson (1934). The spines of the median armature (set 2) of *Rehbachiella* correspond to spines among the anterior set of Eriksson, who did not distinguish between these two different groups. The filtratory setae, named 'Ultimalborsten' by this author correspond to the posterior row (set 3) of *Rehbachiella*. As in the fossil, they comprise a proximal socket, a median part with close-spaced setules and a slowly tapered end with more widely spaced setules. Due to some modification of the endites of modern Branchiopoda – mainly by fusion and compression – these sets may occur at a slightly different position in the different taxa, but they are always present.

As noted above, seizing of food in the viscous regime can be achieved by re-routing water from the surrounding region and actively catching it proximally. Other structures required are the rows of setules that form the sieves for the retention of particles (e.g., Fryer 1987b for daphniid cladocerans; also Nival & Ravera 1979; Crittenden 1981, Figs. 1–6 in particular).

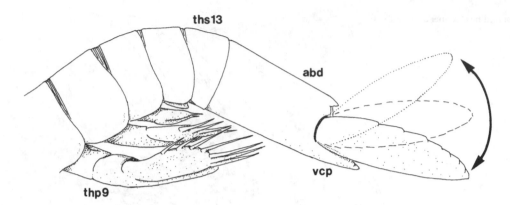

In euanostracan Branchiopoda the chambers between the thoracopods are opened and closed during metachronal beating of the limbs. In consequence, feeding is only possible during swimming. However, a large series of limbs is not essential, as can be seen from Cladocera. Moreover, in *Branchinecta ferox* suction is as effective in the predatory late developmental stages as before the loss of the setules (Fryer 1983), which indicates that even the possibility of inter-limb suction does not imply filtration.

Features indicating a filter-feeding habit are clearly present in the postnaupliar limb apparatus of *Rehbachiella* suggesting a similar feeding and movement activity for its later instars as in Recent Branchiopoda, at least in a general mode. These are in particular:

- a large pliable corm with C-shaped cross-section which form 'sucking chambers' (Fig. 38; for Branchiopoda see, e.g., Cannon 1933; Eriksson 1934; Fryer 1983),

- posteriorly directed endites, with marginal rows of setae forming a close grid for the retention of food (Figs. 33, 38, 49B, 50B; for Branchiopoda see authors listed under previous point),

- setae with regularly spaced setules oriented in characteristic ways ('pectinate setae'; Fig. 35I, J; see Fryer 1983, p. 232, Figs. 92–98, for *Branchinecta ferox* until a body length of about 18 mm, or Schrehardt 1986a, 1987a, b for *Artemia salina*),

- distal parts (exopods) that can be flexed posteriorly during the anterior stroke (re-routing of water into the median capture area (Figs. 21, 39),

- a V-shaped narrowing of the capture area between the limbs from distal to proximal (Fig. 36), and

- a deep sternal invagination which forms a food channel in the thorax for the orally directed food transport (Figs. 4, 33, 36, 49B3; Pls. 16:2; 26:3; 27:4; 29:1; for different Branchiopoda see Cannon 1933, and Fryer 1983 for Euanostraca).

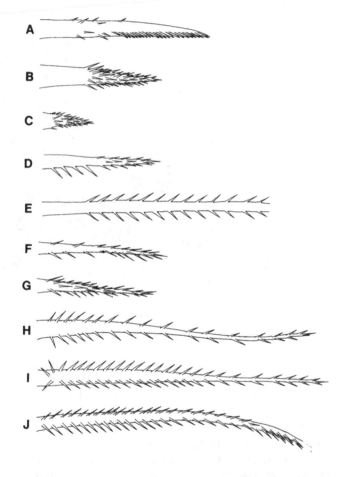

Fig. 35. Selection of setal types (not to scale). □A. Comb spine on median surface of more distal endites of trunk limbs (see also Pl. 16:5). □B, C. Short brush spines of proximal endite of 1st maxilla (Pls. 15:5; 19:6). □D. Proximally pectinated brush spine, occurring in anterior set of enditic setae (UB 55). □E. Exopodal swimming seta with opposing rows of setules (Pls. 22:5; 25:5). □F, G. Setulate spine-like setae of different sizes, occurring on proximal endites of maxillae (Pl. 16:1). □H. Slender, slowly tapered, bipectinate seta with widely spaced setae of advanced larvae or on more distal endites of late larvae (e.g., Pl. 16:7); similar but shorter setae are also on the mandibular basipod (Pl. 13:4) and in the anterior group (1) of the endites (Pl. 15:5; 22:4, 7). □I, J. Filter setae with two rows of densely spaced setules (proximally with 2 μm distance between setules) on endites of 2nd maxilla and thoracopods (e.g., Pl. 14:6; 16:6).

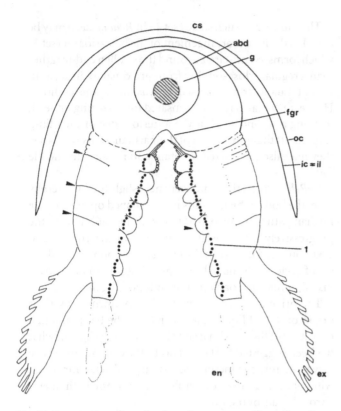

Fig. 36. Cross-section of anterior thoracic segment, as if seen from the anterior; setae of anterior group (1) indicated by dots; one of the proximal setae of posterior row added to show their projection into sternal food groove; dashed circle indicates size of abdomen; short arrows point to segmentation of outer edge of limb; ic/oc = inner and outer surfaces of shield (freely covering the segment).

Assumptions about the apparatus of *Rehbachiella*, even with regard to structures not or no longer present in Recent forms have been facilitated by the careful and detailed functional morphological studies on Branchiopoda by Cannon (1933), Eriksson (1934) and Fryer (various papers). The latter author also included details of ontogenetic changes. However, any functional model of a fossil such as *Rehbachiella* must, of course, remain rather simplified, since (1) various details from the analysis of Recent filter-feeding crustaceans cannot be applied to a fossil, even when fairly well-preserved, and (2) a number of details of the filter habits of Recent crustaceans are still controversial.

Nevertheless, it is assumed that much of what is known about the mode of motion and feeding described for *Artemia salina* by Barlow & Sleigh (1980) can be extrapolated to *Rehbachiella*. During the back swing of the limbs, water is accelerated laterally from the median space (*a* current of Barlow & Sleigh 1980), which gives rise to a current which draws food particles into the median space. With the metachronal beat of the limbs, water is also progressively accelerated in the inter-limb spaces, which is maximal as the limbs complete their back stroke. Adopting this for *Rehbachiella*, water could have been expelled through the filter mesh of the enditic setae of set III, and particles were retained as the inter-limb spaces increase in volume (arrow 1 in Fig. 38). The currents (arrow 2), passing the chambers, joined laterally (arrow 3) to form a posteriorly directed current alongside the tail of the animal (*j* current of Barlow & Sleigh 1980).

As stated for the nauplii, it may be possible that the anterior swing of the limb may have enhanced the re-routing of water from the water body around the animal and inward between the limbs (*v* current of Barlow & Sleigh

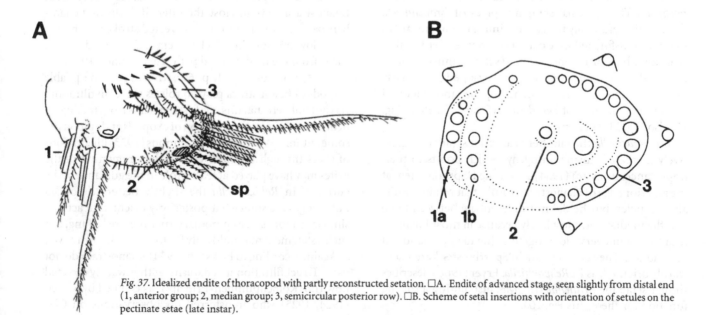

Fig. 37. Idealized endite of thoracopod with partly reconstructed setation. □A. Endite of advanced stage, seen slightly from distal end (1, anterior group; 2, median group; 3, semicircular posterior row). □B. Scheme of setal insertions with orientation of setules on the pectinate setae (late instar).

1980). In consequence, the observations of both the older and the later workers would be correct but refer to different distance levels on the appendages: the anterior swing initiates the inward flow, while the back swing continues the flow at a more proximal level.

A continuous metachronal beat, as is likely with a large number of appendages involved in the apparatus, ensured that nutrients were eventually passed proximally into the sternal food channel, where they were moved towards the mouthparts (arrow 4 in Fig. 38). As in Euanostraca (see Fryer 1983, Pl. 6), the bipectinate proximal setae on the proximal endites of *Rehbachiella* reach far into the sternal food groove curving anteriorly at their extremities (Pl. 16:4; Figs. 33, 36).

According to Eriksson (1934, see also Barlow & Sleigh 1980) an anteriorly directed food transport is confined within the sternitic food channel in Euanostraca, which can be assumed also for *Rehbachiella*. Concerning the nature of this orally directed food transport, Fryer (1983, pp. 300–301) argues convincingly in favour of a major mechanical influence of the posterior row of setae of the proximal endites (set 3). This is in contrast to earlier workers, such as Eriksson (1934, pp. 70–74 and Fig. 2) who suggested a passive transport, stating that all setae of the proximal endites are held in one plane to sieve but cannot push anything forward.

In *Rehbachiella* a single comb spine arises from the enditic crest of the more distal endites (set 2). Its shape and position suggest a grooming or collecting function, scraping off particles from the more posterior limbs. Similar spines occur on the more proximal endites immediately in front of the posterior regular row (set 3), for example in the euanostracan *Branchinecta ferox* (Fryer 1983, Fig. 64). In the Lower Devonian *Lepidocaris*, such spines occur on the more distal endites of the postmaxillulary limbs (Fig. 51A). Hence, it would refer to the median set (set 2) of *Rehbachiella*. The marginal scraping spines of *Branchinecta ferox*, which gradually transforms into a carnivore, are in my view transformed setae of the posterior set (set 3) rather than members of the median set (set 2). Similar coarser marginal scraping devices are also developed particularly on the endopod of various Branchiopoda, and as a distal tuft on the endopod of *Lepidocaris* (Scourfield 1926, Fig. 15; also Fig. 51A herein).

In *Rehbachiella* the number of anterior setae is progressively increased during ontogeny, which points to their importance. Eriksson (1934) mentioned 'Sperrborsten' at the anterior edge of euanostracan limbs that retain unsuitable particles, but he did not differentiate between these and the median set, most likely because in most Euanostraca they stand very close together due to applanation of the endites. The setules of such 'Sperrborsten' are backwardly oriented, as in *Rehbachiella*. Fryer (1966) describes similar setae for *Branchinecta gigas*, also noting a sorting function for these (his Fig. 10).

The number of such setae (set 1) in *Rehbachiella* may be correlated with the arrangement of the posterior set (set 3) which forms a semi-circle around the swollen endite rather than a regular, close-set row from proximal to distal, as in Recent Euanostraca (caused by fusion of the endites). Hence, they may have compensated for a 'leaking' of set 3, forming a second sieve at the anterior edge of each limb (Fig. 38). It the assumption of their articulations is reliable, they may also have acted to retain and remove unsuitable food.

In *Rehbachiella* the limbs were held slightly distolaterally from the body, which results in a V-shaped opening of the median path from proximal to distal (Fig. 36). Due to the progressively more distal orientation of the endites and slight outward flexure of the elongate endopods the whole set of setae might have been spread like a fan and closed distally while approaching the endopods again.

The orientation of the exopods is preformed by their origin on the oblique outer surface of the basipod and a curvature of the proximal end (Fig. 36). In the light of what has been suggested for larval habits, the exopods may have served mainly for locomotion (marginal setae furnished with opposing rows of setules) in rhythm with metachronal beats of the limbs.

Again, their articulation with the corm is indistinct anteriorly but well developed posteriorly (Pls. 14:3; 29:4, 5; Figs. 21, 39). Such partial fixation may have enabled the exopod to move in rhythm with the limb in the back stroke but to flex in the anterior stroke. This flexure during the anterior swing of the limb may have had a similar re-routing effect of the surrounding water core as the exopods of 2nd antenna and mandible in the early larvae. In their posterior position the exopods overlapped each other with the setae reaching over the surface of the subsequent paddles (same figures).

It has been suggested that the foliate exites or epipods of Euanostraca serve to close the inter-limb sucking chambers passively in the anteriorly directed stroke which produces low pressure in the chambers and outward flow of water from the median food path (e.g., Cannon 1933). By contrast, Eriksson (1934, p. 69) stresses that the pliable epipods of Euanostraca play no role as valves in filtration. It is not unlikely that this confusion results from mixing up of the morphology and function of exopods and epipods by some authors. With regard to Cannons (1933) illustrations of slices through the limb corms, the orientation of the exites may have played at least a role in the guidance of the currents. In *Rehbachiella* the slightly better sclerotized outer edges are somewhat posteriorly oriented to act in a similar fashion, since epipodial structures are lacking, but such a statement may hold only for the stages known as yet.

Again, according to Eriksson (1934) Euanostraca do not use a 'Druckfiltration mechanism' with inwardly directed flow (no 'Kolbenpumpen-Prinzip'), as claimed by Storch (1924, 1925) who transferred his observations on Cla-

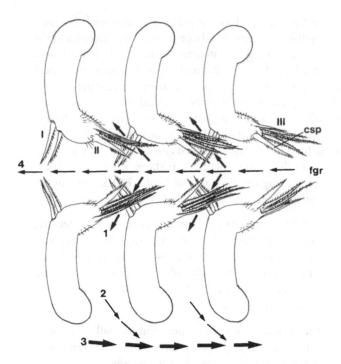

Fig. 38. Part of thoracic filter apparatus, drawn as if cut at the middle of the limb corms to show their orientation; enditic armature partly included (I–III), also with the comb spine (csp); more firmly sclerotized outer edge enhanced; arrows indicate principal water currents: 1, current sucked into inter-limb spaces at effective stroke; 2, outward jet at end of compression phase; 3, lateral current, 4, path of food within the deeply invaginated thoracic sternal food groove.

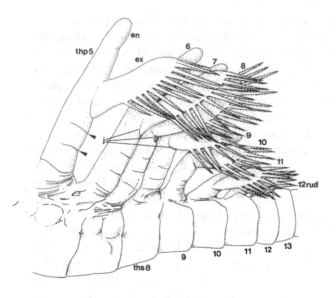

Fig. 39. Outer view of posterior trunk limbs showing overlap of the exopods and their setae, mainly redrawn from UB 87; short arrows point to outer segmentation of the corm and to a paired hump on lateral side of thoracic segments.

docera to the euanostracan apparatus. According to Fryer (1987b, p. 429) and Kohlhage (personal communication, 1989) a large or even enclosing shield is not essential for filtration (as Lauterbach postulated in various papers, e.g., 1974). A lateral current between limbs and shield provides no significant advantages, and filter-feeding Euanostraca are completely devoid of a shield. In *Rehbachiella* the shield, though large, extends no further ventrally than to the insertions of the exopods (Figs. 6, 15, 36) and was apparently not an imperative for feeding.

Mode of life and habitat

The nauplii of *Rehbachiella* were not filter feeders but may have swept in particles ('particle-feeding') while swimming. Appendages, added progressively, became functional gradually, but the naupliar appendages retained their functions and shape over a long period. This implies that naupliar and thoracopodal apparatuses functioned in cooperation, at least for some time, as is reported from Recent Euanostraca (e.g., Fryer 1983, p. 231). In its later stages *Rehbachiella* was probably a swimmer, propelled by the rhythmic metachronal beat of the trunk limb apparatus (including the maxillae) which at the same time was used

for filter-feeding. Again, it may be possible that *Rehbachiella* swam up-side down, as this is the common habit of many small-sized suspension-feeding or swimming microphagous filter-feeding crustaceans.

As can be deduced from extant crustaceans, swimming may have been almost constant in *Rehbachiella* in the search for food, mates, or when escaping from predators. Due to lack of grazing structures on the distal parts of the trunk limbs, it does not seem very likely that *Rehbachiella* scraped off particles from surfaces. Yet, filter feeding may have been only one of the functions of the posterior limbs, since in Crustacea generally the limbs do not usually operate for single purpose only. Even when working as components of a complex apparatus they can act both individually and for multiple function (e.g., grooming and sorting).

Discussion

Affinities of *Rehbachiella*

Position within Crustacea s. str. (= crown-group Crustacea)

The assignment of *orsten* arthropods to Crustacea has recently been questioned in general terms by Lauterbach (1988). This is not the place to respond to all his arguments at length, but it seems necessary to note that Müller & Walossek have never claimed that all components of the fauna are crustaceans, that Lauterbach has never worked on the material himself, and that he has not taken any of the various papers on the *orsten* fauna published since 1983 into acount (cf. his list of references). Hence, his comments on the *orsten* fauna are based solely on theoretical constructs. Moreover, many of his comments and interpretations of functional morphology are inconsistent with the evidence from fossils as well as the literature on Recent and fossil Arthropoda, especially Crustacea.

As a single example, Lauterbach based his theory on his own interpretations of the shape of the anterior appendages of *orsten* forms, neglecting the details. His creation of an early 'crawling' nauplius results simply from misidentification of the appendages from a specimen of the 4th instar of *Bredocaris* by using an unlabelled early SEM micrograph in Müller (1981b). Thus he called the mandible a 2nd antenna (carrying a huge gnathobase), while the developed 1st maxilla turned into a 'still brush-shaped' mandible. This would indeed be a very extraordinary design, if it were correct. By contrast, this larva is not a nauplius but an advanced instar, already possessing well-developed 1st maxillae as well as buds of 2nd maxillae and three thoracopods on the trunk. It is not a crawling larva but well capable of swimming and suspension feeding, as is typical of such larvae living at low Reynolds numbers and can be readily deduced from the various structures required for such life strategy (see above). The particular specimen UB 918 (Müller & Walossek 1988b, Pls. 12:1, 2, 4–8, 13:1, 2, 4, 5, 8, 9, 11) simply lacks the 1st antennae due to preservation, while the next two limbs are somewhat crumpled and distorted distally. Again, this larva has no circular and flat head shield, but an arched one with slightly projecting margins, similar to various other crustacean larvae.

As a matter of fact, the *orsten* assemblages do not represent a homogenous mass of 'stem-group mandibulates', as Lauterbach claimed: besides true chelicerates and *Agnostus*, traditionally understood as trilobite, the material contains a number of true crustaceans and, moreover, arthropods with resemblance to these but also clear differences (see Müller & Walossek 1991 for an overview of the components). These have been recognized now as representatives of the stem lineage of the Crustacea (embracing the Crustacea s. str. or crown-group taxa and its stem lineage; cf. Walossek & Müller 1990). The fossils lack most of the constitutive characters of the crown-group crustaceans but share at least three derived features with the latter:

- a separate 'proximal endite' at the medioproximal edge of all postantennular appendages (Fig. 54B),

- a multi-segmented exopod at least on the 2nd and 3rd head appendages, with the seta arising from the inner edge of the ramus (same figure), and

- non-filamentous locomotory and feeding 1st antennae, with a special setation for this purpose.

Since these characters are missing in any available outgroup they are considered as evolutionary novelties of the crustacean lineage. Ontogenetic stages are known now from all of the animals, including *Martinssonia*. These stem-group crustaceans provide an interesting insight into the progressive development ('additive Typogenese'), modification and completion of the 'crustacean characters' at the crown-group level. Walossek & Müller (1990) went further to present a set of constitutive external characters in the ground plan of Crustacea s. str.. Since these have a bearing on the status of *Rehbachiella*, some of the arguments are considered again here, including:

- the development of a bipartite locomotory and feeding apparatus, made of a naupliar apparatus extending back to the mandibles and a postnaupliar one, with a set of appendages basically adapted to swimming and suspension-feeding and the 1st maxilla used to interact between the two sets,

- the conical telson (as a non-somitic tail end), with terminal anus and a pair of articulate furcal rami serving as steering devices while swimming,

- the ontogeny starting with a nauplius as the most oligomeric type of a feeding larva with the anterior three pairs of cephalic appendages only, and

- the retention of functionality of the naupliar morphology at least until the adult apparatus, which develops gradually during many moults, is functional.

All changes along the stem lineage are assumed to have evolved progressively toward to a more free-swimming mode of life and new feeding habits of both larvae and adults (see also Dahl 1956). The transformation of the originally exclusively sensorial and multi-articulate 1st antennae in the euarthropod plan, as known, e.g., from trilobitomorphs, into locomotory and feeding devices, with sensorial equipment only distally and setation along its posteromedian edge (sweeping), was already initiated in the stem lineage of Crustacea. Interestingly, also *Agnostus pisiformis* shows a certain degree of modification of its 1st antennae, which seems to be an additional indicator of its systematic separation from Trilobita. Again, the distinct 'proximal endite' on all postantennular limb corms of crown-group crustaceans is a retention from the stem-group level.

The evolution of the limb corms of 2nd antenna and mandible and those of the posterior limbs went different ways in line with the divergence of the two apparatuses mentioned above. In the naupliar limbs the 'proximal endite' enlarged to form a distinct portion, the coxa, while the basis (=basipod) remained to carry the two rami. Depending on the feeding state both portions may carry an enditic process (a large grinding plate in the mandible) or at least a seta medially (Fig. 54C–E). The origin of the coxa from a small 'proximal endite', as developed in the stem-group crustaceans, is reflected in the morphogenesis of the limbs (cf. Figs. 4 and 5 in Sanders & Hessler 1964 for *Lightiella incisa*). In consequence, the 'proximal endite' of all the posterior limbs corresponds to the coxa of the naupliar appendages, while the distal part of the postmandibular limb corms corresponds to the single basal portion of the ancestral euarthropod limb type.

The different evolutionary fate of the two portions may be best seen in its extremes: the phyllopodous limb of non-malacostracan Crustacea has a small 'proximal endite' and a large 'sympodite' which is nothing more than the retained but enlarged fleshy basipod carrying the rami (examples in Figs. 46, 47; 54F, G); and postmaxillary limbs of Eumalacostraca, with a well sclerotized, large 'proximal endite' (=coxa) and a basipod of varying prominence (Fig. 43D, F in McLaughlin 1980; also Fig. 48E). In the latter limb type and in all other types where the proximal portion is distinctive, the term 'coxa' has a long tradition. Whenever it is less prominent, however, it is either neglected altogether or provided with a variety of names (compare, e.g., Figs. 3 and 141 in Calman 1909).

A straight line can now be drawn to the exclusively biramous but segmented limbs of all ancient euarthropods (Fig. 54A), in disagreement with theories of a multiramous or even a non-segmented origin of the crustacean limbs. Again, apart from the development of the 'proximal endite', the original basis as well the two rami ('endopod' = 'telopod'; exopod) are simply retained in Crustacea. Additional exites of any kind are considered as novelties of particular crustacean taxa. However, although very useful for ingroup analyses, comparative work on their homology status is still wanting. Criteria for such purpose may be seen in their position on different portions of the corm (coxa, basis), shape, setation, equipment with muscles, and function (e.g., osmoregulation, respiration).

On the other hand, the whole complex of the separate locomotory and feeding apparatus, comprising essential components of the naupliar apparatus to feed the growing larva until the posterior set of limbs, with endites and setation, were functional, is an imperative for the 'last common ancestor' of the 'crown group' (apomorphic to Crustacea s. str.). This basic naupliar apparatus includes:

- locomotory and feeding 1st antennae (from the stem-group level),

- an enhancement of 'proximal endite' and basis of 2nd antennae and mandibles to form two separate portions with one endite each,

- a large, fleshy labrum projecting over a funnel-shaped atrium oris,

- a postoral sternum with paragnaths, which originate from the mandibular sternite, and

- a special setation and particularly setules on all parts concerned with feeding (labrum, atrium oris, sternum, appendages, etc.).

As a further consequence, a large mandibular coxal body, medially drawn out into an obliquely angled grinding plate, characterizes the ground-plan level of Crustacea s. str. (the short term 'mandible' only for the coxa should be avoided for clearness). Again, the 2nd antenna and mandible probably evolved in a similar manner, and primarily, the antennal coxal endite was at least as prominent as that of the mandible. The prominence as well as structural identity of the 2nd antenna with the mandible is still reflected in the morphogenesis of these limbs not only in the Upper Cambrian *Rehbachiella* and *Bredocaris* but also in the various Recent crustaceans (e.g., Benesch 1969, Fig. 24a, b, for the euanostracan *Artemia salina*; Vincx & Heip 1979, Figs. 4 and 5, for the copepod *Canuella perplexa*; Walley 1969, Fig. 1, for the cirriped *Balanus balanoides*).

Prior to this functional level, not even a primordial type of filter-feeding was possible, which also did not necessitate a special larval type, such as the nauplius. All known stem-group representatives still possessed the phylogenetically older trilobitoid 'hypostome', with the mouth exposed at its rear (Walossek & Müller 1990). The basically prominent and fleshy labrum as a free, projecting lobe above the atrium oris and containing musculature and glands (see, e.g., Boxshall 1985, p. 323), characterizes the ground-plan level of Crustacea s. str. Even at this level a crustacean was not filter feeding, since various functional requirements were still missing.

The assumption of a bipartite limb apparatus in the ground plan of Crustacea s. str. is in accordance with Dahl (1976, p. 164) who stated that the 'double feeding mechanism ... is a prerequisite for the existence of autonomous early larvae'. Hence also, the appearance of a nauplius could not have preceded the definition of the anterior locomotory and feeding structures. Accordingly, the natant and feeding nauplius, as the most oligomeric larval type, is considered to be one of the key ground-plan characteristics of Crustacea s. str. (see also Snodgrass 1956). This larva, or better its apparatus, had to support the growing animal until the posterior apparatus became functional (nauplius as the 'locomotive' for the posterior, gradually developing portion). The earliest larvae of *Henningsmoenicaris*, *Goticaris*, *Cambropachycope*, and *Martinssonia* all have four functional pairs of limbs (Walossek & Müller 1990, and further, still unpublished material), and the same is true for *Agnostus* (Müller & Walossek 1987) and possibly all trilobite protaspides.

The 1st maxilla may have already been used to interact between anterior and posterior apparatus, while the 2nd maxilla was still a morphological and functional trunk limb. The design of both the 1st and the 2nd maxilla, however, may not basically have altered much from the type of stem-group representatives once having achieved the ground-plan level of Crustacea s. str.

Prominent 'flagelliform natatory exopods' with inwardly inserting setation occur in Cambropachycopidae, *Martinssonia* (Fig. 54B), and in *Cambrocaris*, a further arthropod from the Upper Cambrian of Poland which could be added to the group of stem-group crustaceans on the basis of these new definitions (Walossek & Szaniawski 1991). In *Henningsmoenicaris* the exopods of the anterior two postantennular appendages have only a few rod-shaped articles with rigid spines; the paddle-shaped exopods of the more posterior limbs develop from the segmented state by fusion during ontogeny (Walossek & Müller 1990). This indicates that changes in the mode of locomotion and life attitudes were progressively evolving early in the stem lineage, but feeding devices in the vicinity of the mouth as well as a differentiated setation were, for example, still missing. An additional derived character closely allied with the new strategies toward the ground-plan level of Crustacea s. str. is the cylindrical telson with articulate furcal rami as steering devices. They are also missing in the stem group or, possibly, initial in *Martinssonia*.

Virtually all components of the bipartite apparatus can be basically recognized in crown-group crustaceans, or at least appear during ontogeny. The further modifications of the structural components of this primordial apparatus are considered to be of great relevance to the recognition of trends within the different lines and, thus, for the evaluation of the phylogeny of Crustacea.

Rehbachiella possesses all the derived characters listed above for the Crustacea s. str., and can be clearly recognized as a representative of the crown-group crustaceans. Evaluating its relationships within this group, its 'primordial' construction seems to dictate a very basic position. This holds particularly for the simple shield, lateral eyes, large labrum and prominent and well-equipped naupliar appendages, the low level of alteration of the 2nd maxilla, and the cylindrical telson with articulate paddle-shaped furcal rami. These features occur in a similar fashion in the basic body plans of all crown-group taxa and are simply symplesiomorphic. On the other hand, at least some of these features must be regarded with much caution, since the description of *Rehbachiella* is based only on larval material. The three naupliar limbs are, for example, already under way to modification in a specific manner, most clearly recognizable in the mandible and its palp. Moreover, this merely demonstrates that *Rehbachiella* still retained much of the ground-plan characteristics of Crustacea s. str., also reflecting its absolute closeness to other early forms assignable to particular crown-group taxa, such as *Bredocaris*.

Exclusion of closer relationships with non-branchiopods

Remipedia. – These cave-living crustaceans, first described by Yager (1981), show remarkably little resemblance to any of the Upper Cambrian stem-group crustaceans and the members of the crown group of Crustacea. Apart from descriptions of new species and aspects of external features, distribution and ecology (Garcia-Valdecasas 1984; Schram *et al.* 1986; Schram & Lewis 1989; Yager 1987a, b, 1989a, b; Yager & Schram 1986), the group is still incompletely known or even misleadingly described. Gonopores, for example, are on the 8th and 15th trunk limbs, but have been described as being located on the 7th and 14th, since Ito & Schram (1988), following Yager (1981), count the maxillipedal segment as belonging to the head.

Other features are confusing: the mandible lacks a palp, which contrasts with that of the basic plan of Malacostraca, but is described with a 'lacinia mobilis' (Schram *et al.* 1986; Schram & Lewis 1989, Fig. 6B, C), known only from Eumalacostraca. The great individual and specific variability of the number of segments is suspect in that the homonomous segmentation of the trunk may be the result of multiplication rather than a primordial type of segmentation. The retention of segmental glands in all head segments save for the antennal one is most probably a primordial feature (Schram & Lewis 1989) and of little importance.

The missing eyes, the small cephalo-thoracic shield, the whole set of anterior appendages back to the 1st trunk limb (= maxilliped; remipedes are carnivores), the locomotory trunk limbs lacking median setation and endites, the reduced shape of the furca, and the reproductive strategy (hermaphrodites with spermatophores; Yager 1989a) are all derived characters. With all this, it is impossible for me to accept this group as the most primordial one of the Crustacea (cf. Yager 1981, 1989a; Schram 1986; Schram & Lewis 1989). Yet, with regard to the character set of Crustacea s. str. (see above), features such as the possession of a hairy labrum (simply covering the mandibular gnathobases), prominent mandibular coxae, large, hairy paragnaths (forming the posterior closure of the feeding chamber around the gnathobases) and the telson with articulate furcal rami underline that Remipedia represent crown-group crustaceans, and the results of rRNA-sequencing, where remipedes come out close to copepods (Abele, personal communication, 1990), are remarkable. Since any description of their ontogeny, which in my view is imperative for understanding these peculiar animals, is still lacking, detailed comparisons with *Rehbachiella* are postponed.

Malacostraca. – The number of synapomorphies of Phyllocarida – restricted to Recent Leptostraca and fossil Archaeostraca – and Eumalacostraca presented by Dahl (1987) convincingly characterize the Malacostraca as a monophylum, and many more may be found in internal features or embryological data (Dahls criticism of Schrams 1986 inclusion of the Phyllocarida within the Phyllopoda is not repeated here; see also Fryer 1987c). The status of particular members within the Eumalacostraca is, however, still subject to controversy (e.g., Burnett & Hessler 1973; Schram 1986; and Dahl 1987 for the hoplocarid problem; Watling 1981, 1983 for Peracarida), but this is not relevant in this context.

Whereas Eumalacostraca have been extensively studied, information regarding the extant Phyllocarida is still scarce and refers mainly to taxonomic descriptions (examples: Thiele 1904; Cannon 1931, also for older references; Hessler & Sanders 1965; Wakabara 1965; Brattegard 1970; Wakabara 1976; Wägele 1983; Hessler 1984; Dahl 1985; and Bowman et al. 1985, also for historical references [more at other places of the text]). General aspects on morphology and status of the Leptostraca have been reported by, e.g., Rolfe (1969, also with fossil record), Kaestner (1967), McLaughlin (1980), Rolfe (1981), Hessler & Schram (1984) and Dahl (1984); information on feeding and other life habits has been provided by Cannon (1927b, 1931) and Linder (1943).

Some uncertainties exist regarding the relationships of fossil Crustacea assigned to the Phyllocarida and which differ to a greater or lesser extent from the Recent Leptostraca (cf. Rolfe 1981; Hessler & Schram 1984). Dahl (1984) pointed to the clear differences in limb morphology and tagmosis of several Cambrian 'phyllocarids' (e.g., Brooks & Caster 1956; Briggs 1977, 1978). He concluded that there is no definite proof of the existence of Cambrian malacostracans, the earliest unequivocal forms being the Archaeostraca which appear in the Ordovician. However, also this grouping which embraces fairly large and well-sclerotized forms (e.g., Broili 1928; Rolfe 1963; Schram & Malzahn 1984; Jux 1985; Feldmann et al. 1986) may be polyphyletic. In particular the detailed description of Bergström et al. (1987) of the Lower Devonian Nahecaris has indicated that Archaeostraca have to be reevaluated carefully. They are obviously not filter feeders, and have walking legs and fewer pleopods than Recent leptostracans and a different telson. They may equally represent a mixture of forms from either the stem lineage of Malacostraca, Phyllocarida, or Eumalacostraca.

Constitutive characters of the ground plan of Malacostraca, which clearly differ from Rehbachiella, are in particular the biramous adult 1st antennae (considered as a derived condition since it develops from a uniramous appendage during late larval development), the mandible developing a unique tripartite 'palp' (see Fig. 51A) after the complete reduction of the naupliar palp consisting of a basipod and two rami, the 1st maxilla lacking basipodal endites (Figs 48C, F, I), the 2nd maxilla with 3–5 endites (Figs. 48D, G, J), trunk limbs basically with a five-segmented endopod (Figs. 48B, E), and a division of the trunk limb set into two functional units (pereiopods and pleopods), which are all considered herein as thoracic. This interpretation assumes a basic division of the crustacean trunk into 14 limb-bearing segments (thorax), one apodous segment (abdominal) and the telson with furcal rami, from which the malacostracan condition developed by splitting the limb series into two units (see also subchapter on tagmosis in the chapter 'Significance of morphological details' below).

The ontogeny of Eumalacostraca again clearly differs from that of Rehbachiella (see chapter on comparative ontogeny, subchapter on Malacostraca). According to Dahl (1987) the epimeric development of Phyllocarida can be derived from the eumalacostracan type but not from any other type developed among Crustacea. Furthermore, in all Malacostraca the 'proximal endite' has gained more prominence in the postmandibular limbs than in most other Crustacea. In the Phyllocarida this enhancement has affected mainly the two maxillae, while in Eumalacostraca it has progressed further with the formation of distinctive sclerotized coxal portions, variously articulating against the basipod, also in postmaxillary limbs.

Cephalocarida. – The small benthic to endobenthic cephalocarids, first described by Sanders (1955), are rare but widely distributed (e.g., Jones 1961; Gooding 1963; Shiino 1965; Hessler & Sanders 1973; Wakabara & Mizoguchi 1976; McLaughlin 1976; Knox & Fenwick 1977; cf. Schram 1986, p. 352, for detailed list with localities). Yet, despite their importance for the phylogeny of Crustacea, little information has been added since the detailed descriptions of external morphology, larval development and aspects of the mode of life by Sanders (1963b) and the meticulous work on skeleto-musculature by Hessler (1964), including comparisons with other crustacean taxa. There are, e.g., notes on their phylogenetic significance (Sanders 1963a), a short note on the reproductive system of Hutchinsoniella macracantha (Hessler et al. 1970) and more or less short descriptions of the larval development of some species (e.g., Gooding 1963; Sanders & Hessler 1964). A brief report on the probable existence of rudimentary compound eyes (Burnett 1981) has been invalidated now by the investigations of Elofsson & Hessler (1990) on the central nervous system, and more interesting new information on the anatomy has been published recently (Elofsson & Hessler 1991; Hessler & Elofsson 1991).

Cephalocarida differ from Rehbachiella most evidently in their tagmosis which comprises nine limb-bearing thoracomeres (last pair modified to minute egg-carriers, usually not mentioned in the descriptions, and segment counted as first abdominal segment) and 10 apodous abdominal segments (excluding the telson), and in the division of the outer ramus of the trunk limbs, which carries a 'pseudepipod' on its outer basis (Fig. 48A; see subchapter on appendages in the chapter 'Significance of morphological details' below). Additional differences from *Rehbachiella* are the retention of a large 2nd antenna, the laterodorsal position of the 1st maxilla with its elongated proximal endite (Fig. 51B), and the whole postnaupliar feeding and locomotory apparatus including components of the limb design. In both maxillae the proximal endite is not a brush, as developed in *Rehbachiella,* while in all postmaxillulary limbs the proximal endite is less pronounced than in the fossil, even more poorly developed in the 2nd maxilla (cf. Sanders 1963b, Fig. 4).

Comparisons between *Rehbachiella* and Cephalocarida are closely allied with assumptions on the status of the latter taxon. The view of a central position of exclusively extant Cephalocarida has been favoured by Sanders (1955; 1963a, b) and has found general acceptance (e.g., Siewing 1960, 1985; Hessler 1964, 1969, 1982a, b; Hessler & Newman 1975; Lauterbach, various papers). However, this has never been founded on an analysis based on the concept of Phylogenetic Systematics. If it can be assumed that features such as the large 2nd antenna with its multi-articulate exopod (>18 annuli), the shape and armature of postmaxillulary limbs, with poorly developed enditic lobes, short setation and most likely five-segmented endopods, the missing sternitic groove in the thorax, and the insertion of the limbs of this recent group, reflect an even more ancestral state than in *Rehbachiella*, it clearly canalizes the search for relationships of the Upper Cambrian fossil: in this form the two antennae undergo progressive reduction during ontogeny, and the numerous details of the postnaupliar limb apparatus indicate a filter-feeding habit, clearly absent in Cephalocarida; again, the endopods of postmaxillulary limbs are only four-segmented in *Rehbachiella.*

Such a statement – appealing as it may seem with regard to the positioning of *Rehbachiella* proposed herein – requires careful consideration, since it implies that in various features Cephalocarida may not have changed over 500 million years. In fact, with respect to the morphology and ontogeny of *Rehbachiella*, the extant Cephalocarida do not appear primitive, but may have become markedly modified from their ancestors, probably in line with adaptation to benthic life in the flocculent zone rather than originating from it. According to Burnett (1981) this may have occurred relatively recently. Such adaptive changes may be seen in (1) the blindness (neither naupliar nor compound eyes developed), (2) the shape of shield and labrum, (3) the anterior position of the antennae, (4) the limb morphology, with specialized rami including the claw-like appearance of the tuft of spine-like setae on the endopods and outwardly directed 'pseudepipods', (5) possibly a modified use of the sucking chambers in accordance with swimming in morphological orientation, (6) the modification of the 9th thoracopods to egg carries, (7) the reduced size of furcal rami, (8) a special reproductive strategy (2 eggs,

hermaphrodites), and (9) the highly specialized nervous system (cf. Elofsson & Hessler 1990).

This list also indicates that this part of the morphology of Cephalocarida cannot be readily transferred to an 'urcrustacean model', nor can it serve to substantiate a central position of Cephalocarida in any phylogenetic scheme of Crustacea. This must be based on synapomorphies, which still have to be worked out.

Presumed synapomorphies. – There are two features that assist in concentrating the search for closer relationships of *Rehbachiella* with Maxillopoda and Branchiopoda.

One is the conspicuous neck organ on the head shield of its early larvae, which occurs, in my view, in its specific design only in two crustacean taxa: in the Maxillopoda this organ is well-developed at least in the Upper Cambrian *Bredocaris*, but relics/modifications thereof occur also in various Recent taxa; among the Branchiopoda such an organ is well-developed and structurally similar in all taxa save for the fossil Lipostraca, where is has not been discovered by Scourfield (1926). So-called 'dorsal organs' occur also on the head of other crustaceans and even Trilobita, but they are considered here to be different in structure as well as function (see subchapter on this organ in the chapter 'Significance of morphological details' below).

The other character concerns the complex filter-feeding apparatus, including the deeply invaginated sternitic food groove in the thorax ('Bauchrinne' according to Eriksson 1934, p. 60) and a special design of the postmandibular limbs (including the maxillae): this apparatus of *Rehbachiella* is, indeed, structurally and functionally related only to that of the Branchiopoda (detailed argument below). Cannon (1927b) has already pointed out the fundamental differences in the thoracic filter apparatuses and mechanism of Phyllocarida and Branchiopoda, which have evolved independently from more primordial types. Again, Dahl (1976) stated that the phyllocarid functional model 'does not seem to provide a good basis for Eumalacostracan evolution and radiation'. As Manton (1977) suggested, filtration has developed independently several times (see also Fryer 1987b). Accordingly, evolution of a filtratory mechanism simply originated from the same basis, i.e. the postnaupliar (postmandibular) apparatus.

In this context it is, however, important to note that the maxillae cannot have become much modified from the trunk limb design at the beginning of crustacean evolution, as can be still seen in Recent Cephalocarida. A similar shape is also reflected in the larval development of Eumalacostraca, Maxillopoda, and Euanostraca, where the 2nd maxilla develops in close contact with the subsequent series. Hence, their different fate within the diverging crustacean lines is imperative for the understanding of the various apparatuses, since these could not have begun their development at the 'thoracic level'. The convergent modification of the maxillae from quite different starting points might, thus, easily be misinterpreted as synapomorphy (see subchapter on maxillulae and maxillae in the chapter 'Significance of morphological details' and under concluding remarks below).

Exclusion of maxillopod relationships. – Support for the recognition of this taxon as a monophylum, established by Dahl (1956), came not least from the description of two new Recent taxa, the Facetotecta and the Tantulocarida (Grygier 1983, 1984; Boxshall & Lincoln 1983, 1987; Boxshall & Huys 1989; Huys 1991), and two Upper Cambrian taxa (Skaracarida: Müller & Walossek 1985b; Orstenocarida: Müller & Walossek 1988b), providing hitherto unknown body plans and structural details. A remarkable step forward to the inclusion also of the Ostracoda was the discovery of living punciid ostracodes that start their ontogeny with an univalved shield (Swanson 1989a, b).

A paedomorphic evolution from more segmented ancestors has variously been suggested for Maxillopoda (e.g., Newman 1983). Indeed, the basically anamorphic but abrupted ontogeny could explain, for example, the retention of a mandible with basipod and rami in the adult. On the other hand, Newman (1983) assumed that Maxillopoda originated from pre-caridoid malacostracans, or better still from a particular stage of a hypothetical larval sequence. In fact, the ontogeny of both Maxillopoda and Eumalacostraca (phyllocarids have epimeric growth) shows no correspondence at any stage. This is even more the case, since the tagmosis still used by Newman as 5–6–5 does not refer to the basic condition in Maxillopoda, recognized now as being 5–7–4 (plus the telson; see in particular Huys 1991; Newman was aware of that but retained the conventional nomenclature; Walossek & Müller 1992). Again, 'pre-caridoid malacostracans' may refer to anywhere in the stem lineage of malacostracans. With regard to the neck organ, the development of the maxilla, and the ontogenetic pattern (below), Maxillopoda are assumed herein to be entomostracans that have branched off by paedomorphosis from a common ancestor with the Branchiopoda (see Fig. 41, char. 1, 2).

The Maxillopoda are clearly distinct from *Rehbachiella* in their tagmosis – 11 trunk segments basically comprising seven limb-bearing thoracomeres and four abdominal segments (plus the telson) – and the specific mode of development of their two lineages (see chapter on Comparative ontogeny, subchapter on Maxillopoda, below). Shared characters in the sense of symplesiomorphies are the head shield, the compound eyes, the feeding and natant nauplius and its appendages, and the basically small size of the 'proximal endite' of postmandibular limbs (ground-plan characters of Maxillopoda). With this, it seems as if the Maxillopoda, apart from the various specialization of its members related in particular to special life habits, have retained much of the ancestral crustacean body plan.

Hence, the relationship of *Rehbachiella* with either the maxillopod lineage or the branchiopod lineage rests largely on the status of the complex character 'filter-feeding apparatus', i.e. whether it is apomorphic only to Branchiopoda or was present already prior to the branching off of Maxillopoda. Due to the possible paedomorphic evolution of the latter only a little of the adult apparatus can be expected to be retained. Possibly the basic postnaupliar locomotory and feeding apparatus was less developed than in *Rehbachiella*, but clarification remains difficult. The Upper Cambrian *Bredocaris*, for example, has a brush limb 1st maxilla with four endites on its corm possibly split into two sets and a reduced exopod. The setal pattern bears a basic resemblance to a larval *Rehbachiella* limb. This is even more apparent in the 2nd maxilla and thoracopods, which have an only slightly enlarged 'proximal endite' and maximally six more endites with simple paired spines. The rami are symmetrical and paddle-shaped, adapted for swimming.

Only the postmandibular limbs of Mystacocarida are similar to the 1st maxilla of *Bredocaris* and the larval limbs of *Rehbachiella* (save for the reduced exopod). In *Dala* (cf. Müller 1983, pp. 94–97, Figs. 1, 2), with probable maxillopod affinities (Müller & Walossek 1988b, p. 30), the thoracopods have a number of endites, yet their setation is much more poorly developed (Fig. 48L; see also Müller 1981a, Fig. 15). In fact, the postmaxillulary limbs of *Bredocaris* are similar only to larval limbs of *Rehbachiella* and Branchiopoda, and in all these, the setal pattern (see also Hessler & Sanders 1966, Fig. 2D–F, for the mystacocarid *Derocheilocaris typica*) can be traced back to the primordial equipment with paired spines, as developed in the stem-group crustaceans (Fig. 5 of Walossek & Müller 1990; Fig. 48K herein for *Martinssonia*), as well as *Agnostus*, and in trilobitoid limb types (examples in Fig. 27 of Müller & Walossek 1987).

As with the specific setal armature of the postmandibular limbs concerned with feeding, a midventral invaginated thoracic food channel seems to be missing primarily in the Maxillopoda which basically have modified their thoracopods progressively for swimming. Development of a cephalo-maxillipedal feeding apparatus, with modification of the 1st thoracopod to a 'maxilliped', is a special feature only of the lineage leading to the Copepoda. In contrast to Schram (1986) and Schram & Lewis (1989), 'cephalic feeding', as in Maxillopoda, is not accepted as a primordial type of feeding. Again, 'thoracic feeding' – in the sense of postmaxillary – as occurring in sessile cirripeds, is clearly derived and cannot have characterized the ground plan of the 'thecostracan' lineage of Maxillopoda.

General features and systematic status of Branchiopoda

The Branchiopoda are used here in the sense of Claus (e.g., 1873), Eriksson (1934), Bate *et al.* (1967), Kaestner (1967),

Tasch (1969), McLaughlin (1980), and Fryer (various papers). They represent a very diverse group of small to moderate-sized crustaceans. Their fossil record reaches, with confidence, down at least to the Silurian, which indicates a long history for this group (Fig. 40; see also Tasch 1963 and 1969 for further records). Ulrich & Bassler (1931) described shell remains from the Cambrian as conchostracan shields, but Müller (1979, 1982) was able to show that these are the shields of phosphatocopines.

A number of fossils of the Middle Cambrian Burgess Shale fauna were discussed by Linder (1945). However, while trying to challenge Störmers ideas on the distinction between trilobites and crustaceans, he went on to indicate affinities of these fossils to Branchiopoda. Their position with respect to Crustacea and even to the Arthropoda is, however, at best unclear (cf. Whittington 1979, also for further references).

Potts & Durning (1980) regard the group as one of the most ancient and, until recently, conservative group of crustaceans. Having most likely originated in the marine environment, they have been exclusively freshwater for most of their existence. Their stasis in external morphology is assumed to be related to adaptation to temporary ponds, which also have not much changed their essential characteristics since (cf. Fryer 1987c). On the other hand, Potts & Durning (1980, p. 475) remark that 'many are physiologically extremely sophisticated'. Thus, particularly those modifications necessary to survive in the new, often extreme, habitats can hardly be traced back to their original design due to lack of intermediate fossil evidence (changes more at the physiological rather than the anatomical level of evolution, according to the above authors).

The number of environments inhabited by the extant species of the Branchiopoda is widespread. They range from the sea (marine cladocerans) up to mountain ranges (examples: the euanostracan *Chirocephalus diaphanus*; Alonso 1985), and from ponds in the Antarctic area (e.g., the euanostracan *Branchinecta gaini*; Jurasz *et al.* 1983) to saline and even hypersaline waters (e.g., *Artemia salina* and certain cladocerans of the genus *Daphniopsis*, according to Sergeev 1990). Life styles are also diverse, though it is presumed here that all are derived from a primordial type of permanently swimming filter feeder (see also Cannon 1933, p. 318). This strategy is considered to be retained basically in three of the Recent branchiopod taxa, the Conchostraca, the Cladocera and the Euanostraca. In the latter group some species may, however, grow to considerable size (about 10 cm) and modify their serial apparatus to become carnivores (Fryer 1966, 1983 for species of *Branchinecta*; increase in size as evolutionary strategy of Branchiopoda, according to Cannon 1933, p. 326). Notostraca and the fossil Kazacharthra as their possible sister group (see below), on the other hand, have most likely moved to a benthic life and have modified their anterior trunk legs, in accordance with enhanced scleroti-

zation of the whole body, flattening, and an increase in size (additionally, multiplication of trunk segments [also polypody in the Notostraca]).

Interestingly, all extant branchiopods have resting eggs which allow survival in ephemeral waters (e.g., Euanostraca, Notostraca and Conchostraca: Alonso & Alcaraz 1984; Mura & Thiery 1986; Euanostraca: Mura *et al.* 1978; Mura 1986, 1991; Thiery & Champeau 1988; Conchostraca: Belk 1970; Martin 1989; Belk 1989). They may even occur together in the same ponds (Alonso 1985). Today only a few forms, all of which are cladocerans, are marine, but these have clearly migrated secondarily back into the sea (cf. Potts & Durning 1980). Remarkable is the morphological and physiological variability of the Branchiopoda, documented by, e.g., sexual dimorphism, seasonal changes of morphology or individual adaptation to availability of food, etc. It is not unlikely that this was already initiated in the marine ancestors (pre-adaptation).

Dahl (1956) has suggested that the different lines of present-day Branchiopoda radiated independently from the sea into freshwater (also Potts & Durning 1980, and seemingly Preuss 1951). The assumed affinities of *Rehbachiella* with the anostracan lineage of the Branchiopoda, as expressed below, would indeed support such an assumption. Again, it has great impact on the status of characters, since even strikingly similar structures of Recent taxa show up as homoplasies (e.g., reduction of 1st and 2nd antennae). Besides apomorphies of particular extant branchiopods, those features of *Rehbachiella* shared with the different taxa can be recognized as symplesiomorphies and retained from the ground plan of Branchiopoda, prior to branching off in to the different lines and prior to their radiation within fresh water habitats.

Details of morphology and ontogeny of the Branchiopoda have been worked out for a long time. Noteworthy in this respect are the studies of Sars and Claus in the last century (cf. Calman 1909 for additional historical references). Considerable information has been added more recently by Eriksson (1934), Linder (1941, 1945, 1952) and particularly Fryer (e.g., 1959, 1963, 1966, 1968, 1974, 1983, 1985, 1987a, b, 1988). Their heterogeneity and mixture of primordial and advanced features has, however, ensured that their monophyly has never been widely accepted, although strongly advocated for example by Calman (1909) and particularly Eriksson (1934). Not least due to the difficulty of working out synapomorphies and to include all presumed members of this group, including such heterogeneous forms as the 'shield-less Euanostraca', the 'multi-legged Notostraca', and the 'bivalved Conchostraca and Cladocera', the hitherto proposed classificatory schemes demonstrate widely diverging opinions.

In the only more phylogenetically oriented attempt made so far, Preuss (1951) criticized the use of plesiomorphies ('Überschätzung ursprünglicher Bauplancharaktere für systematische Belange'), such as numerous segments, many legs, primitive nerve system, nauplius, retention of naupliar eye into the adult. Instead of taking the point, however, his argument was much influenced by the concept of breaking down the group into two isolated components, the Anostraca and the Phyllopoda, rather than to search for synapomorphies.

Recently Fryer (1987c) has summarized and discussed all major attempts of classification in detail, and no repetition is needed here. Fryer also listed the major diagnostic characters of each group and convincingly excluded certain fossil groups, such as a number of fossils from the Middle Cambrian Burgess Shale and the Lower Devonian Acerostraca (e.g., *Vachonia*, assigned to Branchiopoda by Lehmann 1955; name changed to *Vachonisia* by Stürmer & Bergström 1976), now considered as Phyllocarida.

Fryer argued convincingly against the approach of Schram (1986) who placed the branchiopod taxa with the Phyllocarida and the Cephalocarida, which should include living Cephalocarida and fossil Lipostraca, under 'Phyllopoda' (see also the critique of Dahl 1987 from the malacostracan point of view). It has only to be added that the Carboniferous (Lower Pennsylvanian) Enantiopoda (with *Tesnusocaris*, described by Brooks 1955), which have been included by Bate *et al.* (1967) into the Branchiopoda, should be omitted since their status is at best uncertain. Schram (1986) links them with remipedes, but the poor evidence from new collections, provided by Schram & Emerson (1986), has not improved our knowledge of these forms.

All classifications put much weight on special characters, autapomorphies, of a taxon, which leaves in most cases no 'width' for the evolutionary path undergone since it branched off from the sister taxon. Other characters given for taxa are variously clearly symplesiomorphic. Fryer (1987c, p. 357) notes that the 'component subgroups share a constellation of primitive features' and also remarks (p. 367) that the taxonomic units are often ill-defined, if at all (cf. for example Belks 1982 introduction to Branchiopoda, which in many respects contradicts published evidence). However, Fryer himself preferred to draw back from proposing a new phylogenetic scheme but to treat all major groups as distinctive units, stressing their distinctiveness rather than searching for synapomorphies that could help to reconstruct the relationships between the branchiopod taxa. He proposed to challenge the terms 'Phyllopoda' and 'Onychura' and, moreover, suggests that the subordinate taxa of Conchostraca and Cladocera should be raised to ordinal rank (equally to, e.g., Anostraca and Notostraca), so challenging these widely used names.

Indeed, heterogeneity of a group may reflect plasticity and evolutionary success, especially in the light of a long history, while assumption of relationships between taxa rests on shared characters in the sense of common ancestry. Eriksson (1934, p. 31) arguing in favour of the unity of Branchiopoda, stated that differences between its members are large, but 'betreffen die Unterschiede dennoch nur die Ausgestaltung, nicht aber die morphologischen Grundzüge. Wie für die übrige Organisation gilt dies auch für den Bau der Extremitäten'. This is not concrete enough to characterize a taxon as a monophylum, and moreover, does not help much when trying to relate a marine fossil like *Rehbachiella* to such a group. It is, however, an indication that synapomorphies should be sought in the general area of these structures.

In fact, Branchiopoda have differentiated their post-naupliar feeding and locomotory apparatus to a complex filter-feeding system, which in its detailed basic design is unique among Crustacea (cf. Cannon 1927b). This apparatus has been examined in great detail for the various groups in particular by Cannon (1928, 1933), Eriksson (1934) and Fryer (e.g., 1963, 1966, 1968, 1974, 1983, 1985, 1987b, 1988). The gross equipment of the limbs may be plesiomorphic to all Crustacea s. str., but neither the non-filtering Cephalocarida (cf. Sanders 1963b) nor the filter feeders among Malacostraca and Maxillopoda have evolved to a comparable mode of specialization. Again, not only do filter feeders among Phyllocarida and Eumalacostraca, for example, use different limbs and parts of them for filtration, but the mechanisms also are clearly different (for details see chapter 'Comparisons of locomotory and feeding apparatuses' below).

Already Cannon (1927b) recognized that the feeding system of *Nebalia* cannot have originated from the branchiopod type. Again, Sanders (1963a) argued that the functional model of Cephalocarida in a generalized sense can serve as the basis for the two distinctive systems in Malacostraca and Branchiopoda (extant forms have changed the apparatus for different purposes). The latter assumption needs to be proven, but it seems clear that the 'foliaceous limbs' in phyllocarid Malacostraca, Cephalocarida, and Branchiopoda are only superficially similar but do not reflect a common ancestry (no synapomorphy).

The entire complex of the feeding system of Branchiopoda is, thus, considered here as an apomorphy to characterize its monophyletic state (Fig. 41, char. 3). Its structural components (in the ground plan) include:

- a deep thoracic food groove made by a U- to V-shaped invagination of the sternites and representing the posterior continuation of the cephalic sternal food groove (= 'paragnath channel'),

- posteriorly concave limb corms with posteriorly directed basically lobate endites on the limb corm, many in number and with special setation made of three sets, the posterior one of which represents the filtratory set,

- a position of the maxillae and filter limbs at the margin of the filter groove and with the setation of the proximal endites pointing into the groove.

It also includes a basic functional identity at least in:

- the limb-corm type of filtration with backwardly oriented grid,

- the possibility of food currents entering the feeding chamber all along the limb series,

- the formation of sucking chambers between the filter limbs (for features of Phyllocarida and Cephalocarida see the subchapter 'Comparisons' in the chapter on the apparatuses below).

Secondary modification of it within the branchiopod lineages occurs widespread and may lead, for example, to a further complication of the system, such as among cladocerans, or its reduction and even loss, such as in the notostracans and certain cladoceran taxa.

The sister group of the Branchiopoda is seen in the Maxillopoda due to their shared possession of the watchglass-shaped neck organ, surrounded by a ring wall (at least basically, as recognizable in the Upper Cambrian Orstenocarida; Müller & Walossek 1988b; Fig. 41 herein, char. 1). This structure does not occur in a similar design and function in Malacostraca and Cephalocarida.

Relationships within Branchiopoda and possible taxonomic position of REHBACHIELLA

It is not easy to be sure about the affinities of fossils, when the structures they possessed have become largely modified subsequently, particularly of those forms which are near the roots of taxa. Such fossils may document just the first step(s) of modification, but are still far away from the 'end product'. Various 'bridges' might have been 'washed away' in the course of evolution, and, as a consequence, the extant taxa are clearly distinctive but no longer easily recognizable as being related to one another due to lack of clear synapomorphies. A particular example for such a case seems to be the Branchiopoda.

Since a phylogenetic scheme for the Branchiopoda is not available, an attempt is made here to characterize monophyletic units within the group and, with this, to make an assumption on the taxonomic placement for *Rehbachiella*. Current problems with particular lower rank taxa have been discussed by Fryer (1987a, c), and they are not evaluated here in full detail, although some may be moot.

Since the larval neck organ is developed basically in both the Branchiopoda and Maxillopoda, it remains to be solved whether *Rehbachiella* is a representative of the common stem lineage of these two taxa, of the stem group of the Branchiopoda, or already a member of one of its subordinate taxa.

The well-developed postnaupliar feeding apparatus of *Rehbachiella* is structurally as well as functionally of the branchiopod type. Since it is assumed that this apparatus evolved only within the Branchiopoda, this limits the possibilities for its recognition at least as a stem-group

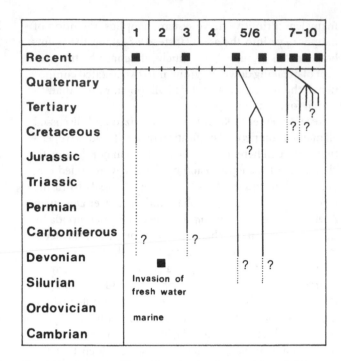

Fig. 40. Fossil record of Branchiopoda, after Bate *et al.* (1967), Schram (1982) and Fryer (1985, 1987c). 1, Euanostraca; 2, Lipostraca; 3, Notostraca; 4, Kazacharthra; 5, 6, Conchostraca with Laevicaudata and Spinicaudata; 7–10, Cladocera with four subtaxa, only Anomopoda reaching possibly into the Cretaceous; ?, questionable fossil records.

representative of this group. Principal shared characters associated with this apparatus are:

- the posteriorly concave limb bases with their backwardly orientated setiferous endites,

- their specific armature and design (also setal sockets with denticles), the deeply invaginated thoracic sternal food channel, and

- the morphogenesis of the whole complex structure.

When reconstructing a phylogenetic scheme with the available data base while searching for possible relationships of *Rehbachiella* within it, it soon became apparent that almost all characters given in the diagnoses for branchiopod taxa are valid only for the extant forms. Recognition of any shared character of the different branchiopod taxa is, more or less, obscured, probably as a result of their long isolated history, which took place in non-marine environments at least since the Devonian. Hence, the available fossil taxa were also taken into account here, since *Lepidocaris* and the Kazacharthra may well serve in estimating the polarity of characters. *Rehbachiella* was also alternatively considered as representative of either of the two branchiopod lineages, and its characters were then tested against the known diagnostic features of the latter, in order to search for synapomorphies and to achieve the best fit.

Even though this procedure may be somewhat subjective, it aided in the recognition of synapomorphic characters of the two major lines of Branchiopoda, the Anostraca and Phyllopoda (An and Ph in Fig. 42), which, in my view, are supportive of their monophyly, and, retrospectively, also helped in the satisfactory inclusion of all three fossil taxa. Already Linder (1945), Preuss (1951, 1957), Dahl (1963), or Bate *et al.* (1967), in particular, have remarked upon the distinctiveness of these lines. Preuss' phylogenetic conclusions were, however, polarized by the assumption that non-filtration represents the plesiomorphic state and that Notostraca accordingly are 'archaic'. His table of 'differential characters' lists exclusively autapomorphies of Euanostraca (= Recent anostracans) and two plesiomorphies, which are set against a whole cluster of shared characters of the subordinate taxa of Phyllopoda. Interestingly, Preuss did not consider the fossil Lipostraca, though stating that they are anostracans and though their morphology has indeed an impact on the bauplan of Anostraca in general. His statements concerning the separate radiation of the different taxa into freshwater are, however, fully acknowledged.

Following Dahl (1963), Tasch (1963), Preuss (1951) and earlier workers, the Anostraca embrace the Recent Euanostraca and the extinct Lipostraca. The fossil record of Euanostraca reaches down with certainty into the Upper Triassic (see Addendum; unnamed silicified fairy shrimp are known from the Miocene, according to Palmer 1957; see also Seiple 1983). A more detailed report of a discovery in the Silurian (Mikulic *et al.* 1985) is still lacking. Tasch (1963) reports a find from the Upper Devonian by Van Straelen (1943, fide Tasch; the publication was not available to me). The Lipostraca are known from the Lower Devonian and comprise the single species *Lepidocaris rhyniensis*.

The Phyllopoda embrace the Recent Notostraca, with a fossil record from the Upper Carboniferous onwards, the Kazacharthra, an extinct group which ranges from about the Upper Triassic to the Lower Jurassic, and the Onychura, following the view of Eriksson (1934) and Preuss (1951, 1957). The latter are considered to include the Conchostraca, with the subtaxa Spinicaudata, known from the Silurian to the Lower Devonian (also from marine deposits, cf. Tasch 1963, 1969), the Laevicaudata, known from the Upper Jurassic onwards (see Addendum), and the Cladocera. However, with regard to the information available, the situation within Onychura remains unsatisfactorily understood (see below).

The assumption of the monophyly of Anostraca and Phyllopoda is founded on the specific expression of a character in the anterior head region.

In the Phyllocarida the lobes of the compound eyes become progressively internalized and eventually embedded into an eye chamber during ontogeny, while migrating towards the dorsal surface (described already by Claus

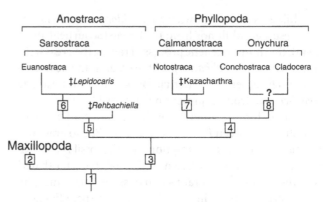

Fig. 41. Presumed relationships within Branchiopoda, including fossil taxa and suggested position of *Rehbachiella*. Selection of major synapomorphic characters of presumed monophyletic units; suggested outgroup (sister taxon) = Maxillopoda; relationships within Onychura embracing Conchostraca and Cladocera cannot be resolved satisfactorily at present, indicated by a '?' (see p. 73). □1. 'Neck organ' on apex of cephalic shield. □2. Tagmosis with 11 trunk segments, i.e. 7 appendiferous thoracomeres and 4 apodous abdominal segments basically, paedomorphosis affecting ontogeny pattern and morphological details (e.g., retention of naupliar mandible), modification of trunk limb apparatus for swimming from primordial type of suspension feeding apparatus. □3. Specialization of postnaupliar feeding apparatus to true filter feeding (incl. thoracic sternitic food channel, etc.). □4. Internalization of compound eyes and shifting of them toward the dorsal surface; naupliar eye composed of four ocelli. □5. Protrusion of forehead region including the compound eyes and reduction of the naupliar neck organ during ontogeny. □6. Reduction of head shield in adults, segmentation of 1st antenna partly reduced, cervical groove on dorsal surface of head, (?) brood pouch formed by 12th pair of thoracopods. □7. Modification of filter-feeding apparatus for bottom dwelling, enhancement of sclerotization, etc. □8. Secondary, eventually bivalved shield, appearing post the naupliar shield after metamorphosis to a pre-bivalved larval stage.

1873; see also Preuss 1951). This process has now been confirmed by the discovery of a pair of slits on the ventral surface of the Kazacharthra (McKenzie *et al.* 1991), documenting, in my view, the original site of the eyes that had migrated inwardly and dorsally (as in Notostraca). A pore may retain the contact with the outside (e.g., Fig. 2 in Preuss 1951; Fig. 41 herein, char. 4). The structural identity in Notostraca, Kazacharthra, Conchostraca, and Cladocera is considered as an indicator that the eye structures do not result from parallel development but developed only once and are apomorphic of the Phyllopoda. Further modification within the group may, however, occur, such as the tendency to fuse the eyes in certain spinicaudate Conchostraca, all laevicaudate Conchostraca, and in the Cladocera.

By contrast, in the Anostraca the corresponding area is progressively raised and slightly separated from the head during ontogeny. In the Recent Euanostraca the process is continued and completed in the postlarval differentiation phase by the gradual formation of eye stalks (cf. Claus 1873, Pls. 2:5, 7; 3:8; 4:11, 13; 5:16; see also Hsü 1933, Figs. 43–48; Valousek 1950, Pl. 1; Baqai 1963, Figs. 5–6A; Bernice 1972, photomicrographs 2–6; Jurasz *et al.* 1983, Fig. 5; Schrehardt 1986a, Fig. 9).

While eyes are not known from *Lepidocaris,* neither from the presumed adult nor from the larvae (Scourfield 1926, 1940), the morphogenetic process of raising the eye is well-documented for *Rehbachiella,* at least until the advanced larvae. The absence of internalization would not exclude *Rehbachiella* from representing a possible stem-group phyllopod, but the progressive change in the other direction shared between *Rehbachiella* and the Euanostraca has led me to favour the assumption that the fossil is a representative of the Anostraca. An additional character shared with the extant Euanostraca – and possible synapomorphy of Anostraca – is seen in the reduction of the naupliar neck organ during ontogeny (not described from *Lepidocaris*; both features are included in character 5 of Fig. 41). In Euanostraca it is replaced by the osmoregulatory distal epipod (misidentified as an exopod by Schrehardt 1987a) during late ontogeny and, more or less, lost (Criel 1991, p. 183).

On the contrary, in all phyllopod taxa the neck organ not only persists into the adult (also occurring in the presence of developed epipods), but may also be more complex in its structure. Information on this organ is, however, still rather uneven, since hitherto studies of this organ have focussed mainly on certain Cladocera (including marine forms), a few Conchostraca, and especially *Artemia salina* (see separate chapter on this organ below). Accordingly, this character requires further clarification by a detailed comparative study of its morphogenetic and functional changes in the different branchiopod taxa, which is still lacking. Again, the status of the presence of a functional neck organ retained into the adult phase is still unclear. Focussing mainly on Cladocera, Potts & Durning (1980) regard it as an advanced character, achieved by paedomorphosis. Notostraca and Conchostraca, however, have this structure also in the adults. Alternatively, the continuing presence of the neck organ up to the adult may represent the plesiomorphic condition.

Additional features supportive of an anostracan relationship of *Rehbachiella* rather than with any other branchiopod group may be seen in the gross design of the slender trunk with its faintly developed tergites and particularly in the growth strategy which is remarkably close to that of the living *Artemia* (see chapter 'Comparative ontogeny'). Again, similarities to the limb apparatus of Euanostraca are greater than to that of any of the phyllopod taxa, but there are also considerable differences in the limb design.

The number of 13 thoracomeres with 12 pairs of limbs of *Rehbachiella* would agree with Euanostraca (genital segments are treated as thoracic, in the sense of Benesch 1969), but it remains difficult to evaluate the status of this feature within the Branchiopoda, because this is much dependent on assumptions about the interpretation of the limb-less posterior trunk region and the tagmosis of Crustacea in general: according to Linder (1945) the number of trunk segments in female laevicaudate Conchostraca is 13, the number of limbs 12 or 13, if the nature of the opercular lamellae as limbs is accepted; what is more, Conchostraca show no segmentation in the post-thorax. Hence, it cannot be excluded that the number of 13 thoracomeres is symplesiomorphic at least to Branchiopoda.

The monophyly of the Phyllopoda can be founded on a further, internal character: the naupliar eye comprises four rather than three ocelli, as in Anostraca and all other Crustacea (cf. Eberhard 1981, p. 24; Schram 1986; Huvard 1990). The conclusion of Paulus (1979) of the basically quadripartite naupliar eye in Crustacea (and also Lauterbach 1986), thus, requires substantiation. Differences in limb musculature between the Anostraca and the Phyllopoda, as has been claimed by Preuss (1951, 1957) are simply based on erroneous homologization of all portions and on erroneous interpretation of the limbs of *Lepidocaris* (see subchapter on Sarsostraca in the chapter 'Anostraca' below).

Accordingly, the use of 'Phyllopoda' in other senses, either as equivalent of 'Branchiopoda' or to enclose all groups save for the Cladocera, is rejected, and the prefix 'Eu' is superfluous. Again, the use of 'Gnathostraca' to enclose Phyllopoda in the above sense and Cephalocarida as a subordinate taxon of the Anostraca (e.g., Dahl 1963; Siewing 1985) should be abandoned: Cephalocarida have been convincingly demonstrated by Sanders (1955, 1963a, b) to be a distinctive monophyletic taxon within Crustacea s. str. This would rather create a paraphylum, not least in the light of the supposed sister group relationships of Branchiopoda and Maxillopoda (see above). The same is true for Lauterbach's (1974 and subsequent articles) 'Palliata' which should embrace all crustaceans with large shields, such as the Phyllocarida (Malacostraca), the Ostracoda (Maxillopoda) and the Conchostraca (Branchiopoda).

Fryer (1987c) has criticized taxa erected to embrace Recent and fossil forms, such as 'Sarsostraca' for Anostraca and Lipostraca and 'Calmanostraca' for Notostraca, Acerostraca and Kazacharthra. According to Ax (1985, 1988, 1989) fossils should not be treated as of equal rank to living taxa but should be placed to the stem lineage of monophyletic units with descendants into the Recent. Willmann (1989b), on the other hand, gives good reasons for fossils to have an equal right in the reconstruction of phylogenetic systems. Once monophyletic units are characterized they should be treated as such, also when the sister group is only a single fossil species. In consequence, 'Sarsostraca', embracing the sister taxa Euanostraca (recent anostracans = Anostraca s.str.) and *Lepidocaris,* is considered as a monophyletic taxon, characterized by the progressive atrophy (or effacement) of the larval shield during ontogeny (besides the fact that *Rehbachiella,* as their possible sister taxon or stem-group form, was marine; Fig. 41, char. 6; characters for subordinate taxa given below).

The Notostraca by far precede Kazacharthra in their fossil record. Both are hypothesised to represent sister groups, sharing at least the design of the anterior pairs of trunk limbs. This was achieved most probably in line with a move to a bottom-dwelling mode of life, loss of the filter-feeding habit, and an enhancement of sclerotization of their common ancestors (Fig. 41, char. 7). For the taxon embracing these two groups, the name 'Calmanostraca' is available, but excluding the 'Acerostraca', which are no branchiopods.

The sister group of the Calmanostraca is seen in the 'Onychura', combining Conchostraca, with the two lineages Spinicaudata and Laevicaudata (in the sense of Linder 1945), and Cladocera. Its monophyly is founded on the unique development of a secondary shield subsequent to the naupliar shield during ontogeny, which eventually becomes bivalved (the larval shield is retained in the form of a 'forehead cover'; see chapter on bivalved shields below; Fig. 41, char. 8). More shared characters have been noted by Preuss (1951), such as: the development of claspers at least on the 1st thoracopods and the claw-shaped furca ('abreptor'; 'absence' in laevicaudate Conchostraca is just a misinterpretation of its 'larvalized' shape; see Linder 1945, in particular his Fig. 7), possibly also the formation of a brood chamber in the dorsal part of the secondary shield.

The situation within the Onychura remains problematical. Fryer (1987c) states that the two conchostracan lines have a separate origin, but without discussing from which branchiopod taxon they should have branched off (possibly also the intention of Martin & Belk 1988, though they interpret with caution by pointing to a possible paedomorphic origin of the Laevicaudata). In this case the name 'Conchostraca' has to be abandoned as referring to a paraphyletic unit – but only given the additional assumption that Cladocera originated from Spinicaudata. In either way the monophyly of 'Onychura' remains untouched.

With regard to sarsostracan Anostraca, *Rehbachiella* exhibits various plesiomorphic characters, indicative of its nature as a stem-group representative of the Anostraca. These are in particular:

- the presence of a large shield (lost during ontogeny in Sarsostraca),

- the well-segmented the 1st antenna (segmented only in the nauplius of Euanostraca, see Fig. 53C and pp. 115–116),

- the two developed pairs of maxillae (see p. 118),

- the segmentation of the endopods of the postmandibular limbs (fused to undivided paddles in Sarsostraca),

- the furca being developed earlier than in Euanostraca (specialized in Lipostraca), and

- the completeness of the larval series.

A corollary of the assumption that *Rehbachiella* represents a stem-group anostracan is that the two branchiopod lineages had branched off already before the Upper Cambrian, in accordance with the assumptions of several authors about a separate radiation of the lines into the non-marine environment. Comparisons of selected morphological characters within the different Branchiopoda and other crustacean taxa, will be given in the next chapters.

If later instars of *Rehbachiella*, i.e. from the post-larval differentiation phase, could be established, it may well be that details in the feeding and locomotory apparatus of *Rehbachiella* give further support to the presumed relationship (besides the remarkable similarity in the setal sockets of the posterior row of filter setae, or the delicate corona of denticles at the sockets of exopodal setae, as in Fig. 4 of Criel & Walgraeve 1989 for *Artemia*). Suitable features could also be the design of the food channel and the specific shape and orientation of the endites in the different postmandibular limbs. Again, more detailed re-study of all major branchiopod taxa with regard to their phylogenetic status is imperative to improve and complete the presented hypothesis.

It would also be interesting to know whether certain similarities between Laevicaudata, Spinicaudata and *Rehbachiella* are more than superficial and thus have a bearing on the status of the Conchostraca among the Phyllopoda. Whereas the fossil record is incomplete, it is worth noting that Spinicaudata precede Notostraca in the geological history and are also known from marine deposits at least in the Carboniferous (cf. Tasch 1963, 1969; Fig. 40 herein). This is in accordance with the remarks of Fryer (1987c, 1988) on the isolated position of the Notostraca due to various modifications of morphology, life habits, feeding structures (see Cannon 1933), and ontogeny (particularly Fryer 1988).

Characters of the two branchiopod lineages

In addition to the above notes on the two major lines, the descriptions of the branchiopod taxa focus on critical characters of the external morphology. Detailed descriptions of the subtaxa, including anatomical and ecological specialities, are widespread in papers and textbooks.

Anostraca

According to the presented scheme, the Anostraca embrace the fossil taxa *Rehbachiella* and *Lepidocaris*, and the extant Euanostraca. *Rehbachiella* may represent the anostracan stem line, sharing the extrusion of the eye region and early reduction of the neck organ with the other members of the group. A plesiomorphic character of *Rehbachiella* is, for example, the presence of a well-developed head shield, which is progressively reduced in size during further evolution of the Anostraca. The monophyly of 'Sarsostraca' to embrace the fossil 'Lipostraca' and the extant 'Euanostraca' is founded, at least, on the lack of a head shield in the adult, the presence of a 'cervical groove' dorsal to the mandibles, and, possibly, the modification at least of the last (12th) pair of thoracopods into egg-carrying devices.

Lipostraca. – The small *Lepidocaris* (Fig. 42G) has been found fairly abundantly in the Lower Devonian (stage uncertain: Siegenian to Eifelian; Fig. 40) Rhynie Cherts of Aberdeenshire, Scotland, a lagerstätte well-known for its plant remains and the first report of a terrestrial insect (spring tail). Embedding in a chert matrix has resulted in a similar exceptional bodily preservation, as in the *orsten*, with setation also still in place and with a number of ontogenetic stages. Thanks to Scourfield (1926, 1940), various details are known and have found their way into the textbooks. Remarks are confined here to special features of this fossil and to inconsistencies in the original description.

Lepidocaris has 17 postcephalic trunk segments plus an elongate telson with a unique double furca and articulated lateral outgrowths (Scourfield 1926 counted the head as the first somite and telson as the last to achieve the number of 19, known from Euanostraca). Unable to recognize the 2nd maxillae, Scourfield (1926, p. 164) believed them to be represented by a small structure behind the 1st maxillae. It is remarkable that there is no space between the 1st maxillae and his 1st thoracopod (Scourfield's Fig. 32), in particular in the males, which even have a prominent, clasper-shaped 1st maxilla (his Figs. 24, 25, 29). Furthermore, the first trunk limb is noted as having a cephalic insertion (Scourfield 1926, p. 165 and Pl. 22:3, 5), while there is at least one more pair of trunk limbs than in Euanostraca. Lastly, the structure to which Scourfield was referring is remarkably similar to the openings of the maxillary glands.

Hence, Schram's (1986) interpretation that this limb is the trunk-limb shaped 2nd maxilla is convincing. In consequence, *Lepidocaris* had not only well-developed 1st maxillae, at least in the males, but also 2nd maxillae with the shape of a trunk-limb, 13 thoracomeres, four apodous abdominal segments and a cylindrical telson. Twelve postmaxillary body segments had flattened limbs in the males, while the females modified the last two pairs. Characteristic are the lack of eyes and the modifications of the postmaxillulary limb apparatus for scraping (anteriorly) and swimming (posteriorly). A head shield is missing in the presumed adult, but is present in earlier larval stages (Scourfield 1940, Fig. 3: approximately TS8 according to *Rehbachiella* stages).

The feeble 1st antenna is composed of three podomeres. The 2nd antenna is prominent, and its corm is distinctly subdivided into a coxa and basipod. The endopod is two-segmented, while the exopodal annuli are fused to few segments (Scourfield 1926, Pl. 23:1, 2). The adult mandible lacks the basipod and rami, present in larvae (Scourfield 1940, Fig. 5). As in Euanostraca, a 'cervical groove' demarcates the position of the mandibles on the dorsal surface of the head.

No traces of epipodial structures were recognized by Scourfield on any of the postmandibular limbs, which are essentially biramous. However, as in *Rehbachiella*, which also lacks epipods up to the latest instar, the outer edges of the limbs are bi- or tripartite. Of the 1st maxilla the setiferous proximal endite is present in the females (Scourfield 1926, Fig. 14), similar to that of Recent Euanostraca, while the limb is a clasper in the males (Scourfield 1926, Pl. 23:6). The postmaxillulary limb apparatus is divided into three sets of limbs (Fig. 46F–H):

- 2nd maxilla (redefined) and anterior two thoracopods compact, with a corm carrying a large proximal endite and five smaller endites; flattened one-segmented endopod with scraper setae; exopod smaller and leaf-shaped (Fig. 50A),

- 3rd–5th limbs more slender; median armature of corms as in the 1st set but less developed; rami symmetrical, outer edge of limbs with incisions,

- 6th–10th limbs as before but with rudimentary armature medially.

The symmetrical rami of the natatory posterior limbs resemble more maxillopod than euanostracan limbs, but a similar design can be seen, e.g., in the 5th thoracopod of the juvenile euanostracan *Branchinecta paludosa* (Linder 1941, Fig. 12d; Fig. 46E herein) or the last (71st) limb of the notostracan *Lepidurus lynchi* (Linder 1952, Fig. 23; Fig. 47F herein). Again, symmetry of rami occurs also in the 2nd antenna of conchostracans.

In clear contrast to Recent Euanostraca, two pairs of thoracopods form the egg pouch: the 11th pair is an 'egg pouch cover' and is still limb-shaped, while the 12th forms the egg pouch itself. The existence of a 13th pair of rudimentary limbs posterior to the egg pouch in females has been supposed by Scourfield (1926, pp. 169–170 and Fig. 22) but remains uncertain. (Even if it had existed, this would not contrast with the euanostracan morphology, since in these a 13th pair is present but is not externalized. It remains at developmental stage 2 according to Benesch 1969, p. 352. It is also not involved in the formation of the brood pouch.) Features unknown from any other Branchiopoda are the developed pleural scales of the anterior thoracomeres and the peculiar telson: according to Scourfield (1926, pp. 160, 161, e.g., Figs. 7, 8), it terminates in four rod-shaped extensions, most likely a derived state, and comparable to the rod-shaped furcal rami of other small Crustacea, such as Cephalocarida, Copepoda, *Bredocaris*, or Remipedia.

Few details are known of its development as yet, which is seemingly anamorphic; metamorphic changes, as Scourfield (1926, p. 184) claimed, are not apparent. *Lepidocaris* may have lived in fresh water, probably with enriched content of silica (Scourfield 1926, p. 154; recent research confirms the assumptions of a fresh-water environment, according to Clarkson, personal communication, 1990).

Due to the assumed sister-group relationships of Lipostraca with Euanostraca any affinities with Cephalocarida, as stated by Schram (1986, pp. 340–343), must be rejected. Also Fryer (1987c, p. 364) notes that 'of the characters listed in the text to show the similarity of Lipostraca and Brachypoda, several are erroneous, some are dubious, and others in fact emphasize the branchiopod nature of the Lipostraca'. In fact, Schram's list contains various plesiomorphies, such as the shield, natatory antenna, proximal gnathite [endite], 'thoracoform' 2nd maxilla, and anamorphic ontogeny, as well as errors, such as the biramous mandibular palp (in fact only the basipod and two-segmented endopod), polyramous trunk limbs (essentially biramous), and a horseshoe-shaped cephalon (actually the shield, which is missing in adult *Lepidocaris*). On the other hand, Schram did not consider the egg pouch and other characters (e.g., larval design and morphogenesis of limbs) that link *Lepidocaris* with Euanostraca (he prefers to relate the egg chamber of Lipostraca to the minute egg carriers of Cephalocarida, in fact comprising the 9th pair of thoracopods) or the strikingly different rami of trunk limbs of Cephalocarida, but refers to size or lack of eyes as 'shared structures' between these two taxa.

Euanostraca. – Euanostracans show, in many respects, a very conservative design, an impression not least influenced by their serial homology of postmaxillary limbs (Fig. 42H). Most species are microphagous filter-feeders, while size increase to more than 10 cm and morphogenetic changes of the filter apparatus may also permit carnivorous habits (Fryer 1966 for *Branchinecta gigas* and 1983 for *B. ferox*).

The head is shorter than in Lipostraca, most probably in the course of a further compression of the maxillary segments, which are still visible by their feebly developed boundaries in early larvae (Fig. 53D for *Artemia franciscana*). The larval neck organ is very large and takes up most of the larval shield (Figs. 44B), the posterior margin of which, however, is clearly recognizable in the nauplius immediately after hatching (Fig. 53B).

The trunk comprises 19 segments plus a cylindrical telson with a terminal anus and leaf-shaped to elongate furcal rami. The 12th and 13th trunk segments are the genital segments. They are well separated at least in *Branchinella* species (Thamnocephalidae), according to Geddes (1981), but variously show a tendency to fusion. Since the segments are modified thoracomeres, according to Benesch (1969, pp. 401, 439), the segmentation of the trunk includes 13 thoracomeres, 12 of them bearing appendages, and an apodous abdomen which consists of six segments and the telson. Higher numbers of segments in certain genera clearly represent the derived state (e.g., *Polyartemia*).

The 1st antenna shows no segmentation in adults, but segmentation is at least recognizable, e.g., as a faint subdivision in *Branchinecta occidentalis* (Heath 1924), or as folds and muscles in *Artemia salina* (Benesch 1969, Fig. 23). In larval, possibly also later, stages a division into two portions is still recognizable in *Chirocephalus*, which has very long 1st antennae (see, e.g., Oehmichen 1921; Valousek 1950; both for *C. grubei*). Likewise the nauplii of *Artemia franciscana* (Fig. 53C) show a clear segmentation pattern enhanced by rows of denticles. This indicates that the missing segmentation of later stages is simply due to effacement rather than a primary loss (see section on the 1st antennae in the chapter 'Significance of morphological details' below).

The 2nd antennae are large, dominant feeding and locomotory organs up to late larvae, and eventually become largely modified and different in the sexes, particularly in Thamnocephalidae (e.g., Geddes 1981; Belk & Pereira 1982). In the males they become clasper organs. The larval mandible is, as in other crustacean groups, biramous, according to Benesch (1969) who also studied the musculature (particularly his Fig. 24b; see also Baid 1967; Bernice 1972). During the postlarval differentiation phase the palp completely atrophies. The 1st maxillae are present with their prominent 'proximal endite' (= 'gnathobase' in Fryer 1987c); the 2nd maxillae are smaller but similar (in both these are not the limbs themselves).

The corms of the foliaceous thoracopods bear a series of 5–6 flattened and vertically oriented endites (six, according to Fryer 1983, because the 'proximal endite' may be composed of two elements; see also McLaughlin 1980, Fig. 5D). Endopod and exopod are undivided (not one-segmented) and flattened (Figs. 46A–E). These develop progressively from lobate protrusions, as in *Rehbachiella* or *Lepidocaris*, and become applanate eventually (particularly Nourisson 1959, Fig. 1–6; Bernice 1972, Fig. 1–7; Schrehardt 1987a, Figs. 24, 27, 28). In the early stages the two rami are clearly developed, while the exites on the outer side develop gradually. The prominence of the rami may vary between the different species, from the endopod being the larger ramus to the reverse condition (Figs. 46A–D).

Up to two fleshy epipods arise more proximally from the outer edge of the corm. Of these, the distal one serves as an osmoregulatory organ, after the neck organ has become atrophied (see particularly Croghan 1958b). The more proximal one, which can subdivide into two leaf-shaped portions (e.g., Geddes 1981), may be used for respiration (Schrehardt 1987a).

Evolutionary trends of Euanostraca affected various internal features, such as the loss of the dorsal wall of the heart, as well as external ones, such as the complete loss of the shield during ontogeny, the enhancement of sexual dimorphism, and adaptation to extreme habitats (e.g., hyper-saline environments). Plesiomorphic is the persistence of derivatives of segmental glands in all postantennular head and thoracic segments, probably even in the segment of the 1st antenna (Benesch 1969, pp. 433–435, list on p. 436; also Warren 1938 and Fränsemeier 1939). This feature has frequently been overlooked but is of significance for phylogeny.

Recognition of *Lepidocaris* as well as *Rehbachiella* as anostracans is of relevance for the status of several features of the extant Euanostraca. For example, while a shield, as recognizable in *Rehbachiella*, must have characterized the ground plan of Anostraca, it had dwindled progressively during further evolution. Both Euanostraca and *Lepidocaris* lack the head shield (synapomorphy of Sarsostraca); however, this is valid only for adult stages, since not only the larvae of *Lepidocaris* have a prominent shield (see above), but also the nauplius of *Artemia franciscana* immediately after hatching (cf. Rafiee et. al. 1986, Fig. 4E; incorporated in Figs. 28A and 53B herein). With this, a shield must still have been present, at least in early larval stages, in the ground plan of Sarsostraca, while further reduction occurred in Euanostraca, probably in line with the enhancement of the naupliar neck organ (increase in the number of cells) and compression of the head region (cf. Fig. 53D). The neck organ is reduced before the adult state, but remnants of it are retained in some extant species as a small structure of unknown function on the apex of the head (see chapter on this organ below). Such a neck organ is not known from *Lepidocaris*, but early larval stages were not found by Scourfield (1926, 1940). In *Rehbachiella* this organ is exclusively restricted to the early larvae and becomes effaced already after delineation of four thoracomeres.

Again, the 1st antenna is well-segmented, and even more clearly recognizable in the larvae (Scourfield 1940, Fig. 4). Remnants of segmentation occur also in Euanostraca (see above). Hence, segmented 1st antennae were present in the ground plan of Anostraca, as known from *Rehbachiella*, as well as the ground plan level of Sarsostraca (plesiomorphy). In Euanostraca the 1st maxillae consist only of the proximal endite. According to Fryer (1983), however, these play a vital role in transport as food scrapers, being well-equipped with pusher-spines and setae. Taking the well-developed 1st maxilla in male *Lepidocaris* into account, their definite reduction to the proximal endite took place in the stem line of Euanostraca. On the other hand, the 1st maxilla of *Rehbachiella* is already much shorter than the subsequent limbs. Moreover, of the four specialized endites on the corm, the proximal endite is by far the largest and most important element (the only portion retained in female *Lepidocaris* and in both sexes of Euanostraca). The plesiomorphic trunk-limb shape of the 2nd maxilla of

Lepidocaris indicates that this state was retained for longer within the Anostraca, and this limb was reduced eventually within the Euanostraca.

A significant structure shared between Euanostraca and *Lepidocaris* is the modification of the posterior thoracopods of females to form an egg pouch. At present it is, however, difficult to interpret this as a valid synapomorphy: in Euanostraca the 12th–13th thoracomeres may fuse with one another, but only the 12th pair of limbs is modified into plates that form the pouch; in *Lepidocaris* the 11th–12th limbs form this structure (see above). Hence, specialized reproductive features, such as in Recent Euanostraca, may not have been ground-plan characters of Anostraca for two reasons:

- in the males of *Lepidocaris* the last pair(s) of thoracopods is(are) not modified as reproductive aids as in the females, indicative of only partial completion of the reproductive strategies of Recent Euanostraca,

- in Recent euanostracans the specialization of the 12th and 13th segments does not occur in the larval phase, i.e. before delineation of the abdominal segments starts, while it is presumed that only the larval sequence of *Rehbachiella* prior to this phase is documented, though at least the 11th limb is already achieving a typical limb shape (ds3b–4a; 12th is a bud).

This indicates that reproductive features may have evolved since the Cambrian in the Anostraca, also implying that they developed independently within the different branchiopod lines. As in *Rehbachiella*, in *Bredocaris* the last thoracic segment has ordinary thoracopods, while they are reproductive aids in Recent members of Maxillopoda. Hence, in general terms a more basic level of reproductive strategies in Crustacea may have existed in the Early Palaeozoic.

Specific to *Lepidocaris* are in particular the morphological changes in the life style away from the primordial mode of filter feeding, including: blindness, modification of the postmandibular limbs series by a splitting into an anterior feeding portion and a posterior set of progressively more exclusively natatory limbs, the clasper-shaped 1st maxilla in the males, the egg-pouch cover of 11th thoracopod in the females, and the unique furcal rami. Apparently already foreseeing the status of these fossils as suggested herein, Eriksson (1934, pp. 89–97) concluded that *Lepidocaris* cannot be the 'urtype' of an anostracan but that its development 'ging von primitiven – jedoch, nach allem zu beurteilen, sicherlich nicht von sehr primitiven – Anostraken aus'. The large natatory 2nd antenna seems to be a retention of the basic plan, as found in *Rehbachiella* (larval), however, fusion of the ring-shaped exopod articles, fewer endopodal segments, fairly short setation, and a bundle of setae on one of the portions of the corm (brush function?) cast doubt on their primitive status (personal observations of the type material).

The various constitutive characters of the Euanostraca are well-described and need no repetition. A possible autapomorphy of *Rehbachiella* may be the enlargement of the spine-bearing ventrocaudal processes.

Phyllopoda

The Phyllopoda are distinguished from Anostraca particularly by their internalization of the eye region. They comprise the Calmanostraca, with the Notostraca and the exclusively extinct Kazacharthra, and the Onychura, with the laevicaudate and spinicaudate Conchostraca, and the Cladocera (but see below), all with extant representatives.

Calmanostraca. – The Notostraca have been well-described particularly by Linder (1952), Longhurst (1955) and Fryer (1988; including morphogenetic changes of the limb apparatus), and so only a few additional remarks need to be made. For example, the internalized compound eye (e.g., Claus 1873, Pl. 7:5b, c) has shifted dorsally into close contact with the persistent neck organ (e.g., Linder 1952, Fig. 19 , or several species; Longhurst 1955, Fig. 4) and in close contact with the naupliar eye, which also lies dorsally underneath the integument (Eberhard 1981, Fig. 9, after Nowikoff; regrettably, in Siewing's [1960] widely adopted body plan of Notostraca [his Fig. 18A] this organ is termed 'naupliar eye'). The pit anterior to the compound eye (same figure) is the opening of the eye chamber.

The position of the compound eyes can already be recognized in Mesozoic representatives of this group (Chen 1985). It should be added that only in notostracans is the naupliar eye enclosed in a 'pocket' together with the compound eye, while in the other members of the phyllopod lineage the naupliar eye is in front of or ventral to the eye chamber (Eberhard 1981, p. 16; character not known in detail from Kazacharthra).

The naupliar structures become largely modified during development, and progressively the head and much of the trunk become enclosed in a large, well-sclerotized, univalved shield. The two pairs of antennae are much reduced in the adult state. This is particularly true for the 2nd which, however, retains its typical crustacean shape until the late stages (e.g., Claus 1873, Pl. 8:5; Fig. 45G herein). The mandible lacks the palp in the adult, as in other Branchiopoda (unclear whether it is biramous in early larvae). The 1st maxillae are present with their large proximal endites, carrying rigid spines (Fryer 1988, Figs. 100, 101), while the 2nd one is much smaller (e.g., Longhurst 1955, Fig. 11).

The trunk limb series is not homonomous as, e.g., in Euanostraca (compare Fig. 47D–F), which led McLaughlin (1980, p. 9) to term the anterior set back to the 11th pair of trunk limbs (egg carrier in females) 'thoracic' and the posterior set 'abdominal'. The 1st trunk limb is very long and may extend far beyond the shield. Its proximal endite is separate and large, the subsequent four endites are elongate, or, better, their median surfaces are extremely drawn out. Its developmental path can be recognized along the limb series as well as in the morphogenesis of the limbs. The endopod is fairly small, while the exopod is leaf-shaped and of varying size in the different species. Notostraca have one hose-shaped epipod located proximal to the exopod (Fig. 47D, E) similar to the elongated epipod of Conchostraca.

The subsequent trunk limbs are smaller and progressively change into a more phyllopodium-like shape (Fryer 1988, Figs. 6–11; also Figs. 47E, F herein). Up to six pairs may occur on each of the subsequent 12–17 trunk segments (e.g., Linder 1952, Fig. 20; 'polypody', see Siewing 1985, p. 885). The number of apodous abdominal segments is variable, and annulation of the body is not correlated with the number of segments, since the body rings can be spirals (Linder 1952, Figs. 3–7). In the posterior limbs the rows of setae on the lobate endites become more distinctive and referable to the three sets of *Rehbachiella* (Fig. 48F). The proximal endites are vertically oriented, and it seems as if the posterior row (set 3) of *Rehbachiella* points anteriorly (Fryer 1988, Figs. 29, 118). The set of trunk limbs is not used for filtration, but for feeding on all kinds of food available to a bottom dweller (also carnivorous).

The labrum is made of a sclerotic plate, superficially resembling the 'hypostome' of trilobites. It is regarded herein as a special structure associated with life at the bottom. Further adaptive features in this way can be seen in the flat shield, the size and orientation of the limbs, and the sclerotization of the trunk with modification of the supraanal flap and 'cerci-like' furca (counteractors of the shield in analogy to horseshoe crabs, according to Eriksson 1934, pp. 234–235). The lack of correlation between trunk segmentation and limbs and the high infra-specific variation (Bushnell & Byron 1979) demonstrate that the multilegged trunk must be regarded as a derived and not as a primitive feature, as claimed by Lauterbach (1986).

Of the large number of features of Notostraca, the modification of the whole series of appendages away from filter feeding, enhanced sclerotization, applanation of the wide shield, sclerotized labrum, absence of a food groove, and a telson with slender furcal rami (most probably all in line with adapting to a bottom dwelling life) are examples of features that also are possessed by the Upper Triassic to Lower Jurassic Kazacharthra (Novojilev 1957, 1959; Chen & Zhou 1985; McKenzie *et al.* 1991), a diverse group but known only from deposits of Kazachstan and South China (Fig. 42A; more data in Schram 1986, pp. 360–363; and Fryer 1987c; synapomorphies of Calmanostraca).

The trunk limbs of Kazacharthra show the same basic design as the anterior trunk limbs of Notostraca, save for the lack of epipods (compare Figs. 47B and 47D). Kazacharthra, however, have a lower number of thoracopods than Notostraca, possibly 11, but a higher number of apodous trunk segments, while the latter have multiplied their trunk segments (the last segments are even spiral-shaped), the anterior ones carrying more than one pair each (up to six pairs; the total number may reach 71 pairs).

Other features of interest in Kazacharthra are:

- the well-segmented 1st antenna, with up to 15 annules,

- the well-developed 2nd antennae, with a three- to four-segmented endopod and a 10–15-segmented exopod,

- mandibles with strong coxal gnathobases and missing palp,

- the maxillulae represented only by the proximal endite (maxillae obscured on all specimens available to McKenzie *et al.* 1991),

- smaller 1st thoracopods than the subsequent ones,

- the 2nd to 6th legs are larger than the remaining set, and

- the proximal endite being much elongated (see Fig. 47B).

With regard to these characters the Kazacharthra are not considered as representatives of the stem-group of Notostraca, but as their sister group, which is in accordance with their later appearance in the fossil record.

Various authors have pointed to the similarities of Notostraca and Conchostraca (e.g., Martin & Belk 1988; Siewing 1985, p. 890). Besides the conclusion that Calmanostraca represent the sister group of Onychura and not of Conchostraca alone, shared details of the appendage morphology would simply refer to their common ancestry (characters of the ground plan of Phyllopoda). This is especially true of the definite shape of the exopod and the hose-shaped single epipod with its position in close contact with the outer edge of the exopod (compare Figs. 46J–L and 47A). This may be supported by the musculature system. Preuss (1951, 1957) observed muscles which split into portions that run into the proximal extension of the exopod and into the epipod (his 'pseudepipod'). Corresponding muscles are missing in the distal epipod of Euanostraca which also inserts in a more proximal position than in Phyllopoda (Preuss 1951, Fig. 1; Preuss 1957, Fig. 12; also Benesch 1969, Fig. 12).

The conservative design of the Notostraca since their first appearance in the fossil record, with a large shield, large number of segments, and serial homology, has obviously polarized the view to consider these as 'archaic'. In the light of the important new discoveries of the morphology of Kazacharthra by McKenzie *et al.* (1991) as well as the characters of *Rehbachiella* and the tagmosis of Conchostraca, this can be interpreted in a different way. In consequence, the appearance of phyllopodous, endite-bearing corms in the posterior limbs of Notostraca resembling larval limbs of other Branchiopoda is interpreted as a relic of an ancestral morphology (cf. Fig. 44 of Eriksson 1934; Fig 47D–F herein). The derived status of the anterior thoracopods can also be deduced from their morphogenesis (e.g., Claus 1873, Pls. 6–8). The presence of a well-developed 2nd antenna in Kazacharthra (McKenzie *et al.* 1991) as well as in the Onychura indicates that the small size of this appendage in Notostraca represents the apomorphic state. In contrast to adult Onychura, the adult 1st antenna is segmented in both the Notostraca and Kazacharthra. The derived status of the notostracan mode of life has been described in great detail by Eriksson (1934, pp. 231–254), but this work has to be expanded to Calmanostraca in general.

The shields of certain Notostraca are remarkably similar in all aspects of outline to that of *Rehbachiella* (compare Pl. 28 herein with Pl. 2:1, 2 of Linder 1952 for *Lepidurus packardi*). Besides the significant size difference, the shields of Kazacharthra are more moderate relative to the body, covering more or less only the limb-bearing trunk segments (the number given by Novojilev 1959 has proven to be wrong, see above). This seems to indicate that the notostracan shield, which covers most of the trunk, is secondarily enlarged.

Onychura. – As with other branchiopods, the members of this group show a mixture of primordial features and highly modified ones. Yet, they share the unique development of a secondary dorsal shield during ontogeny, which becomes more or less bivalve eventually. This shield (see also the subchapter on bivalve shields in the chapter 'Cephalic shields and carapaces') is no continuation of the naupliar shield but originates at the rear of the head or in the anterior trunk region, while the naupliar shield becomes more or less effaced or remains as a sclerotic forehead cover. The internalized compound eye and naupliar eye with four cups, features shared with Notostraca and synapomorphic for phyllopods, have been mentioned above.

Conchostraca, with shell sizes up to about 2 cm, show a conservative design in a regular segmentation and seriality of trunk limbs, but as in Notostraca the number is variously modified, particularly since the insertion of the posterior limbs, varying from 10 to 32 pairs, does not correspond to the segmentation in the posterior limbs of Spinicaudata (Linder 1945, Fig. 7, obtained from Sars; McLaughlin 1980, Fig. 7). A detailed summary of the distinctive characters has been provided by Linder (1945) and Fryer (1987c), hence remarks are confined to selected characters of interest for the comparisons with *Rehbachiella* and other Branchiopoda.

Even in recent textbooks (e.g., Siewing 1985) it has been overlooked that Conchostraca comprise two very distinctive groups, the Laevicaudata and the Spinicaudata, although already Linder (1945) has pointed to the separate status of these groups in detail. Recently, Fryer (1987c) listed a large number of distinctive features (without weighing them), such as:

- the missing 'growth lines' in Laevicaudata (moulted) and differences in the formation of the 'hinge' (true hinge in Laevicaudata),

- the fixation of trunk segments in Laevicaudata but their variable number in Spinicaudata,

- the large telson with a pair of claw-like furcal rami in Spinicaudata but feebly developed rear in Laevicaudata (furca not missing).

Fryer's list emphasizes the distinctiveness of the two groups, while demonstrating that for each character the plesiomorphic state can be found in the one or the other group. In other words, differences are only in the degree of development of the characters, and a common ancestor of both can be found by summing up the plesiomorphic states. For evaluating relationships, synapomorphies should be present among these characters. Spinicaudata, for example, have modified the first two pairs of trunk limbs as claspers in the males. Laevicaudata have modified only the 1st to a clasper, or right or left limb of the 2nd pair, or modified the 2nd pair slightly. This difference is more than small, and the status of having two pairs modified to claspers is most likely the apomorphic state. Hence, the character itself – claspers at least on the 1st trunk segment – is considered here as synapomorphic, since it is missing in the outgroup.

Similar examples can be found throughout Fryer's list (e.g., heart with four or three ostia), which emphasizes the close relationships of both groups. These characters are either missing or more primordial in their degree of development within the other phyllopod taxa and, hence, can serve only to support a monophyly of Conchostraca. Another character shared by both groups is the ontogeny. Fryer (1987c) correctly noted that the nauplius of Spinicaudata has neither a head shield nor the 'cruciform' head as developed in Laevicaudata (compare Figs. 44D and F). Again, these are only specializations in one or the other direction: 'loss of shield' and 'cruciform head' both represent the apomorphic states of 'presence of shield' and 'ordinary head'. However, the transformation to the so-called 'heilophora' or pre-bivalve larva in both taxa (Figs. 44F, G) is not known from other phyllopods and may be a useful character to unite both taxa, though being somewhat more pronounced and metamorphosis-like in Laevicaudata.

Little is known as yet about the ontogeny of Laevicaudata, but it is likely that this group has simply shortened the sequence even more than has the Spinicaudata. Further shared similarities lie in the appendage morphology. For example, in both groups the two rami of the 2nd antenna are of the same shape, being multi-segmented and flagel-late. Similarities can also be found in the shape of the trunk limbs.

Obtained from the ground plan of Onychura, in both the Laevicaudata and Spinicaudata a bivalved shield is progressively developed as a new structure behind the naupliar shield after a metamorphic change to the 'heilophora'. The morphogenesis (and allometry) of the originally univalved shield follows different paths in the two lineages, leading to different hinge structures and different outline. Moreover, Spinicaudata do not shed their shield, which leads to characteristic 'growth lines' (e.g., Linder 1945, pp. 3–5).

As in Spinicaudata, the globous shield of Laevicaudata encloses the animal completely, according to Martin & Belk (1988, particularly their SEM-picture, Fig. 1b), in contrast to the view of Siewing (1985, p. 890). The area of fusion with the body is unclear. According to Linder's (1945) illustration (his Fig. 6, obtained from Sars), it seems to be connected to the anterior trunk region since the first two limbs are inserted below. Again, the maxillary region seems to be completely free from the shield and anterior to its area of fusion. This would be in accordance with Strength & Sissom's (1975) observation that the shield of Spinicaudata grows out from the first trunk segment.

The anterior head portion – or complete head – is free from the secondary shield and separately moveable but also well-sclerotized. Linder (1945, pp. 5, 6) in remarking upon the differences in the head of the two groups, states, however, that although the larval differences are great, differences between the head of adults are only a matter of degree.

The lobes of the compound eye are closely set together in most Spinicaudata (save for *Cyclestheria*) and are always fused in the Laevicaudata (apomorphic status of the eye). In both they are positioned internally (not visible under SEM) and in close contact to the neck organ. As in Notostraca the eye chamber is connected with the dorsal surface (Eberhard 1981, Fig. 79 after Nowikoff; Martin & Belk 1988, Figs. 1c, d, 2d, e). In Conchostraca the naupliar eye is separate from the eye chamber, anteroventrally in Laevicaudata (Eberhard 1981, Fig. 13 after Nowikoff) and more anteriorly in Spinicaudata.

The 1st antenna may be of considerable length in some species of Spinicaudata (Battish 1981 for *Leptestheria* sp., *Caenestheriella ludhianata* and *Ocyzicus dhilloni*), but it is unknown whether their annulations reflect a former segmentation or are secondary. The 1st antennae of Laevicaudata are shorter but at least two-segmented (Martin *et al.* 1986). This clearly indicates that, as in Anostraca and Notostraca, the 1st antenna was still segmented in the ground plan of Conchostraca.

A distinctive feature of the two lineages is the trunk segmentation. For Laevicaudata, Linder (1945) counted 11 segments in the males and 13 in the females, the last segment bearing the 'opercular lamellae'. It may be possible that these are specialized limbs, although Linder found neither limb muscles nor a ganglion in this segment. Spinicaudata always have more limb-bearing segments, though the number varies considerably. Again, in the posterior region the correlation between segmentation and limbs is lost (Linder 1945), indicating the advanced status of this feature. Abdominal segments are not delineated in either group, while the furca is differently developed: in Spinicaudata it is firmly sclerotized and armed with spines, while it is feeble in the Laevicaudata, and similar in appearance to larval furcae of the former group (Linder 1945, p. 10 and Fig. 7a for a laevicaudate and 7d for a larval spinicaudate).

The distinctive morphology of Laevicaudata (Fig. 42D) and Spinicaudata (Fig. 42E) indicates indeed two clearly separate taxa: the Spinicaudata which comprise a number of subtaxa and show polypody, and the Laevicaudata which embrace the single family Lynceidae and with the impression of a slightly immature design (Linder 1945; Martin & Belk 1988). The larvae are relatively dissimilar, but this is not unexpected in distinctive lineages: for example, the hatching larva of Laevicaudata is almost completely encircled by a flat shield (Fig. 44F), while the shield is small in early larval Spinicaudata (Figs. 44D, G, 45A, B).

As to the Cladocera, Potts & During (1980) note that these may represent the youngest offshoot of the Branchiopoda, as far as can be stated from the known fossil record (Fig. 40) that reaches only into the Lower Tertiary (Fryer, personal communication, 1991, hints to yet unpublished ephippia from the Lower Cretaceous of the USSR). The report of Permian Cladocera – or better their ephippia – from eastern Kazakhstan (Smirnov 1970) would substantially extend this record, but the data presented are not convincing. According to Fryer (1987c, p. 366) this is also the case with records of remains from the Cretaceous other than ephippia.

The advanced state of the small-sized cladocerans (mostly only up to 3 mm long; Fig. 42C) has frequently been mentioned. Hence, any similarities in detail with *Rehbachiella* are most likely nothing more than symplesiomorphies. Derived features of Cladocera, taken as a whole, are in the eye morphology with fusion of compound eye lobes and the greater distance of the naupliar eye from the brain (probably due to the curvature of the head). Evolutionary trends of the four cladoceran taxa are recognizable, for example, in a varying degree of reduction of internal and external features (e.g., tagmosis, organs), and modification of reproductive strategies. Infraspecific variability in some groups may affect the setation size and pattern (e.g., Crittenden 1981), seasonal changes of setation (e.g., Korinek *et al.* 1986).

According to Belk (1982) or Siewing (1985, p. 886) these features may have been achieved by paedomorphic evolution, while Fryer (1987b, c) favoured independent development. He proposed to treat the four distinctive cladoceran subtaxa Ctenopoda, Anomopoda, Onychopoda and Haplopoda separately (see also Fryer 1987a). His re-definitions of these groups represent a mixture of characters that are partly found also in other Branchiopoda or crustaceans (e.g., 'short head', '1st antennae tubular and uniramous', 'labrum large') as well as autapomorphies. These substantiate monophyly of the different taxa and corroborate the long known distinctive status of each group within the Cladocera, but give no clues for the relationships between them (by searching for synapomorphies).

The status of the Cladocera remains unresolved until more conclusive evidence for the relationships between the four cladoceran taxa and their origin is available. This leaves us also with the uncertainty about the Conchostraca. It may even be possible that either Laevicaudata or Spinicaudata gave rise to certain cladoceran taxa. This would, indeed, not only challenge the monophyly of Cladocera but also, and going even farther than Fryer (1987c), make each of the four groups of cladocerans sister groups of different conchostracans. The assumption that the Conchostraca are monophyletic is favoured here, but more conclusive statements, however, require further detailed studies of the two conchostracan groups – in particular the ontogeny of Laevicaudata – and a re-study of the relationships of the four cladoceran taxa.

Comparative ontogeny

Remarks

Because in a series of developmental stages none can fail to survive, the long set of *Rehbachiella* larvae (Fig. 5) may indicate that environmental as well as biological factors (e.g., food availability, predator pressure) were favourable enough to guarantee durability of any individual instar. Many if not all modern crustaceans, however, have substantially modified their series, usually by considerable acceleration. This has affected in particular the early larval phase, including the naupliar and postnaupliar stages. Most effort is put into the phase of postlarval differentiation, i.e. when the postnaupliar feeding and locomotory apparatus is already, at least partially, functional.

Where naupliar stages are retained, these are mostly passed through rapidly, and in many cases the earliest instars do not feed. The spinicaudate conchostracan *Eulimnadia texana*, for example, may complete the whole life cycle within seven days (Strength & Sissom 1975); in *Paradiaptomus greeni*, a calanoid copepod that occurs in turbid rain pools immediately after the first monsoon showers, adults appear on the 9th day (Rama Devi & Ranga Reddy 1989). According to Izawa (1975) the parasitic copepod *Colobomatus pupa* does not grow during its five non-feeding naupliar stages.

Major strategies of Crustacea that modify the ontogeny pattern are:

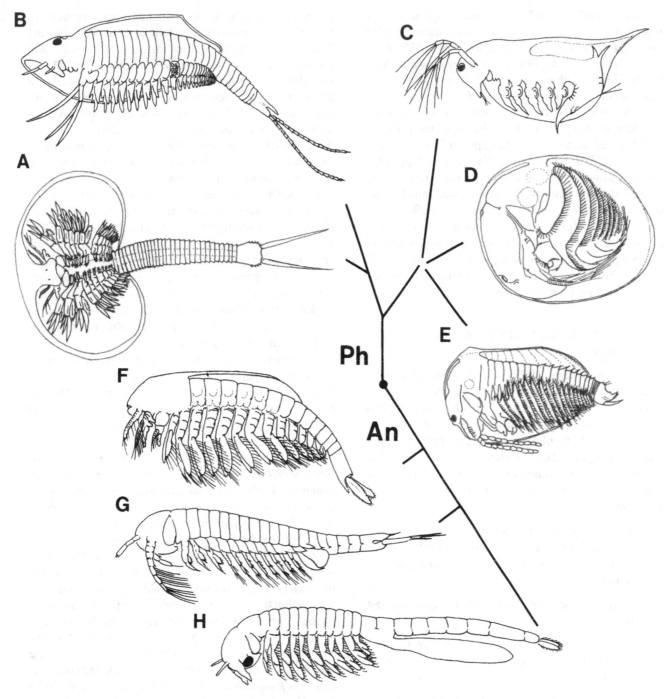

Fig. 42. Body plans of Recent and fossil Branchiopoda; lateral view, with left side of head shield omitted when necessary, save for Kazacharthra which are in ventral view; direction changed in some cases in this and subsequent figures for consistency; *An,* Anostraca; *Ph,* Phyllopoda. □A. Kazacharthra (basically after Novojilev 1959, Fig. 4B, but largely remodelled according to new evidence of Chen & Zhou 1985 and McKenzie *et al.* 1991). □B. Notostraca (after Calman 1909, Fig. 16). □C. Cladocera (modified from Kaestner 1967, Fig. 745). D–E. Conchostraca. □D. Laevicaudata (after Martin *et al.* 1986, Figs. 2, 3). □E. Spinicaudata (after Calman 1909, Fig. 17). □F. *Rehbachiella kinnekullensis* at stage TS13. □G. Lipostraca (*Lepidocaris rhyniensis*) (modified from Scourfield 1926, Pl. 22:3; limbs drawn as in other figures in his paper). □H. Euanostraca (after Hsü 1933, Pl. 1).

- development of group-characteristic sets of larvae ('phases'),

- metamorphosis-like jumps by suppression of stages, or even of complete phases (no true metamorphosis, according to Snodgrass 1956),

- development of lecithotrophic larvae,

- delay of external delineation of body segments and/or appendages,

- partial to complete development within the egg (leading to epimeric ontogeny).

The term 'nauplius' has been variously applied to early crustacean larvae regardless of their often quite different developmental level. For example, Anderson (1967) noted that the hatching 'nauplius' of Conchostraca may have already laid down the anlagen of the two maxillary segments and of the anterior eight trunk segments, the proliferation zone and the telson internally, though externally only the three naupliar limbs can be seen. Hence, even a newly hatched larva with only three pairs of limbs may not be a true nauplius (in the sense of Kaestner 1967, pp. 921–928; also named 'orthonauplius'). In subsequent stages also limb development may be suppressed, though other features obviously progress. A further modification is when structures of posterior segments appear first in development (e.g., mystacocarids). Far-reaching suppression of the larval phase occurs, for example, in leptostracan Phyllocarida (epimorphy of Nebaliacea; cf. Linder 1943), which in this respect are by no means primordial.

Development of REHBACHIELLA *as a comparative reference series*

The sequence of stages of *Rehbachiella* as found is interpreted herein to embrace the larval phase. Assuming that no stages were skipped, the series of *Rehbachiella* – with as many as 30 instars, i.e. presumably 29 moults from the nauplius, until 13 trunk segments are completely developed – is more gradual than in any other known Crustacea, and there are no rapid transitions in external morphology. In this respect, an attempt is made to use this pattern of *Rehbachiella*:

- for comparison with a selection of those representatives of the different crustacean subclasses that seemingly reflect the typical and most gradual sequence within a taxon,

- to search for general developmental strategies amongst Crustacea, and

- to evaluate how much and in which direction the development of a particular group could have been modified.

Anderson (1967, p. 48) remarked that 'during crustacean development, segments are added progressively to the trunk from a growth zone lying immediately in front of the telson'. Since embryological details are not available for *Rehbachiella*, identification of the sequence has had to be restricted to the external expression of segmentation and appearance of appendages during the phase of postembryonic growth.

Weisz (1946) pointed to the existence of a basic pattern underlying the development of Recent euanostracan Branchiopoda (cf. also Table 1 in Weisz 1947). He recognized that segment formation in *Artemia salina* occurs very gradually and in three steps: *a* at the cellular level, and *b* and

c as visible externally. Within each interphase between two moults, constrictions appear on the larval trunk. Following Benesch (1969), the two genital segments of *Artemia* – and Euanostraca in general – (trunk segments 12 and 13) are considered as thoracomeres. Only the 12th one carries a pair of modified appendages, while the next one does not appear externally. Hence, *Artemia* would theoretically require 26 instars to delineate all 13 thoracomeres. Taking the constrictions as equivalent to the appearance of an incipient segment in *Rehbachiella*, this is exactly the same number recognized in the fossil.

The benefit of the use of the segmental pattern is the possibility of correlating instars at the same number of segments and to compare their structural equipment. This demands significant reference points, such as the appearance of the first postmaxillary segment, developed trunk segments in general, to some degree also the orthonauplius (but see above). Even if moulting stages are missed (e.g., Anderson 1967 missed 9 out of 19 instars of *Artemia salina*; see discussion in Fryer 1983, pp. 331–340), the sequence remains consistent in general.

Since the time scale of moults (duration) and development of structures ('biochronism' according to Weisz 1946, 1947, who also criticized the use of moults [1947, p. 87] as 'anthropocentric practice of clocking') is purely relative, comparisons that are based on segment increase are seen as the only appropriate way in which to relate ontogenetic patterns of different crustaceans to one another. Moreover, segmentation generally precedes the appearance of other external structures on a segment (in particular the limbs). In *Rehbachiella* the appearance of limb anlagen is delayed by 1–2 stages (= apodous thoracomeres between limb-bearing one and telson; see Fig. 43A), while to achieve a functional state requires about 5–10 stages (save for the 1st maxilla which is faster). Accordingly, a larva may be considerably older in terms of segment segregation than in terms of limb development as a result of this delay. With this, the description of the complete sequence of more than 20 stages for the euanostracan *Branchinecta ferox* by Fryer (1983, Fig. 27), recording moult stages and limb development, precluded comparison with *Rehbachiella*, since no reference points could be found between the nauplius and the adult to correlate the two sequences.

On the other hand, even in highly modified sequences with suppression of many stages, segment stages can still be correlated by their segmentation – as long as they show delineated segments. Because it is common among Crustacea to reduce or efface external body segmentation, preference is given here to those forms that seem to be representative for a particular taxon, i.e. exhibit a more or less regular segmental pattern and have not modified it too greatly. As contrasting examples a few ontogenetic patterns of members of the thecostracan lineage of Maxillopoda have been selected.

Some further difficulties in staging appear in the first larval phase, in which the maxillary segments are laid down. In *Rehbachiella* these are budded off regularly. Though no break occurs in development, a 'naupliar phase' can be distinguished from a 'thoracic phase' by the incipient appearance of the first thoracomere (TS1i). In modern crustaceans, however, no such distinction is evident.

Comparisons focus on the general pattern, not least because comparatively few descriptions of larval sequences refer to the segmental pattern, and data on segment state and limb shape have been found to be rather incomplete or often not described in detail. Since it has proved necessary to generalize and to 'adjust' data from other authors, there may be some discrepancy from original descriptions. I take the responsibility for such inaccuracies, though believing that they do not greatly influence the general trends in a particular developmental pattern.

Ontogeny of several groups, such as the Euanostraca, the Mystacocarida, or the Cephalocarida, shows significant interspecific variation in the number of moulting stages as well as in the appearance of external features (examples included in Table 5). Finally, environmental factors such as temperature or salinity may also lead to considerable infraspecific modification of a larval cycle, such as in Euanostraca (e.g., Weisz 1946; Hentschel 1967, 1968) or Conchostraca (e.g., Mattox & Velardo 1950). Among Branchiopoda, detailed comparisons with the phyllopod Notostraca and Conchostraca cannot as yet be made, due to paucity of data to be incorporated adequately. The highly derived Cladocera are not considered herein.

Comparisons

The data have been compiled in two different ways. One scheme (Fig. 43) is a slight modification from that of Sanders (1963b, Fig. 27) and refers to moult stages. It enhances particularly the increase in the appearance of segments and appendages, and the degree of delay between their formation. Development of appendages can be seen along the rows. Jumps in segment increment are also clearly recognizable (marked by arrows).

For simplification and adjustment of the data base, limb development has been divided into two steps only. For the 'undeveloped' state (hollow circle) no difference is made between rudimentary and highly reduced appendages. In consequence, the small 2nd to 5th trunk limbs of Mystacocarida are treated as 'indefinite'. This also refers to modified genital appendages of Euanostraca (12th limb = thp12), Copepoda (thp7) and Cephalocarida (thp9). These never gain a completely segmented state, in contrast to, e.g., the 2nd antenna of Euanostraca which changes eventually into a reproductive aid in the postlarval differentiation phase. For Cephalocarida the first appearance of the 9th thoracopod and its further developmental

path are mentioned neither by Sanders (1963b) for *Hutchinsoniella macracantha* nor by Sanders & Hessler (1964) for *Lightiella incisa*.

While the first appearance of limbs is clear in *Rehbachiella*, this is less so in other branchiopods, particularly *Artemia*. A limb bud is considered here as visible externally when representing a spine (e.g., 1st maxilla) or a single or bifid hump (2nd maxilla, thoracopods), which would correspond approximately to Benesch's (1969) stage 3. For *Rehbachiella* the developmental stage approaching ds5 (cf. Fig. 27) is considered as the earliest stage of functionality (filled circle). This would approximately correspond to stage 6–7 in Benesch's staging. I have attempted to find a similar level of development for the other crustaceans also. Though the maxillae remain small in Euanostraca, they are treated as functional when they start to function as brushes, according to Benesch (1969) and Fryer (1983). The status of the naupliar appendages has been taken as developed from the beginning, though, in some cases, the mandibles may be still somewhat underdeveloped (e.g., *Rehbachiella*, Euanostraca).

The second mode (Table 4) correlates the '*Rehbachiella* stages' along the X-axis with the same segmental equipment of other crustaceans. This illuminates particularly the degree of abbreviation of a sequence, recognizable as gaps between moults, and the position of the phases retained or even expanded.

In most cases, the larval trunk buds off segments without having a delineated posterior segment behind the budding zone. Even when so, it remains difficult to distinguish between the telson alone and the telson with one abdominal segment fused to it, or even the complete incipient abdominal portion. Accordingly, the rear is demarcated in Fig. 43 either by a stippled line, regardless of whether it is effaced or incomplete, or by a straight line when clearly set off. Only in *Artemia* and Cephalocarida the telson is clear (T in Figs. 43B, F), while in Copepoda, Mystacocarida and Eumalacostraca the last portion may have included at least the last abdominal segment and/or the telson (Figs. 43C–E).

Results

Branchiopoda. – Postembryonic development has been described mainly from a limited number of euanostracans. Important accounts of various aspects of growth are from Claus (1873) for *Branchipus (Chirocephalus) stagnalis*; Claus (1886) for *Artemia salina*; Oehmichen (1921) for *Branchipus (Chirocephalus) grubei*; Heath (1924) for *A. salina* and *Branchinecta occidentalis*; Cannon (1927a) for *Chirocephalus diaphanus*; Hsü (1933) for *Chirocephalus nankinensis*; Weisz (1946, 1947) for *A. salina*; Valousek (1950) for *Chirocephalopsis grubei*; Pai (1958) for *Streptocephalus dichotomus*; Nourisson (1959) for *Chirocephalus stagnalis*; Gilchrist (1960) for *A. salina*; Baqai (1963) for *A.*

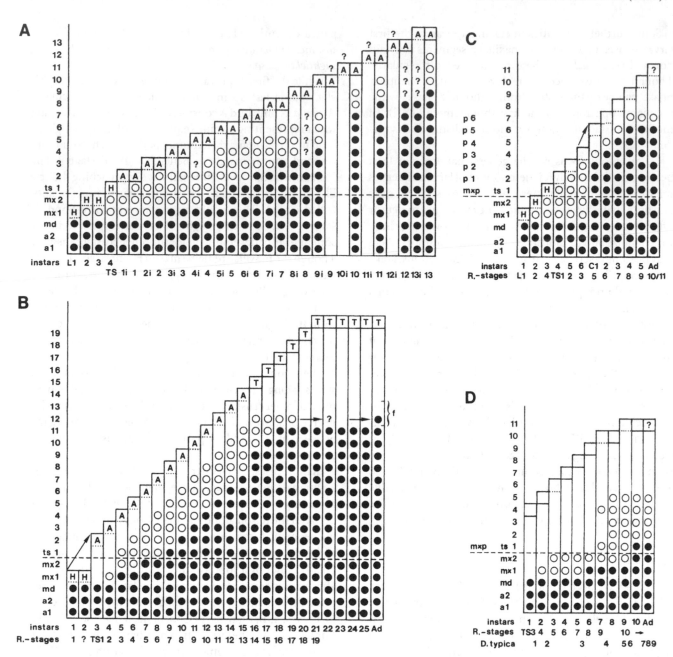

Fig. 43. Life cycles of selected crustaceans. Filled circles = developed and functional appendages; hollow circles = incipient, ill-developed, vestigial; dashed lines = boundary between cephalon and appendiferous trunk region (='thorax'). Abbreviations for last body portion: H, undivided larval hind body; A, undivided abdomen; T, telson; missing or incomplete delineation indicated by dotted line; ?, unclear status. Arrows indicate 'jumps' in segment addition. □A. *Rehbachiella kinnekullensis*, both series combined; missing stages marked by '?'. □B. *Artemia salina* (data from Benesch 1969); morphogenesis of 12th thoracopod unclear, becoming a genital limb in both sexes; 12th and 13th thoracomeres fused (f). □C. *Drescheriella glacialis* (from Dahms 1987a and personal communication, 1990); current terms of limbs on Y-axis; 7th thoracopod (P6) modified to penis in males; unclear whether or not the telson is fused with a further, 11th trunk segment (for ground plan of Maxillopoda see pp. 87–88). □D. *Derocheilocaris remanei* (from Delamare Debouttevelle 1954); status of last trunk segment unclear (stages of *D. typica* added). □E. *Macropetasma africanum* (from Cockcroft 1985 and personal communication, 1990); leg = 'pereiopods', pp = 'pleopods'. □F. *Hutchinsoniella macracantha* (from Sanders 1963b); morphogenesis of small 9th thoracopod unknown, becoming an egg carrier in females (marked by an asterix).

salina and *Streptocephalus seali*; Anderson (1967), Baid (1967), Hentschel (1967, 1968) and Benesch (1969), all for *A. salina*; Bernice (1972) for *Streptocephalus dichotomus*; Amat (1980) for *Artemia* sp.; Fryer (1983) for *Branchinecta ferox*; Jurasz *et al.* (1983) for *Branchinecta gaini*; and Schre-

hardt (1986a, b, 1987a) for *A. salina*. The quality and completeness of data presented, however, is very diverse, and even for the single species *Artemia salina* descriptions vary considerably between authors (see also Fryer 1983, pp. 331–340).

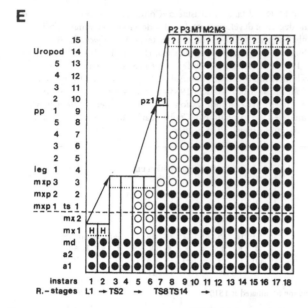

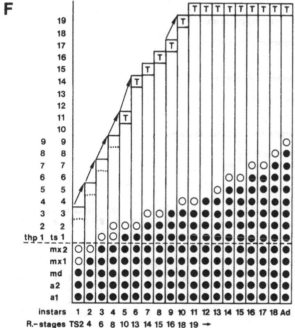

Fig. 43 (continued).

As mentioned above, the number of moults varies inter-specifically to a great extent in Euanostraca. Long sequences occur in *Artemia salina*, with 25 stages (Benesch 1969), and *Branchinecta ferox* with 21–22 (Fryer 1983). In general, development starts with an 'orthonauplius' (e.g., *A.* and *Branchinecta occidentalis*). It may, however, range as far as a larva with six delineated trunk segments, as in *Chirocephalus grubei* (cf. Benesch 1969, p. 350). Fryer (1983, pp. 331–336) recognized euanostracan development as being more anamorphic than has been stated by earlier workers, and comparable to the anameric development of Cephalocarida.

The detailed description by Benesch (1969) permits a detailed comparison with the development of *Artemia salina*. In this species the first four instars have only the three naupliar appendages. Three postmandibular segments appear at the 2nd moult (=TS1, according to *Rehbachiella* stages), and one more at the third moult (TS3) together with the limb bud of the 1st maxilla. Two more buds appear at TS4, and beyond this stage addition of segments and limbs is very regular throughout the larval phase (since the maxillae do not develop much further, they are considered as 'definite' at TS3, resp. 5; see Fig. 43B). Delineation of the apodous abdominal segments does not occur before the postlarval phase (>TS13). Addition of limbs is terminated at the 16th instar (=TS14; Fig. 43B). At stage TS13, the *Artemia* larva has 7–8 developed postcephalic limbs and four more rudiments. Limb development continues into the postlarval differentiation phase, with the modification also of the genital appendages of the 12th segment, and is eventually completed (data mainly from Benesch 1969). While limbs are added, two apodous segments are retained throughout, and it takes 3–4 moults, respectively 6–8 *Rehbachiella* stages, to define a limb.

As compared to the fossil, some concentration in the sequence, with phases of more rapid change, occurs in *Artemia* exclusively in the early part of development (Table 4). In terms of segment increase and addition of limbs, development is basically strictly anamorphic. Rapid jumps are not apparent, which is in accordance with the observations on *Branchinecta ferox* by Fryer (1983, pp. 233, 332–334, Fig. 27; also Benesch 1969), who also discusses the contrasting results of Anderson in detail.

Instar 11 (>TS10) of *Branchinecta ferox* still uses only its naupliar appendages while all postmandibular limbs are rudimentary. Beyond this stage, about one limb becomes functional per moult progressively. As compared to the limb development of *Artemia* and *Rehbachiella* this is recognized here as a considerable delay and a derived condition, in contrast to Fryer (1988). According to Benesch (1969) the 13th segment develops limb anlagen only up to his developmental stage 2, i.e. prior to externalization.

The similarity between euanostracan growth and that of *Rehbachiella* is particularly reflected in the size increment of *Artemia* (gross mode also similar in other species). In Figures 22–23, the data of Weisz (1946) have been included from TS1 to TS13. The overall size difference between these two species is very small, and an *Artemia* larva at stage TS13 is about 2 mm long, when *Rehbachiella* is approximately 1,7 mm long. A 'lag phase' is missing in *Artemia* (Fig. 22A, C, 23A, D), but this can be correlated with the shortness of the head relative to that of *Rehbachiella* (reduced size of the segments of the maxillae). The difference is even almost nil in the growth of the thorax, which is not affected by the changes in the head. Here, *Artemia* differs from *Rehbachi-*

Table 4. Comparisons of ontogenetic patterns between *Rehbachiella kinnekullensis* and selected taxa of Crustacea. Columns refer to larval stages of *Rehbachiella*, line 1 = instar, line 2 = head appendages (no differentiation made in the degree of development), line 3 = developed postmaxillary limbs, line 4 = rudimentary to not fully functional postnaupliar limbs (maxillae and trunk limbs); uncertain data in brackets (see text). A. *Rehbachiella*. B. *Artemia salina* (Branchiopoda, Euanostraca, data from Benesch 1969). C–F. Maxillopoda. C. *Drescheriella glacialis* (Copepoda, from Dahms 1987a and pers. comm. 1990). D. *Derocheilocaris remanei*, (Mystacocarida, from Delamare Deboutteville 1954 and Hessler & Sanders 1966). E. Cirripedia (Thecostraca, generalized from different authors; Cya and Cyb = cypris stage prior and after attachment). F. *Bredocaris admirabilis* (Orstenocarida, from Müller & Walossek 1988b). G–H. Cephalocarida. G. *Hutchinsoniella macracantha*. H. *Lightiella incisa* (from Sanders 1963b and Hessler & Sanders 1964). J. *Macropetasma africanum* (Eumalacostraca, Eucarida, Decapoda, from Cockcroft 1985 and personal communication 1990). Arrows demarcate span when positioning of stages not exactly possible (see text); asterix notes: first appearance of modified 9th trunk limb of Cephalocarida unknown; abbreviations: L = larva, TS trunk segment stage, PZ1–3 = protozoea stages of *Macropetasma*, My1–3 = mysis stages, PL = postlarva.

```
     LS          TS
     1 2 3 4  1i 1 2i 2 3i 3 4i 4 5i 5 6i 6 7i 7 8i 8 9i 9 10i 10 11i 11 12i 12 13i 13       Further segmentation unknown
A    ■ ■ ■ ■   ■ ■ ■ ■ ■ ■ ■ ■ ■ ■ ■ ■ ■ ■ ■ ■  +  ■  ?   ?   ■  ?   ■  ?  ■  ■
     3 3 3 3   3 3 3 3 4 4 4 4 5 5 5 5 5 5 5 5  ?  5      5      5      5  5  5
     - - - -   - - - - - - - - - 1 1 2 2 3 3  ?  4      7      7      ?  ?  9
     - 1 1 2   2 3 3 3 3 3 3 4 3 3 3 3 3 3  ?  3      2      2      ?  ?  3

                                                                              13 14 15 16 17 18 19  Segmentation terminated
B    ←■ ■→     ■   ■   ■   ■   ■   ■   ■   ■   ■   ■   ■      ■      ■       ■  ■  ■  ■  ■  ■  ■  ■  ■  ■  ■ Ad
       3 3   [3] [4] [4] [5] [5] [5] [5] [5] [5] [5]   [5]    [5]    [5]→ → → → → → → → → →[5]
       - -     -   -   -   -   -   1   2   3   2      6      7      8  9 10 11 11 11 11

C    ■ ■       ■   ■       ■   ■       cop  ■   ■   ■   ■     Ad ... Segmentation terminated at TS10
     3 3       3   3       3   3       5   5   5   5   5     5
     - -       -   -       -   -       3   4   5   6   6     6
     - 1       2   3       4   5       1   1   1   1   1     1 ... Last limb modified to penis in males

D              ■   ■       ■   ■       ■   ■   ■   ■   ■   ■  Ad ... Segmentation terminated
               3   3       3   3       3   4   4   4   4   5
               ■   ■       ■   ■       ■   ■   ■   ■   ■   1
               -   1       2   2       2   1   2   6   6   6  4 ... Last four limbs vestigial

E    ■↔■←→■  ■  ■↔■  ■←———→■   Cya Cyb Ad   Segmentation terminated at TS6/7
       3 3     3  3  3  3          3  5  5  5
       - -     -  -  -  -          -  6  6  6
       - -     1  7  7  7          8  7  1  -   Last limb does not become developed after attachment

F    ■↔■←———→■  ■↔■←———————→        Ad   Segmentation terminated at TS7
        3     4  4  4  4            5    Abdominal segment suppressed?
        -     -  -  -  -            7
        1     2  3  4  5            -

     1 2 3 4 1i 1 2i 2 3i 3 4i 4 5i 5 6i 6 7i 7 8i 8 9i 9 10i 10 11i 11 12i 12 13i 13 14 15 16 17 18 19  Segmentation terminated
G           ■        ■        ■        ■        ■        ■  ■  ■  ■        ■  ■  ■  ■  ■  ■  ■  ■ Ad
            3        4        5        5        5        5  5  5  5        5  5  5  5  5  5  5  5
            -        -        -        -        1        1  2  2  3        3  3  4  4  5  5  6  6  7  8
            2        1        1        2        1        1  1  1  -        1  1  -  1  1  1  1  1  1*

H                                      ■                 ■  ■  ■  ■        ■  ■  ■  ■  ■  ■  ■ Ad
                                       4                 5  5  5  5        5  5  5  5  5  5  5
                                       -                 -  1  2  2        3  4  4  5  5  6  6  7
                                       1                 1  1  -  1        1  -  1  -  1  -  1  ?*

     1 2 3 4 1i 1 2i 2 Same segm.  6i 6 7i 7 8i 8 9i 9 10i 10 11i 11 12i 12 13i 13 14 Segmentation terminated
I    ■ ■        ■ ■ ■ ■            ■                     ■  ■     ■  ■  ■     ■ ■ ■ ■ ■ ■ ■ Ad
     3 3        3 3 3 3            5                     5  5     5  5  5     5 —————— 5
     - -        - - - -            2                     2  2     9  9  14    14 ————— 14
     1          6 7 5 5            -                        -     —————— -
                        PZ1                          PZ2, 3      My1–3      Pl1–6
```

ella only in so far that the special development of the two genital segments, 12 and 13, causes a cessation in the growth of trunk (arrow in Fig. 22B, 23B), not recognizable for *Rehbachiella*. The hatching nauplius of *Artemia* is about 350 μm long, which is more than twice as long as the *Rehbachiella* nauplius. Hence, it may be possible that it is more advanced than it seems to be externally (←?→ in Table 4), which may be deduced from an emerging 'prenauplius' (Fig. 53A).

Delayed development relative to *Rehbachiella* is apparent in the process of development of the trunk (including the limbs), described in detail by Benesch (1969) and of the furca. The latter structure develops in a similar manner to that of *Rehbachiella*, but its growth is very slow during the larval phase and definition postponed far into the postlarval phase (e.g., Baqai 1963, Figs. 54–62; Bernice 1972, Fig. 11).

The development of *Lepidocaris* is only incompletely known. Yet it shows that the trunk limbs are much delayed relative to segment increment and to *Rehbachiella* (compare Scourfields 1940, Fig. 2, about TS8, and Fig. 7B herein for the same stage of *Rehbachiella*). This is also true for the furcal development: at a TS9 of *Lepidocaris* only one seta has yet appeared on the incipient ramus.

The ontogeny of the other branchiopod taxa is much less fully documented. As a whole, all Branchiopoda of the phyllopod lineage obviously show a tendency to shorten the sequence, but based on different strategies. A brief summary is given by Linder (1945) for Notostraca as well as the Laevicaudata and Spinicaudata (referring to Sars). Aspects of notostracan development have been described mainly by Claus (1873) for *Triops (Apus) cancriformis* and, more recently, by Fryer (1988) for *T. cancriformis* and *Lepidurus arcticus*. In both the sequence is not given in its entirety, so that it could not be included here in detail.

Fryer (1988) described the development of *Triops longicaudatus* and *Lepidurus arcticus* in detail and noted the occurrence of nauplii, while, according to Benesch (1969) the Notostraca mainly hatch as a larva already possessing six delineated trunk segments (see Fig. 45D, E; Benesch 1969, p. 350). Fairly quickly it moults to a much more advanced larva, about 2 mm long, which already possesses a number of trunk segments and limbs. At the 4th moult more than 21 segments are budded off and at least eight limbs are present in *Triops longicaudatus* (Fryer 1988, Fig. 115). The difficulty of determining later stages prevented Fryer from continuing the description beyond his 7th instar, about 3 mm in length. While the early larval phase is accelerated (Fryer 1988, p. 90), though still anameric, a large number of moults occur in the postlarval differentiation phase (>TS13). As far as it can be established from the fragmentary growth data, the length increase in the early phase is significantly different from and much slower than in Euanostraca and *Rehbachiella*.

Information on conchostracan development is also still scarce. Generally, only important aspects of development or few significant instars have been mentioned (e.g., Lereboullet 1866; Cannon 1924; Gurney 1926; Botnariuc 1947, 1948). Anderson (1967) described the development of *Limnadia stanleyana* in detail but rejected Weisz's sequencing after segment stages, and his description thus cannot be included here. Bishop (1968) studied postlarval growth of the same species mainly by monitoring the increase in the number of growth rings on the shield. Conchostraca may hatch as nauplii (Benesch 1969, p. 350, but see above). Besides this, their development (Anderson 1967 for the spinicaudate *Limnadia stanleyana*) also shows considerable compression of the early larval phase, probably associated with adaption to their special habitats which requires short life cycles (cf. Potts & Durning 1980). Another speciality of both the laevicaudate and the spinicaudate Conchostraca is the metamorphic change to a pre-

bivalved larva, or 'heilophora stage' (Gurney 1926 for the laevicaudate *Lynceus (Limnetis) gouldi*; Linder 1945; Fig. 44F herein). As in the Notostraca the length increase, on the other hand, is very low at the beginning. The change to the heilophora is in accordance with a phase of delayed growth.

Maxillopoda. – As compared to *Rehbachiella* stages, all maxillopod taxa complete their development prior to stage TS13 (examples in Table 4). Segment addition is basically terminated at a maximum of 11 trunk segments, while in Mystacocarida and Copepoda, taken as examples here, the maximum number of developed segments is ten. Of the seven pairs of thoracopods in the ground plan of Maxillopoda, the last pair has become modified as copulatory aids in the males of Copepoda and members of the thecostracan lineage, most likely as a result of parallel evolution (cf. Müller & Walossek 1988b; Boxshall & Huys 1989; Huys 1991).

Copepoda exhibit the most complete sequence among the members of Maxillopoda. The reproductive stage is reached after a maximum of 11 instars: six nauplii/metanauplii and, after a metamorphosis-like change, five more instars, the copepodids (e.g., Vincx & Heip 1979). The data are taken from the development of the harpacticoid *Drescheriella glacialis*, completely documented by Dahms (1987a, also personal communication, 1990). As compared to *Rehbachiella*, only one larva is missing in the naupliar phase (Table 4; unclear whether it is the L2 or L3 stage). Another jump equivalent to four *Rehbachiella* stages takes place at the moult to the copepodid phase (co in Table 4). The appendages increase in number gradually (Fig. 43C), but remain as anlagen until the 1st copepodid stage (= TS5). At this stage the anterior three thoracopods become functional simultaneously. Segment addition and limb development both continue very gradually during the copepodid phase until TS8 when limb addition is terminated. Only in this phase, a transient apodous segment appears. Abdominal segmentation is completed within the last two moults. The 3rd to 7th thoracopods (P2 – P6) become functional over a single moult each, corresponding to two *Rehbachiella* stages. With this, development of Copepoda reflects most of the early larval phase of *Rehbachiella*, but is completed more rapidly until an equivalent of TS10 in terms of differentiation.

Mystacocarida have a similar number of instars (10 at most) as the copepods, but this is not greatly relevant, since their mode of development differs considerably from these and from *Rehbachiella*. *Derocheilocaris remanei* (data from Delamare Deboutteville 1954; Sanders 1963b; and Hessler & Sanders 1966) hatches with three pairs of appendages, but its segmentation is already that of a stage TS3 (Fig. 43D). As compared to *Rehbachiella*, the complete early phase seems to be skipped (Table 4).

Further growth is characterized by a regular increase in the number of segments (even delay between the 7th and 8th moult), which is terminated at TS11(the last possibly including the telson, according to Huys 1991). Over two more moults no segments are added. Limb development shows remarkable delay. Head limb development is not completed before the last immature stage. Again, the first trunk limb, the maxilliped, appears at TS9, together with the buds of the 2nd and 5th limbs; all five trunk limbs are present by the next stage, but the posterior four do not gain any structure throughout.

Development of *Derocheilocaris typica* (cf. Hessler & Sanders 1966) is even more accelerated, while, after the appearance of all trunk segments, it takes four more moults to differentiate into the adult state (bottom line in Fig. 43D). The large number of apodous segments between the last appendage and the trunk end and the protracted delay of limb development and reduction in their number are recognized as clear indicators of the adaptation of Mystacocarida to life in the interstitial environment, together with the reduction of the early larval stages.

In addition, two members of the thecostracan Maxillopoda are included: the Cirripedia and *Bredocaris*. In Cirripedia, though highly modified due to their sessile life style, the larval shape and habit is remarkably conservative. The series consists of maximally six nauplii/metanauplii and one 'cypris stage', which transforms into the adult after attachment (e.g., Bassindale 1936; Costlow & Bookhout 1957, 1958; Crisp 1962; Anderson 1965; Walley 1969; Dalley 1984; Achituv 1986; Moyse 1987; Egan & Anderson 1988, 1989).

Staging of the larvae is difficult because segmentation is completely effaced in the naupliar stages; moreover, the 1st maxilla remains as a bud throughout. The 2nd maxilla and the six thoracopods develop on the larval trunk but in most cases do not appear externally, except for spines indicative of their progressive internal development (e.g., Figs. 7, 8, 12, 13 in Egan & Anderson 1988). Buds, as in *Bredocaris*, are recognizable in the lepadomorph barnacle *Ibla quadrivalvis* (Anderson 1987, Figs. 1f, 5). Workers on cirriped development traditionally do not put much weight on a detailed description of the increasing number of spines. Moyse (1987) only briefly mentioned the correlation for a stage IV metanauplius, which permits this instar to be correlated at least with a TS6 stage. All of the five early stages are not definitely assignable at present, and are placed arbitrarily somewhere between nauplius and this stage (←?→ in Table 4).

Reduction has also greatly affected the posterior part of the trunk. Segmentation is terminated at TS6 externally, but there are indications that a 7th thoracomere is laid down internally in Cirripedia also (Fig. 14a in Walley 1969 for a cypris prior to attachment) which does not bud off later (* in Table 4). As in Copepoda, the thoracopods are functional earlier than in *Rehbachiella*, but in striking

contrast to Copepoda, they appear as well-developed limbs simultaneously at the moult to the cypris, which, hence, clearly does not correlate with the 1st copepodid (see Table 4 and below).

Development of *Bredocaris* (cf. Müller & Walossek 1988b) starts with a metanauplius which can be correlated with a stage L3 of *Rehbachiella* by its limb development. Accordingly, the first two stages are skipped. In further contrast to *Rehbachiella*, but as in other thecostracan Maxillopoda, only the maxillulary segment subsequently coalesces with the head during the larval phase. The maxillary segment remains on the hind body for four more stages, as is indicated by the close contact of its limb to the set of trunk limbs buds, progressively increasing from one to four. All these limbs remain as buds throughout. The precise staging of the 'nauplii' with *Rehbachiella* stages meets the same problems as in the cirripeds, since the segmentation of the trunk is also effaced (←?→ in Table 4), and external expression of posterior trunk limbs suppressed, also resulting in a somewhat different placement of cirriped and *Bredocaris* 'naupliar' stages.

Table 4 confirms the assumptions of Müller & Walossek (1988b) that the subsequent 'cypris phase' is most likely completely skipped in *Bredocaris*. At the metamorphosis-like jump to the adult, which to some degree looks as an 'adultized' cypris, the maxillary segment and all seven thoracomeres appear simultaneously, with feeble segmentation and well-developed limbs, while the postthorax is undivided. Accordingly, the segment pattern remains uncertain and cannot be clearly correlated (Table 4). The substantial reduction of segmentation in the thorax, the abdomen, the thoracopods, and the missing articulation of furcal rami in accordance with accelerated development is seen as an indicator of a special life strategy of this fossil, probably beneath the substrate-water interface.

Both the Cirripedia and *Bredocaris* retain parts of the early larval phase but quickly terminate segmentation. Their pattern accentuates again the distinctiveness of mystacocaridan development. It is only superficially similar to that of Copepoda and clearly set off by its delay of limb differentiation and the metamorphosis-like jump to the 'cypris' respectively adult, with simultaneous development of 6/7 thoracopods at about TS7/8. Again, it is evident that the 'cypris' cannot be regarded as the developmental equivalent of the 'copepodid'.

This detailed agreement in the ontogeny pattern strongly supports the assumptions that *Bredocaris* is a representative of the Thecostraca s. str. rather than a representative of their stem group.

Malacostraca. – Phyllocarida have no early larval stages ('epimorphic development' after McLaughlin 1980). According to Linder (1943, citing older references), two moults occur within the breeding chamber; these stages already have pleopods at least as anlagen. This phase is followed by up to six free-living mancoid and juvenile stages until the reproductive stage is reached. Since these stages already resemble adults, ontogeny of Phyllocarida is reduced to the end of the postlarval differentiation phase, when segmentation is completed.

Among the Eumalacostraca only Euphausiacea and Decapoda have retained a true nauplius stage. Development has been described in various species, mainly from those reared in the laboratory (for example euphausiids: Boden 1950; Mauchline 1971; Knight 1973, 1975; brachyurans: Fielder *et al.* 1979; Greenwood & Fielder 1979, 1980, 1984; penaeids: Fielder *et al.* 1975; Cockcroft 1985; atyids: Salman 1987; palaemonids and alpheids: Gurney 1938). Schminke (1981), describing the adaptational strategies of Bathynellacea, compared these with developmental patterns among several eumalacostracan groups.

The penaeid *Macropetasma africanum* has been chosen here with regard to its complete documentation by Cockcroft (1985, also personal communication, 1990). Its long larval sequence suggests a rather unmodified and typical postembryonic pattern among Eumalacostraca. Since penaeid nauplii are non-feeding, their morphology is characterized by missing feeding structures on all naupliar appendages and a poorly developed labrum accordingly (which in this respect is very similar to corresponding non-feeding nauplii of various non-malacostracan crustaceans or the non-feeding Upper Cambrian type-A nauplii; see Müller & Walossek 1986b and Walossek & Müller 1989).

Development of *Macropetasma* starts with two nauplii having three pairs of appendages and an unsegmented hind body. The next two stages have no more limbs, but segmentation is increased to partial delineation of four more segments, the two segments of the maxillae and two of the trunk (corresponding to TS2). While further external delineation is delayed for two more stages, the maxillae and two trunk limbs appear as anlagen. The impression of a stagnation between the 5th and 6th larvae, as suggested by Fig. 43E, is slightly misleading since it results merely from the restriction herein chosen of dividing limb development into two steps only. In fact, the 6th 'nauplius' has already more advanced postmandibular limbs with both rami developed.

Further development passes over rapidly to continue within the later larval phase. Protozoea I (= TS8), comprises all thoracomeres, developed maxilla, the anterior two trunk limbs, and buds of the 3rd, protozoea II has all segments and buds back to the 8th trunk segment, and protozoea III has eight developed trunk limbs and anlagen of the uropods. At mysis I the uropods are functional, and buds of the remaining pleopods are present. Mysis II and III are complete. These instars are followed by six more postlarvae (Fig. 43E). The biggest jump in segment formation occurs in the protozoea phase, while the final number is reached at the 2nd protozoea. It remains unclear whether another segment still remains with the telson, as has been

variously suggested, since phyllocarids have one more trunk segment ('?' in Fig. 43E).

Ontogeny of *Macropetasma*, and possibly the eumalacostracans as a whole, is characterized by substantial jumps in segment formation, and in the appearance of limbs and their development, but a lack of continuously delayed appearance of limbs in front of the budding zone. Together with the juveniles, ontogeny comprises a total of 18 instars until the adult is reached. Interestingly, the 'gaps' between trunk end and last limb, indicating apodous segments, are completely filled with appendages during successive stages. Again, no apodous segments remain referable to abdominal segments of other crustacean subclasses, in other words a phase of segment formation of limb-less abdominal segments, as in Maxillopoda and Branchiopoda, is missing.

As compared to *Rehbachiella*, the ontogeny of *Macropetasma* is quite different in its shortened early larval phase, with sets of instars, and the occurrence of the majority of stages within the postlarval differentiation phase, beyond the 30 stages of the fossil. This subdivision into a naupliar, a protozoeal, a mysis, and a juvenile phase in *Macropetasma*, readily distinguishable in Table 4, is clearly distinct from all other crustacean subclasses. Again, with regard to limb differentiation in *Rehbachiella*, it is noteworthy that *Macropetasma* has no more than two developed limbs at a stage corresponding to TS14. A limb may be fully functional after two moults in general, but this cannot be correlated with *Rehbachiella* stages.

Cephalocarida. – Cephalocaridan development has been claimed by Sanders (1963b) to reflect the ancestral state among living Crustacea. According to this author segment formation as well as limb development is very gradual, and generally one pair of limbs appears every second moult. It is true that the larval sequence is long, but it is no longer than in the Branchiopoda or the Malacostraca. In fact, Sander's Fig. 27, as well as the schemes presented herein (Fig. 43F and Table 4), reveal a different picture of the developmental pattern of Cephalocarida. Ontogeny of *Hutchinsoniella macracantha* (data from Sanders 1963b) starts with a larva with all head appendages and two more trunk segments, i.e. at stage TS2, while *Lightiella* hatches with already as many as seven developed trunk segments (= TS7). As compared to *Rehbachiella* staging, seven stages are skipped in *Hutchinsoniella* and 17 in *Lightiella*.

As has already been recognized by Fryer (1983, p. 335), segment increment shows steps varying from 1–3 per moult (= 2–6 *Rehbachiella* stages). The number of moults between appearances of new limbs varies from one to three (corresponding to up to 12 *Rehbachiella* stages), which is by no means an even increase by two, as stated by Sanders (1963b). Development of a limb is also variable from one to three moulting steps (corresponding to a range of 2–12 *Rehbachiella* stages). Again, the maximum number of limbs is achieved very late, revealing a considerable delay as measured against segment formation. For example, the TS13 stage (6th instar) has no more than one developed and one rudimentary trunk limb.

When segment increment is terminated (TS19; 11th instar), still only three developed trunk limbs and a rudimentary one are present. Nine pairs of trunk limbs develop, of which eight become developed eventually (seven in other species). The last pair remains small throughout and finally becomes modified into egg carriers. Since Sanders (1963b) did not mention it for the earlier stages, it is not clear when this limb first appears (* in Table 4 and Fig. 43F).

Table 4 emphasizes how far the early developmental phase is abbreviated in Cephalocarida: it takes just six stages to get to the TS13 stage and only two in *Lightiella incisa* (data from Sanders & Hessler 1964). As in Malacostraca, most of the moults occur in the late phase, 8 out of 18 after termination of segment increase in *Hutchinsoniella* and 7 out of 12 in *Lightiella*.

The number of apodous posterior trunk segments rapidly increases, resulting in a large discrepancy between segment formation and appearance of limbs. Remarkably, the first apodous abdominal segments already appear very early during development, if not present from the beginning. From illustrations of *Hutchinsoniella* by Sanders (1963b) and *Lightiella* by Sanders & Hessler (1964) it becomes apparent that maturation of abdominal segments parallels and even precedes that of the thoracomeres.

Conclusions

Both modes of comparison, following either the moult cycle or segment increment, suggest that the highly variable design of ontogenetic patterns among Crustacea is indeed underlain by a common basic strategy, i.e. an origin from a regular series which starts with a true nauplius. The early phase seems to be best reflected in the very gradual developmental pattern of *Rehbachiella*. Beyond the present sequence, further development may have been similar to that of Recent Euanostraca. Hence, assuming that the anamorphic series of *Rehbachiella* represents much of the plesiomorphic state among Crustacea s. str., application of its stages as a standard reference measure helps recognizing distinctive strategies of the different crustacean taxa in modifying particular portions of the developmental series.

Virtually all Recent taxa seem to have reduced the first step in the external delineation of trunk segments, termed 'incipient' in the fossil. It remains unclear, however, whether this occurred independently or must be taken as an argument against a position of *Rehbachiella* within Branchiopoda. Further studies on comparative ontogeny are required for clarification of this unresolved issue.

When comparing ontogenetic patterns with reference to segment staging, it becomes obvious that a large number of moults does not necessarily imply a primitive mode of

development for the higher-rank taxa. It is useful, however, to evaluate the character state within the particular taxa. What is of importance is the schedule of moulting in relation to growth. External segment formation terminates at different levels in all crustacean groups, while moulting and differentiation can be continued in a specific manner until the adult state is reached. Moulting may even continue throughout life. Again, segmentation may be terminated quickly, as in the Maxillopoda, while the sequence up to this state is a very gradual one, reflecting much of the primordial type of ontogeny.

With the exception of the Mystacocarida and the Cephalocarida, development basically starts with a true nauplius or at least at a stage close to it. In most Recent groups there is a tendency to accelerate the early larval sequence, while the major moulting period and differentiation occurs within the postlarval developmental phase. This is taken to the extreme in the Malacostraca and the Cephalocarida, but from very different starting points (Table 4). Euanostraca and Copepoda exhibit a very gradual sequence, exactly 50% of *Rehbachiella* instars, but this is only numerical. Relative to *Rehbachiella*, the Copepoda have skipped one metanauplius and one thoracic stage until TS10 (when neglecting the intermediate stages), *Artemia* has skipped two metanauplii, as can be seen below, where instars of the two early phases (left column: until appearance of the 1st thoracomere; right column: until given segment number) are listed for selected stages of trunk development:

	TS5	TS8	TS10	TS13	
Branchiopoda					
Rehbachiella	4+10	4+16	4+20	4+26	
Artemia	2+5	2+8	2+10	2+13	
Maxillopoda					
Drescheriella	3+4	3+7	3+9	—	terminated
Derocheilocaris	0+3	0+6	0+9	—	terminated
Cirripedia sp.	2+4	—	—	—	not completed
Bredocaris	1+4	—	—	—	not completed
Cephalocarida					
Hutchinsoniella	(0+2)	0+4	0+5	0+6	TS5 not represented
Malacostraca: Decapoda					
Macropetasma	(2+4)	2+5	no more moults until TS14		
	TS5 not represented				

Lumping of stages to sets is clearly an apomorphic state of ontogeny and a particular feature of Eumalacostraca, which may have up to five distinctive sets (Table 4). These sets are clearly not correlated with similar phases of any other crustacean taxa. In the 'copepodid phase' Copepoda reflect the primordial state among Maxillopoda, with a regular increase of segments as well as appendages. Hence, simultaneous appearance of thoracopods in the thecostracan Maxillopoda and in the Eumalacostraca is merely the result of parallel evolution.

An important strategy of development is the different speed of limb appearance and development relative to segment formation. This can be seen in particular in the number of limbs and their achievement of functionality at corresponding segment stages. While *Rehbachiella*, Euanostraca and Maxillopoda have similar numbers of limbs, for example at TS8 (11–12), Cephalocarida have seven at most, and *Macropetasma*, as a representative of eumalacostracans, eight. Of the trunk limbs, six to seven are functional in Maxillopoda, three or four in *Rehbachiella*, two in *Artemia*, two in *Macropetasma* and none in the Cephalocarida. The list below shows the number of functional postmaxillary limbs and anlagen (in brackets) at selected numbers of developed segments, and the maximum number reached eventually:

	TS5	TS8	TS10	TS13		maximum
Branchiopoda						
Rehbachiella	1(3)	?3(3)	7(2)	8–9(4–3)		12
Artemia	0(3)	2(4)	4(4)	7(4)		12
Maxillopoda						
Drescheriella	3(1)	6(1)	6(1)	—	6 (groundplan: 7)	
Derocheilocaris	0(0)	0(1)	0(5)	— 1+4 buds (groundplan: 7)		
Cirripedia sp.	internal buds until cypris			6 (groundplan: 7)		
Bredocaris	buds until supposed adult					7
Cephalocarida						
Hutchinsoniella	—	0(2)	1(1)	1(1)		9
Malacostraca: Decapoda						
Macropetasma	—	2(1)	—	2(6)[TS14]		14

Similar differences can be seen in the appearance as well as in the achievement of functionality (or reduction) of the two pairs of maxillae (Table 4). For example, in the Copepoda the appearance of the maxillae and anterior two trunk limbs gradually progresses as in *Rehbachiella*, but faster than in all other forms with which it is compared (see above); development of the maxillae is, however, delayed until the beginning of the copepodid phase. At TS5 even more thoracopods are functional than in *Rehbachiella* as a result of simultaneous development of the maxillae and three thoracopods. This advantage is kept until TS8 when the development of appendages is terminated (in the sense of the categories used). This process of rapid achievement of functional limbs differs from that of *Rehbachiella* and Euanostraca, which may result from a condensation of the later larval phase with inclusion of elements of the postlarval differentiation phase ('adultization').

In Cirripedia and *Bredocaris*, as representatives of the Thecostraca s. str. (Ascothoracida, Cirripedia, Facetotecta; see also Grygier 1984), the head is completed at about TS6–7, which, again, indicates that the 'cypris phase' and the 'copepodid phase' are not developmental homologues. This can also be seen in the position of the simultaneous development of the thoracopods (2nd maxillae reduced in extant thecostracans). Accordingly, the generalizing term 'podid phase', as proposed by Newman (1983) camouflages such a striking difference.

Delay of the maxillae takes longer in Euanostraca than in *Rehbachiella*, and functionality is not reached before about TS5 (difficult, since these limbs are very reduced; brackets

in Table 4). In *Hutchinsoniella* both maxillae appear simultaneously at TS2 and are progressively incorporated until TS6 (two moults = eight *Rehbachiella* stages). In *Lightiella* the maxillae are developed by the 2nd instar which corresponds to a TS13 stage. In *Macropetasma*, these limbs appear together with the first two thoracopods somewhere after TS2 and are functional at TS8.

At TS13, the largest stage of *Rehbachiella*, 12 thoracopods are present, of which 8–9 are just about fully developed, and in *Artemia* 11 thoracopods, seven of which are developed. By contrast, Cephalocarida have at the most two thoracopods at this stage, merely one being functional. A corresponding stage is not present in *Macropetasma*; at TS14, eight thoracopods are present, but only two developed.

Differences in the termination of segment increase and in the formation and number of apodous posterior trunk segments are also remarkable. In my view, they have an important bearing on the understanding of the tagmosis of Crustacea in general. Malacostraca terminate at TS15 (Phyllocarida) and 14 (Eumalacostraca), respectively. The last trunk segment is apodous in the Phyllocarida. Despite the possibility that this segment is included in the caudal end, a free transitional segment does not appear in Eumalacostraca at any stage during development (Fig. 43E). The sets of apodous segments in the hind body are always accomplished by the next step with the same number of limbs.

Branchiopoda and Cephalocarida both terminate segment addition at TS19. In detail, this number shows up as a composite of two tagmata which in fact develop quite differently in both groups. In Branchiopoda the apodous abdominal segments are not delineated before the postlarval phase (enhanced in Fig. 43B by shading), while the transient ones (most regularly 2) develop into thoracomeres with a delay of generally two stages (= four *Rehbachiella* stages). As in Branchiopoda, the apodous trunk segments referring to the abdomen develop after completion of the thorax in Copepoda (after TS8; shading in Fig. 43C).

In sharp contrast, the thoracomeres and the limb-less abdominal segments develop at least in parallel in Cephalocarida (see above and shaded area in Fig. 43F). This may point to the existence of two separate proliferation zones in this group, a unique feature among Crustacea. As a further consequence, it cannot be excluded that, with this subdivision of the budding zone in front of the telson, two separate evolutionary pathways could have led to the specific number of segments found in this taxon, neither of which reflects the primordial equipment of the ancestors of Recent Cephalocarida.

There is some remote similarity to the pattern shown in Mystacocarida, and this is also evident in the high interspecific variability of these two groups which in aspects of their morphology appear very conservative. This simply indi-

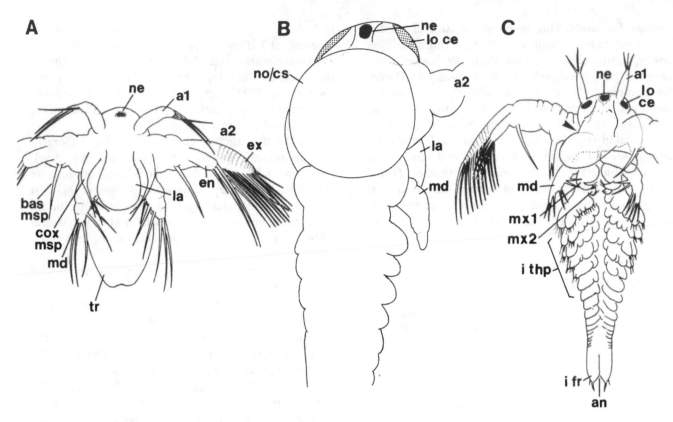

Fig. 44. Selected larval types of Recent Branchiopoda; ventral views save for Fig. B; setation shortened and simplified for clarity in some cases in this and the subsequent figures (not scaled). A–C. Euanostraca. □A. Nauplius of *Branchinecta ferox*, Euanostraca (after Fryer 1983, Fig. 1). □B. Larval *Branchipus torvicornis*, 0,75 mm long, with huge neck organ and developing compound eye (after Claus 1873, Pl. 2:5). □C. Advanced larva of *Branchipus stagnalis*, 1.2 mm long; arrow points to bend on labrum (after Claus 1873, Pl. 2:7). D–F. Conchostraca. □D. Nauplius of spinicaudate *Imnadia voitestii* (after Botnariuc 1947, Fig. 23). □E. Advanced metanauplius of spinicaudate *Eoleptestheria variabilis* with initial secondary shield (after Botnariuc 1947, Fig. 14). □F. Advanced larva prior to moulting to the corresponding heilophora stage of the laevicaudate *Lynceus gouldi*; left side ventral view, right side dorsal view: larva with all-enclosing naupliar shield, new shield is already recognizable below the cuticle (arrows; modified from Gurney 1926, Figs. 1, 2). □G. Heilophora larva of *Leptestheria intermedia* with enlarged but still univalved secondary shield (after Botnariuc 1948, Fig. 1).

cates, however, that both groups share a similar life strategy, supporting the assumption, expressed also elsewhere in this paper, that Cephalocarida are fairly derived meiofaunal forms. In detail, mystacocarids skip the earliest larval stages and subsequently develop gradually, with reduction of trunk limbs (also in number), terminate quickly, and adultize within a few more moults without segment increase (Fig. 43D). Cephalocarids skip the earliest stages and develop in jumps of roughly four *Rehbachiella* stages; limb development is remarkably delayed but gradually completed in an extended phase post TS13 (also Fig. 43F).

Two more characteristics of developmental strategies among crustaceans are noteworthy, both, however, being more relevant for subordinate taxa. External delineation of segments may be effaced or much delayed though internally the segments are already not only segregated but also differentiated. This can be best seen in thecostracan Maxillopoda. Again, species dependent variability is considerable among all groups, but does not greatly affect the general trends of the particular taxon.

In summary, the ontogeny of *Rehbachiella* seemingly has more in common with euanostracan development than with that of the other crustacean subclasses, not least in the light of the theoretical approach of Weisz (1946). This similarity is particularly true for the phase between TS7 and TS13, while in the early phase, *Artemia* exhibits a considerable delay in limb formation and differentiation. Slight numerical discrepancies may even be due to the difficulties in correlating the different developmental stages of limb formation. This overall similarity might, however, merely indicate that both the Euanostraca and *Rehbachiella* document much of the primordial anamorphic pattern of development of Crustacea s. str., if it were not for the striking similarity in growth increment (see above).

Maxillopoda are clearly set apart by their early termination of segment addition basically at TS11 (see also Huys 1991), also reflected in a typically small size of all representatives. Besides this, they show up as the one crustacean subclass which has retained, in the Copepoda, a very complete larval development in the early phase. In the light of rapid completion of development in the copepodid phase,

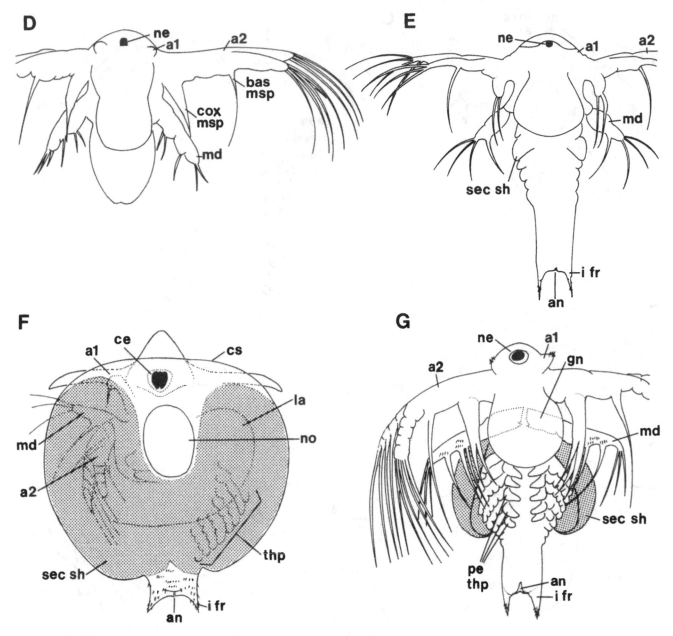

Fig. 44 (continued).

its last stage, TS10, might be better correlated with about a TS8 stage of *Rehbachiella*, since the addition of the last abdominal segments may already refer to the here highly condensed postlarval phase. Again, functionality of limbs is achieved much earlier than in *Rehbachiella*, which supports the assumption of a paedomorphic origin of the whole group. According to Westheide (1987) precocious sexual maturity reached still at a larval level of shape leads to a mixture of larval and new adult characters in a paedomorphic group.

Developmental patterns of Cephalocarida and *Rehbachiella* are quite distinctive, showing differences in general strategy as well as in the fate of particular details. It is

not the gradual development of limbs itself, but the obvious delay relative to the development of the segment, that shows up as a highly derived character of Cephalocarida. Moreover, the simultaneous delineation of thoracic and abdominal segments has no parallel in any other Crustacea. It not only supports Fryer (1983, p. 336) who stated that 'it can no longer be claimed that the Cephalocarida shows a more primitive pattern of development than the Euanostraca' but points even further. Such a strategy cannot represent the primordial pattern from which other types had originated. Again, in terms of moults the individual limbs become functional more rapidly in Cephalocarida than in Euanostraca (along the rows in Fig. 43). This

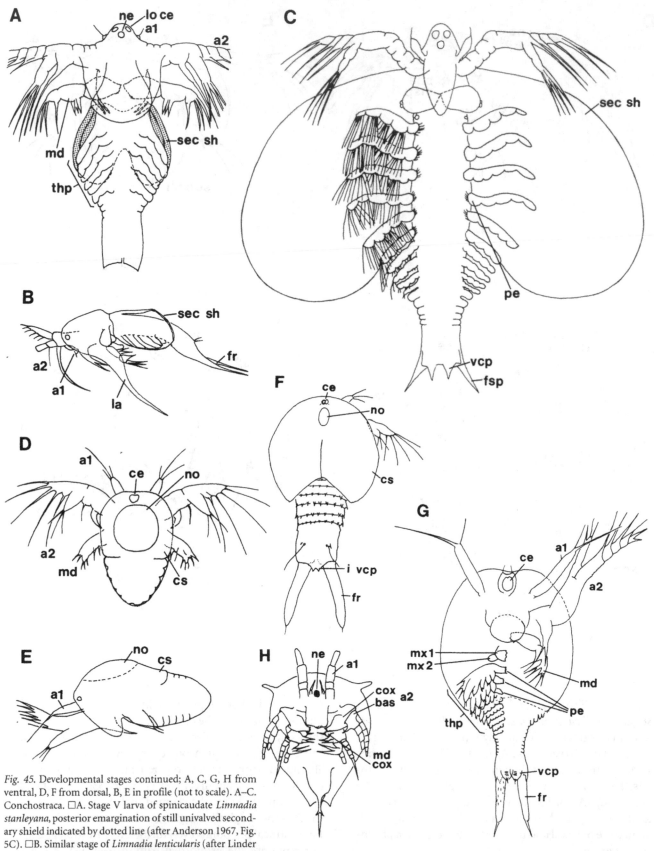

Fig. 45. Developmental stages continued; A, C, G, H from ventral, D, F from dorsal, B, E in profile (not to scale). A–C. Conchostraca. □A. Stage V larva of spinicaudate *Limnadia stanleyana*, posterior emargination of still univalved secondary shield indicated by dotted line (after Anderson 1967, Fig. 5C). □B. Similar stage of *Limnadia lenticularis* (after Linder 1945 [from Sars], Fig. 2b). □C. Stage VI larva of □L. *stanleyana* after metamorphosis to bivalved stage (after Anderson 1967, Fig. 8). D–G. Notostraca. □D. Hatching stage of *Triops cancriformis* (after Longhurst 1955, Fig. 13b). □E. Same stage (after Claus 1873, Pl. 6:1c). □F. Neonatus larva of *Lepidurus arcticus*; note the shifted neck organ (after Longhurst 1955, Fig. 13C). □G. Advanced larva (4th stage) of *Triops cancriformis* (after Claus 1873, Pl. 7:4). □H. Nauplius of the cirriped *Balanus balanoides*; note the huge antennal coxa (modified from Walley 1969, her Fig. 1).

may be another novelty in the evolution of this group achieved in line with adaptation to a bottom mode of life, which necessitates earlier functionality of limbs than in swimming forms.

A description of remipedan development is still lacking. Until this is available, the pattern of *Rehbachiella* is considered to approach the basic pattern of Crustacea s. str., i.e. without distinctive phases or loss of stages ('jumps').

Following strictly an anameric mode of development, *Rehbachiella* shows no significant differences to other crustaceans. The exception is the two-step development of thoracomeres, which cannot satisfactorily be explained at present. Except for the theoretical approach of Weisz (1946), there is nothing comparable mentioned for other Crustacea, so it may even be a peculiarity of *Rehbachiella* rather than the primitive mode that it appears to be. In either way, such a mode of ontogeny was very likely not developed prior to the crown-group level of evolution. This type, starting with an orthonauplius, is thus considered as a synapomorphy of the crown-group crustaceans, the Crustacea s. str., which also implies that the special naupliar feeding and locomotory appendages were present by the same level (cf. Walossek & Müller 1990). During further evolution each crustacean lineage has created its specific ontogeny.

Comparisons of locomotory and feeding apparatuses

General remarks

According to Dahl (1956) 'the most important single selective factor' of crustacean evolution is probably the mode of feeding. It should be added that locomotion is closely coupled thereto. Since each group evolved special aids exclusively dependent on its ground plan characters, synapomorphies may show up particularly in the characteristic feeding and locomotory structures. If these structures are preserved in fossils they are helpful to clarify the systematic position.

Since the postnaupliar limb apparatus of *Rehbachiella* obviously developed progressively toward swimming and filtration, comparisons focus on these life habits. The term 'filter apparatus' should only be applied to apparatuses that are specially equipped for filtration. The broad generalization of the term to cover all kinds of feeding apparatuses where a series of appendages is involved (e.g., Lauterbach 1974 and subsequent papers) disguises the fact that the functional system 'filter apparatus' is a highly complex one which operates only when various demands are fulfilled (see also Dahl 1976, p. 164).

In his pioneering studies published around 1930, Cannon worked out the morphology, mechanisms and differences in locomotion and feeding systems of the various

Crustacea. According to this author (1927b) the mysidacean and nebaliacean limb systems are similar to one another. Whereas this supports the unity of Malacostraca it also shows that filtration developed more than once even within this group from more primordial types. Foliaceous limbs, as in Nebaliacea among the Phyllocarida, should be secondarily developed and only superficially resemble the phyllopodous limb types of other Crustacea. It is not unlikely that they developed in accordance with compression of the whole anterior malacostracan limb series enclosed within the large shield.

Moreover, Cannon recognized that the mechanisms of the branchiopod type of apparatus are quite distinctive in morphology as well as the mechanics of filtration. This led him to refute a possible derivation of the nebaliacean type therefrom. Not all of Cannons conclusions 'survived', but at least the fact that filtration as a mechanism evolved independently several times among Crustacea has been given further support (e.g., Fryer 1987b, p. 427). Indeed, filter feeding can be achieved by the use of quite different structures, and the apparatuses and/or their modifications found among subtaxa may thus not be directly homologous to one another. In its special design, however, it can definitely serve to characterize particular groups. This is even more apparent when 'filtration' occurs at a different size scale and in different environments: most of the Recent Phyllocarida (Nebaliacea) are benthic mud-dwellers, while Euanostraca are permanent swimmers.

The necessity of a large shield for filtration has been claimed by Lauterbach in various papers, but such assumptions neglect the fact that this mode of feeding, as a function, does require currents as an essential element (cf. Fryer 1987b) but not a shield. Euanostraca, which are amongst the most effective filter feeders (Fryer 1987b) lack a shield. Again, euphausiids as well as *Rehbachiella* have a shield, but in both the slender filter aids extend beyond the shield. Notostraca have a large shield but do not filter-feed. Nebaliacea also have a large shield but, according to Cannon (1927b), were not filter feeders originally. Their foliate limbs suggest different functions (respiration?) having enforced the presence of a prominent shield.

Seriality, moreover, is no strong argument, since reduction of the number of limbs does not negate a function 'filter feeding', as can be seen in Cladocera and Ostracoda. The complicated and distinctive mechanisms of cladoceran filtration and other modes of feeding, described in much detail by Fryer (e.g., 1963, 1968, 1985, and 1987b as a summary) as well as the types of maxillary/maxillipedal filter feeding among Maxillopoda (also referred to as 'cephalo-maxillipedal feeding'; cf. e.g., Koehl & Strickler 1981; Boxshall 1985) are not considered herein, since they clearly represent derived states.

In accordance with Cannon (1927b) 'filtration' in general is recognized as a derived mode of feeding, and not at all primordial to Crustacea s. str. Secondary modification

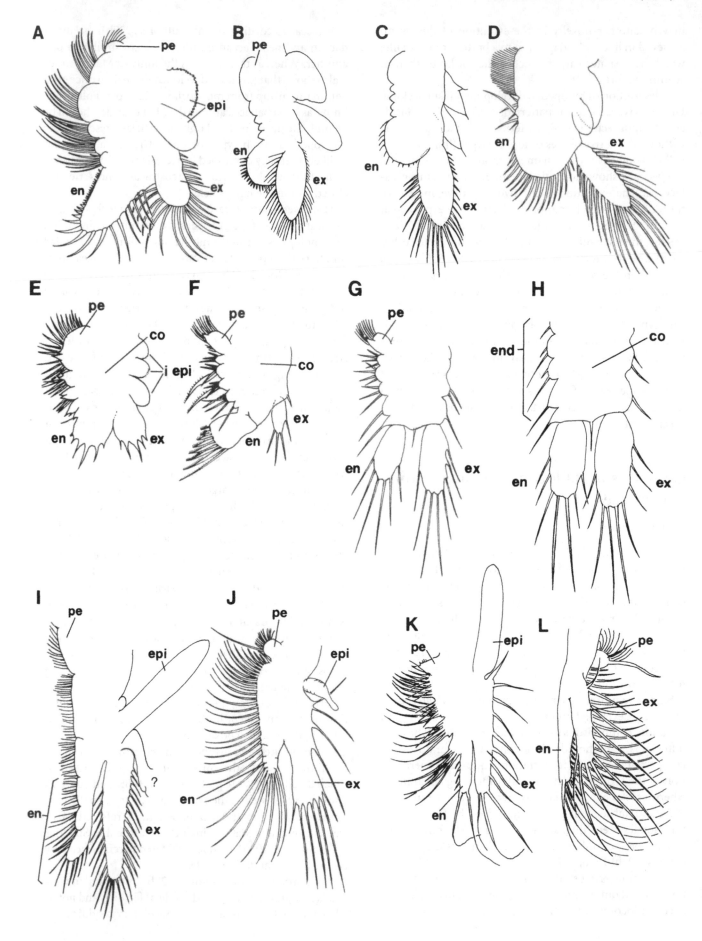

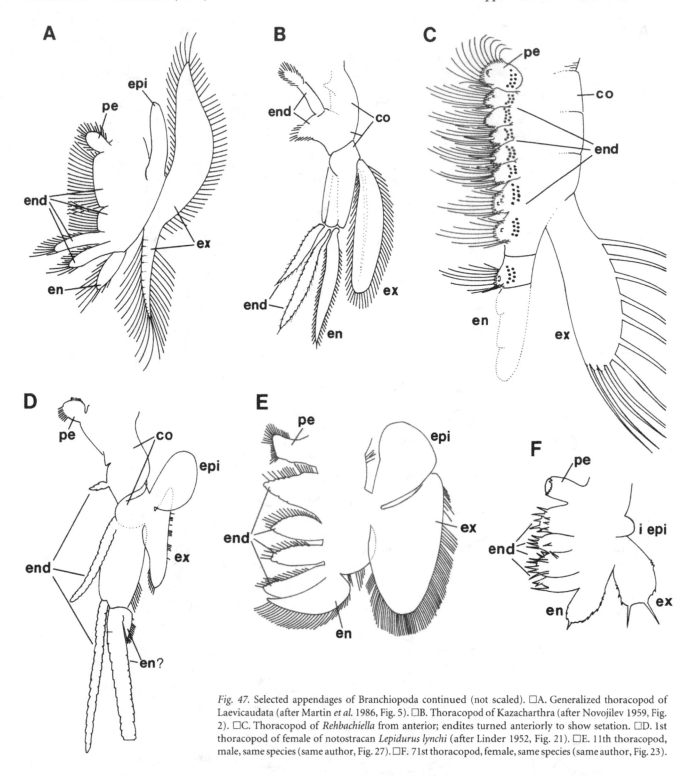

Fig. 47. Selected appendages of Branchiopoda continued (not scaled). □A. Generalized thoracopod of Laevicaudata (after Martin *et al.* 1986, Fig. 5). □B. Thoracopod of Kazacharthra (after Novojilev 1959, Fig. 2). □C. Thoracopod of *Rehbachiella* from anterior; endites turned anteriorly to show setation. □D. 1st thoracopod of female of notostracan *Lepidurus lynchi* (after Linder 1952, Fig. 21). □E. 11th thoracopod, male, same species (same author, Fig. 27). □F. 71st thoracopod, female, same species (same author, Fig. 23).

Fig. 46 (opposite page). Selected appendages of Branchiopoda (not to scale); all appendages oriented in the same direction in this and the subsequent figures. A–E. Euanostraca. □A. First thoracopod of male *Branchinecta gaini* (after Jurasz *et al.* 1983, Fig. 7a). □B. Thoracopod of male *Branchipus (Streptocephalus) stagnalis* (after Claus 1873, Pl. 5:17). □C. 11th thoracopod of *Tanymastix stagnalis* (after Eriksson 1934, Fig. 4). □D. 9th thoracopod of female of *Parartemia zietziana*, median setation omitted (from Linder 1941, Fig. 24d). □E. 5th thoracopod of juvenile, 2 mm long *Branchinecta paludosa* (from Linder 1941, Fig. 12d). F–H. Thoracopods of *Lepidocaris rhyniensis*. □F. One of the anterior scraper limbs. □G. One of the median set of limbs. □H. One of the exclusively locomotory posterior limbs with symmetrical rami (from Scourfield 1926, Pl. 23:7, 9, 10). I–L. Onychura. □I. Thoracopod of 3 mm long spinicaudate *Eocyzicus dhilloni* (from Battish 1981, Fig. 46). □J. Thoracopod of spinicaudate *Sida crystallina* (from Eriksson 1934, Fig. 20). □K. Thoracopod of larval spinicaudate *Cyzicus tetracercus* (from Botnariuc 1947, Fig. 12). □L. First thoracopod of cladoceran *Holopedium gibberum*, lateral view (from Eriksson 1934, Fig. 29).

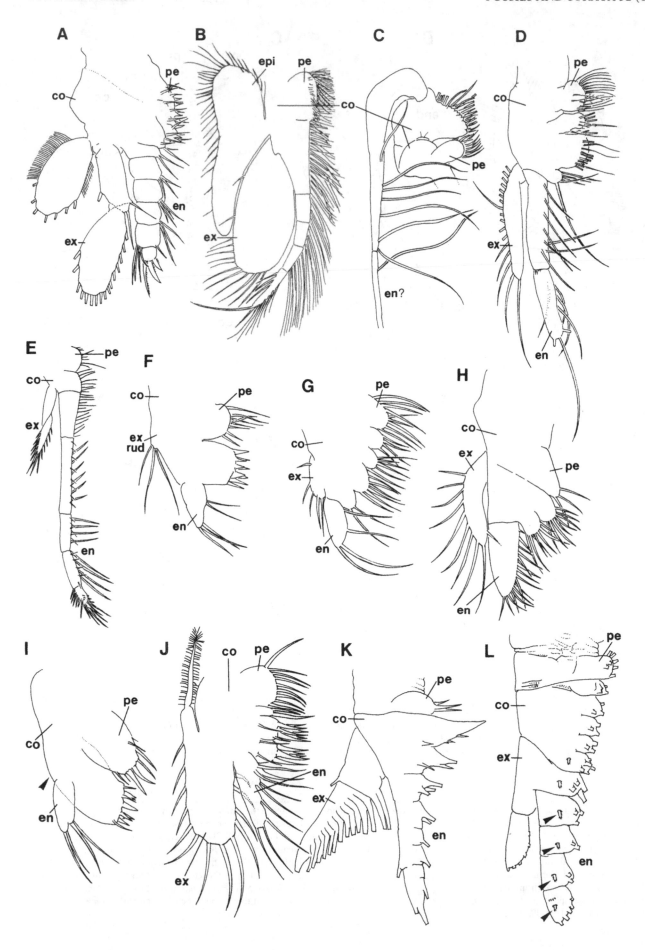

occurs for example in the development of special sensory structures adapted for the perception of food at different Reynolds numbers and at different distances from the body. In the light of recent studies of life habits of small crustaceans by high speed cinematography, the usefulness of the term 'filtration' has been variously questioned (see also chapter on 'Functional morphology and life habits'). Fryer (1987b), however, gives good reasons for retaining this widely used term. In particular he points to various studies showing that filter grids are true filters (sieves) and not solid walls, as claimed by Koehl & Strickler (1981).

Possession of a filter apparatus does not preclude other feeding habits. Particles spanning a wide size range can for example be trapped merely by modification of the filter setules and their spacing. Thus, food may consist of smallest algae as well as of small animals. Active capturing of food or scavenging is also possible. Particles may be scraped off distally (grazing) and filtered after accumulation proximally (e.g., Fryer 1987b for various cladocerans and the Devonian anostracan *Lepidocaris*).

Cannon (1927b) found two different mechanisms in the mud-dwelling *Nebalia*, one for feeding on large particles and one for filtration. Filtration of most large and small-sized forms is basically (sessile forms neglected) coupled with permanent motion (swimming), often, if not mostly, done in an upside-down orientation (examples: Copepoda and Euanostraca). Likewise size dependent, the trunk limb apparatuses used for filtration vary widely, since different filtratory aids are required.

With the discovery of the Cephalocarida in the early 1950's a new crustacean body plan became known and was claimed to serve as the centre from which all crustacean plans could be derived (Sanders 1963a). This together with the features of the ground plan of Crustacea s. str. (cf. Walossek & Müller 1990) has led to the elucidation of a number of structural and functional elements of filter-feeding apparatuses of known crustaceans for comparison with *Rehbachiella*. Our initial intention of evaluating the

suggested affinities of the fossil to Branchiopoda inevitably has led also to considering the possible origin of the branchiopod type of filter feeding and the status of Cephalocarida, today having a serial limb apparatus but not being filter feeders (Sanders 1963a, b).

The bipartite locomotory and feeding apparatus

Basically common to all Crustacea s. str. are the structures of the naupliar apparatus, according to Walossek & Müller (1990): the postlabral feeding chamber bordered by the large labrum anteriorly, which projects over the atrium oris, the paragnaths as outgrowths of the mandibular sternite, the paragnath channel between the lobes, a sternal setation, and an enhanced mandibular proximal endite to form a coxa with obliquely angled grinding plate. In virtually all Crustacea s. str. its cutting edge is differentiated into a anterior 'pars molaris' and a denticulate 'pars incisivus'.

The principal differences occur in the retention of the mandibular palp and the use of the succeeding appendages, the maxillae. A palp in the adult state is developed only in Malacostraca (Phyllocarida as well as Eumalacostraca) and basically in Maxillopoda. In the latter, this retention of the basipod and two rami is assumed to be closely linked with abbreviated growth by neoteny. In the former, a uniramous three-segmented palp reappears after complete atrophy during later larval development (e.g., Knight 1975, Fig. 7) and developed into a prominent structure, serving for various purposes in the feeding process, such as grooming, capturing and pushing in particles (e.g., Hamner 1988, p. 160 for euphausiids).

Its presence in both the Eumalacostraca and the Phyllocarida (e.g., Kensley 1976, Fig. 2c; Quddusi & Nasima 1989, Fig. 2C) is considered as a further apomorphy of Malacostraca (Fig. 51C, D). All other crustaceans, which lack such a palp, use different appendages for these purposes. *Rehbachiella* possibly uses its 1st maxilla, while euanostracans use their 1st trunk limb, since both maxillae are reduced to their 'proximal endites' only.

As mentioned earlier, the consideration of the stepwise modification of the anterior postmandibular limbs into 'maxillae' is imperative for any comparisons of limb apparatuses and the understanding not only of the diversification of the systems but also the synapomorphies in their structural design. Basically, the 2nd maxilla was a morphological and functional trunk limb. This can be deduced from Recent Cephalocarida (Sanders 1963a, b), but also from its trunk-limb shape in early Maxillopoda (Müller & Walossek 1988b) and early anostracan Branchiopoda (Schram 1986, and herein).

In euphausiids the 2nd maxilla has at least 4–5 enditic lobes (Mauchline 1967, e.g., Fig. 7; Knight 1975, Figs. 8–9; Fig. 48G herein). Mauchline (1967, p. 9) claims that in *Bentheuphausia amblyops*, which has three-segmented endopods on both maxillae, the 2nd maxilla 'looks more

Fig. 48. Selection of appendages from other Crustacea (not to scale). □A. Second maxilla of cephalocarid *Sandersiella acuminata* (from Ito 1989a, Fig. 3). B–D. Leptostraca. □B. 1st thoracopod of *Speonebalia cannoni* (from Bowman et al. 1985, Fig. 2a). □C. 1st maxilla of *Nebalia marerubri* (from Wägele 1983, Fig. 12). □D. 2nd maxilla of same species (same author, Fig. 13). E–G. Euphausiacea. □E. Generalized 2nd thoracopod (Mauchline 1967, Fig. 3b). □F. 1st maxilla of *Thysanoessa raschi*, furcilia VI (from same author, Fig. 16c). □G. 2nd maxilla of same species, furcilia VII–VIII (from same author, Fig. 16f). □H. 2nd maxilla of mysid *Mysidopsis furca* (from Bowman 1957, Fig. 1C). I, J. Decapoda. □I. 1st maxilla of *Caridina babaulti basrensis*, stage III (from Salman 1987, Fig. 3d; arrow points to position of exopod reduced in this species). □J. 2nd maxilla of same species and stage (from same author, Fig. 3e). □K. Postantennular limb of Upper Cambrian *Martinssonia elongata* (from Müller & Walossek, 1986a, Fig. 4). □L. Thoracopod of Upper Cambrian *Dala peilertae*; arrows point to sensory bristles (redrawn from Müller 1983, Fig. 1D).

like a limb than a mouthpart'. The same can also be stated for the shape of the 2nd maxilla of leptostracan Phyllocarida, when compared with that of *Rehbachiella* (compare Figs. 48D and Fig. 33 for *Rehbachiella*). Again, according to Walossek & Müller (1990), the 1st maxilla also may not have altered much from the limb design known from the Upper Cambrian stem-group crustaceans

In consequence, the starting point of development of the postlarval limb apparatus of Crustacea s. str. should have been a modification of the 1st maxilla as an additional 'mouthpart' – which still does not imply that it had already changed its shape to any extent. All further steps, the modification of the 2nd maxilla and subsequent limbs, should belong to evolutionary paths inside the different major branches. Hence, 'thoracic apparatuses', if existing, should have evolved even later – and by convergence. Comparison of these alone neglects the fact that by this stage the various taxa had already become separated. Specialities characterizing secondary paths and different evolution are, however, valuable for ingroup analyses. These include, for example, a further addition of 'mouthparts' – parallel in the various groups (copepod lineage of Maxillopoda; Remipedia; Eumalacostraca).

Comparisons

Detailed information on the limb apparatus of *Nebalia*, as a representative of Recent Phyllocarida, has been provided by Cannon (1927b, 1931), and critically reviewed by Linder (1943). Since then, little has been added, but taxonomic descriptions indicate that habitats, life styles and morphology may be rather variable among leptostracans. The majority seem to live in deeper waters (e.g., Linder 1943; Hessler & Sanders 1965), but they have been captured also from hydrothermal vents (Hessler 1984) or in caves (Bowman *et al.* 1985).

Of filter-feeding apparatuses among Eumalacostraca, the feeding structures of euphausiids are chosen here, described, e.g., by Mauchline (1967) and Hamner (1988) who has studied the mechanism of the krill, *Euphausia superba*, by high-speed cinematography. Filter feeders among the euphausiids can be readily distinguished from others by their specific armature. Branchiopod apparatuses have been studied in detail by Eriksson (1934), Barlow & Sleigh (1980), and Fryer (various papers, 1985 also for *Lepidocaris*). A list of the various feeding apparatuses of Branchiopoda is given by Storch (1925), emphasizing the variety of modifications that are possible with a serial feeding apparatus. Among these the euanostracan apparatus is considered to be close to the basic type and is chosen here as representative for Branchiopoda.

A selection of parameters of postnaupliar limb apparatuses, as listed, has aided in estimating the affinities of the *Rehbachiella* apparatus (no hierarchy intended):

1. insertion of limbs below the sternal level or above it

2. design of the sternal food path, being shallow, elevated, or deeply invaginated ('channel')

3. importance of the proximal endite/portion of the maxillae and orientation of its setation

4. number of enditic lobes on either of the two maxillae

5. importance of the proximal endite/portion in the posterior limbs

6. presence or absence of enditic lobes, and, if present, whether they are knob-like or applanate, fused, or turned against the axis of the limb

7. armature of inner edge into sets of spines/setae

8. direction of endites anteriorly or posteriorly

9. orientation of setation and specific function in accordance with endite orientation

10. direction of the food current into the inter-limb capture area

11. shape of inter-limb food path, being open or closed posteriorly

12. mode of orally directed food transport, either on the sternal surface or within a sternitic channel

13. degree of rigidity of the limb corms to act as sucking devices

14. presence of sucking chambers between the limbs

15. mode of metachronal beat

16. portion(s) of a limb that produces the motion and food currents

17. part(s) of the limb that are responsible for filtration

18. use of endopods and exopods

19. compression of the limb series concerned with filtration or serial decrease in the size of the segments

20. increasing or decreasing size of proximal limb portions of the anterior postmandibular limbs (mx1 – thp1, 2)

Malacostraca. – The major morphological and functional differences between the *Rehbachiella* and the malacostracan type are exemplified in Figs. 49 and 50. In filter-feeding Malacostraca – nebaliaceans (Cannon 1927b) as well as euphausiids (Hamner 1988) – food enters from the front (Fig. 49A$_1$). All eight anterior thoracopods may be involved, but one or two of the posterior limbs may not be used – this is species-dependent. In *Nebalia* the complete limb apparatus is much compressed (see Fig. 46 in Hessler 1964) and enclosed within the large head shield. With the movements of the endopods, in both the phyllocarids and euphausiids food particles are transferred to the more

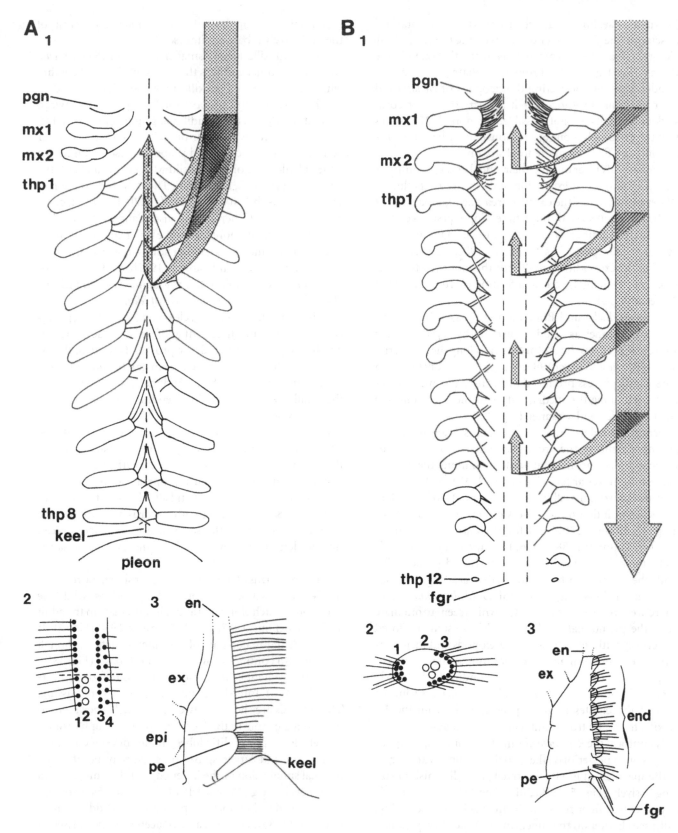

Fig. 49. Schemes of filter apparatuses, to show orientation of limbs, endites and setation; arrows indicate water currents into capture area. □A. 1, horizontal section through apparatus of phyllocarid Malacostraca (Leptostraca); anterior region stretched, 'x' indicates position of mouth, dashed line indicates midventral keel (modified from Cannon 1927b, Figs. 1, 2; see also Hamner 1988, Fig. 3, for Euphausiacea); 2, arrangement of setae on inner edge of thoracopod (1–4); 3, cross-section to show position of thoracopod, setation, and shape of sternite (after Claus 1889, his Pl. 8:8 and photographs kindly provided by E. Dahl). □B. 1 apparatus of *Rehbachiella*; dashed double line indicates invaginated midventral food groove; 2 and 3 corresponding to A (see also Fryer 1983, Figs. 141–145, for Recent Euanostraca).

proximal setae. Finally, in nebaliaceans the horizontal grid of setae on the proximal endite shovels the food anteriorly along the sternal surface, which is distinctly keeled in the thorax (see Fig. 49A₃). From the foliate design of the slender, ventrally projecting thoracopods of *Nebalia* and their position close together it appears that inter-corm sucking currents are not possible. Cannon (1927b) states that the endopods alone are responsible for both the production and filtering of the food stream.

Euphausiids filter outside their shield and use compression pumping, with expansion and closure of the limbs. Water brought into the capture area during expansion of the basket is squeezed through the endopodal sieves during closure (see Hamner 1988, Fig. 1; and Fig. 3 for morphology of thoracopods). As Hamner (1988, p. 161) states 'neither the exopodites [small] nor the mouthparts [maxillae and maxillipeds] create a feeding current outside of the feeding basket'.

In *Rehbachiella* and Branchiopoda (based on the euanostracan model, as described in detail by Fryer 1983) the food current enters the capture area all along the limb series (Fig. 49B₁; see also Fryer 1983, p. 231). In conjunction with the sucking process produced by the limb corms, particles are passed along the median edges of the limbs into the deep sternitic food channel (Fig. 49B₃). It is there that transport toward the mouth occurs, most likely by sweeping movements of the proximal endites. Since the limbs move in a metachronal beat while swimming, the sucking maximum moves anteriorly (Fryer 1983, p. 300).

In malacostracans all medial surfaces of postmaxillary limbs are pointing more or less orally, arranged in a V-shape (Cannon 1927b, Fig. 1 for *Nebalia*) or almost in an oval (see Hamner 1988, Fig. 6, for euphausiids). Accordingly, their filter setae insert anteriorly and point orally, while the rows of comb setae are in the back (Fig. 49A₂; see also Hessler 1984, Figs. 3B–J for *Dahlella caldariensis*). There are three to four sets of setae, which seem to originate from the primordial double row. The system is closed posteriorly with the last pair of pereiopods and their 'closure setae' which bear anteriorly pointing setules in front of the pleopods (Hammer 1988; corresponding to the 'Sperrborsten' in branchiopods, see Eriksson 1934). Again, the sizes of the endites increase progressively from the 1st maxilla to the 1st trunk limb (Figs. 49A₁, 50A).

By contrast, the *Rehbachiella* and the branchiopod apparatus is open posteriorly, the 'proximal' endites are largest in the maxillae, at least primitively, and diminish in size progressively (Figs. 5, 49B₁, 50B). The endites are posteriorly oriented, filter setae are in the back row (set 3; Fig. 49B₂; see chapter on the function of the adult apparatus), while the closure setae form the anterior set (set 1; also Fryer 1966, Fig. 2 for the carnivore *Branchinecta gigas*; two rows with backwardly pointing setules in *Rehbachiella*). Such arrangement is exactly the reverse of that of *Nebalia*.

Again, a third group of spines is developed on the crests of the primitively lobate endites (set 2).

The 1st maxilla is only smaller in size in *Rehbachiella*, while its proximal endite is the largest of the set. Together with that of the 2nd maxilla they most likely acted as brushing devices to transport the food particles between the cutting edges of the mandibular grinding plates (cf. Fig. 33). This accords with other Branchiopoda but is in clear contrast to the function and shape of these two limbs in either phyllocarid or euphausiid Malacostraca (Fig. 51C, D; see also Figs. 48C, D and F–J). The chamber function of the maxillae is shown by Cannon (1927b, Fig. 4b–d, combined in Fig. 51C herein; see also Cannon 1931). The vertically oriented, more blade-like inner edge of the basipod of the 1st maxilla, lacking endites, is even capable of mastication in *Nebalia* (Cannon 1927b, p. 364) and euphausiids, recognizable by the rigid spines (e.g., Mauchline 1967, Fig. 5, for *Thysanopoda tricuspidata*).

This arrangement is also in clear contrast to the cephalocaridan apparatus, where neither of the maxillae operate as brushes (Fig. 51B). Instead, the proximal endite of the 1st maxilla is elongated to reach from far outside into the paragnath channel, while the 2nd maxilla is essentially as the trunk limbs, its proximal endite being even more poorly developed.

Differences between the two malacostracan apparatuses, besides the striking discrepancy in size, are particularly in the use of the limb parts concerned with filtration: the secondary setules of euphausiids operate at the same level as the setules of nebaliaceans and all other crustaceans (see Hamner 1988, Figs. 4, 5). In other words, the forces of the viscous regime affect the setae in non-eumalacostracan filter feeders, while they are at the setule level in euphausiids.

The limb corms of the latter are rather short, more or less divided into two portions but lacking endites, while the endopod is much elongated and five-articulate in the adult state (Mauchline 1967, Fig. 3; Hamner 1988, Fig. 3). It represents the major part of the filter structure, though setation also continues towards the limb basis (Fig. 48E). Setal arrangement bears a resemblance to that of Leptostraca (Fig. 48B). Inter-limb sucking chambers are missing. The exopods are short and two-segmented. They seem to play no major role in the feeding process of euphausiids.

Well-developed enditic lobes are not developed on the long corm, but larger spines or setae may point to an original subdivision of the inner edge of the limb corm in some species (e.g., Barnard 1914, Pl. 39, for *Nebalia capensis*). The distal endopodal podomere is paddle-shaped (e.g., Dahl 1985). The large and foliaceous exopod arises, as in all other Crustacea, from the sloping outer margin of the basipod portion which is indistinctly set off from the slender endopod. It may be valve-like or slender and leaf-shaped, thus appearing much as in Branchiopoda (Fig.

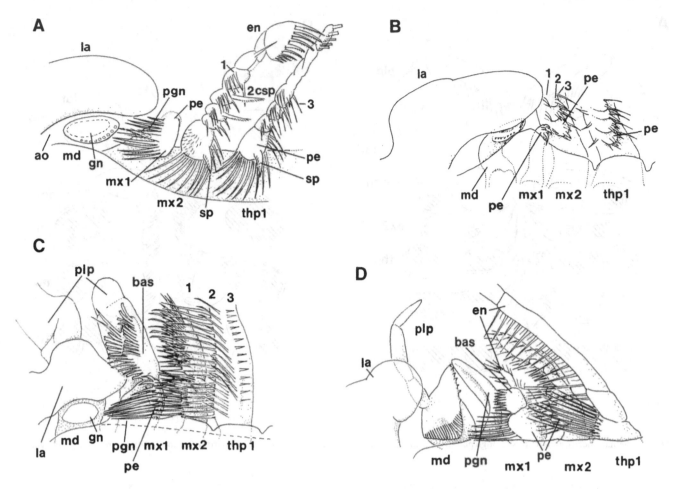

Fig. 50. Semi-sagittal views to show arrangement of appendages and post-labral feeding chambers of selected crustaceans, slightly schematic. □A. Reconstruction of *Lepidocaris* (modified from Fryer 1985, his Fig. 3), direction changed in accord with the other drawings. □B. Same area for Cephalocarida (modified from Sanders 1963b, his Fig. 4, and Hessler 1964, his Fig. 2). □C. Same view for *Nebalia* (combination of Figs. 4b–d of Cannon 1927b), setation of limbs simplified; area slightly stretched in long axis of the body for clarity. □D. Same view for Euphausiacea (modified from Mauchline 1967, Fig. 1); area slightly condensed, because stretched in original picture.

129:1, 2, in Brooks *et al.* 1969; Wakabara 1976, Fig. 2C, for *Paranebalia fortunata*; Clark 1932, Pl. 2:14, for a posterior limb of *Nebaliella caboti*), which becomes even more apparent when comparing immature leptostracans (e.g., Vader 1973, Fig. 1C, for *Nebalia typhlops*).

Due to enclosure within the shield, participation in locomotion, as in Branchiopoda and most likely also *Rehbachiella*, is not possible in Nebaliacea. The insertion of their epipod is in most descriptions elegantly obscured but seems to arise from the outer edge of the proximal limb portion. Cannon (1927b) assumes that the foliate exopod and epipod act as valves during the filter process (respiratory function assumed but never clarified in detail; see also Pillai 1959, Fig. 9, for the 1st thoracopod of *Nebalia longicornis*).

Cephalocarida. – Sanders (1963b, p. 9) remarks that these are not filterers but non-selective deposit feeders. This is evident from the whole organization of the limb apparatus and its equipment with setae, as checked against the above

list. Again, no anterior currents were detected between the limbs, and food transport is effected mechanically. This seems to contrast with the statement that the fleshy limb corms form sucking chambers. The reason may lie in a special benthic mode of life of the Cephalocarida. Sanders (1963a) notes that Cephalocarida move with their ventral surface down, in contrast to, e.g., filter-feeding Branchiopoda, which indicates already some differences in the use of the limbs. The animals oscillate rapidly with their limbs, and it is not unlikely that their metachronal beats serve not only to collect food but also to use the sucking system for different purposes, for example to adhere to the bottom while gliding over it (Strickler, personal communication, 1989). Such a habit would indeed be an interesting evolutionary adaptation for life in the flocculent layer.

The setation resembles that of *Rehbachiella*, in particular with regard to its arrangement in three sets (Fig. 51B). The number of the fairly short rigid spines and spine-like setae, however, is much lower (Sanders 1963b, Figs. 3, 4, 25, 26; Ito 1989a, Figs. 3, 4). Pronounced endites with pectinate

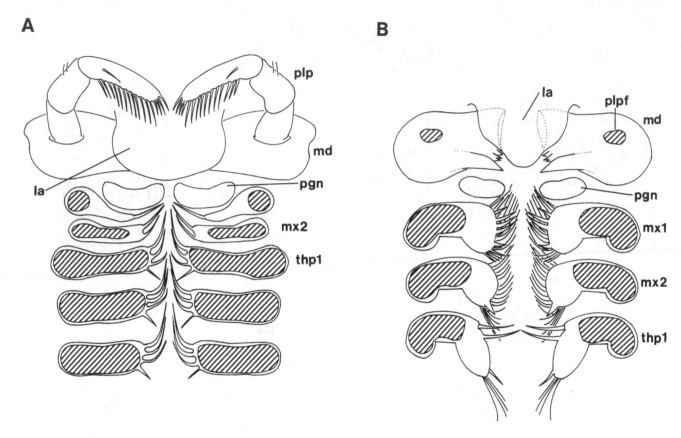

Fig. 51. Schematic horizontal sections through postoral regions. □A. Phyllocaridan Malacostraca, modified from Cannon 1927b, Fig. 6. □B. *Rehbachiella*.

setae forming filter grids for retention of small-sized food are lacking, as well as closure setae. More striking are the slight elevation rather than invagination of the postcephalic food path (Sanders 1963b, Fig. 3), the outward insertion of the postmandibular limbs (also Fig. 51B) and poorly developed proximal endites.

Another striking difference from *Rehbachiella* and Branchiopoda is the median armature of the 1st maxilla, which develops only on the proximal endite which is much elongated and spinose distally. The three incipient endites of the basipod are reduced during ontogeny (Sanders 1963b, Fig. 17). Only at the level of the 1st maxillae is the cephalic sternum deeply invaginated between the paragnaths, as can be seen in various other crustaceans, and the long endite of the 1st maxilla is bent down into this cleft (Sanders 1963b, Fig. 6).

A further contrasting feature of Cephalocarida is their five- or six-segmented, cylindrical endopod with its distal tuft of rigid spines forming a 'claw'. The outgrowths on the outer edge of the corm have been referred to as 'exopod' and 'pseudepipod' by Sanders (1963b), but while the nature of the exopod is beyond question, that of the remaining part is unclear (see subchapter on postmaxillary limbs in the chapter 'Significance of morphological details'). From their outward orientation and setation it can be

assumed that both are involved in locomotion (held laterally while the animal moves close to the surface).

While the differences between the two functional systems of Branchiopoda/*Rehbachiella* and Malacostraca are apparent, the question remains whether the cephalocaridan apparatus could serve as a basis for both, one of them or neither of them. The short setation does not support filter feeding in Recent Cephalocarida but would not necessarily exclude the possibility that their ancestors fed in this way. Their unspecialized trunk-limb shaped 2nd maxilla is a necessity for an ancestral form, but again this is no indicator of closer affinities to any particular group since the examples now known show that this status is retained from the ground plan of Crustacea s. str. in the different crustacean lineages.

Interestingly, Cephalocarida share structures with both systems. At first sight this would seem to be an elegant confirmation of its basic position. However, the characters are exclusive, which demands an interpretation of their character state. If, for example, the invaginated sternitic food groove represents the derived state, as favoured herein, Cephalocarida and Malacostraca share a symplesiomorphy. The position of the limbs lateral or laterodorsal to the sternal region would point in the same direction. On the other hand, Phyllocarida and Cephalocarida both have compressed their thoracic region (anterior part in phyllo-carids), for which no explanation can be given at present.

Shared structures with Branchiopoda are, for example, the posteriorly open capture area and missing specialization of the 1st thoracopod (both probably symplesiomorphic), the lost mandibular palp (status unclear), fleshy limb corms for limb base feeding and sucking chambers between them, four endites on the 1st maxilla, including the proximal one (also in *Rehbachiella*), at least in the larval limbs, and at least six on all posterior limbs.

Presuming that, besides the enhancement of the proximal endite and the peculiar design of the endopod, the malacostracan 1st maxilla lacking any sub-division of the basipod into lobate endites (even biting abilities) reflects the plesiomorphic design among Crustacea s. str., the latter features at least leave the possibility open that the branchiopod type of locomotion and feeding had indeed evolved from the apparatus of cephalocaridan ancestors, as has been suggested by Sanders (1963b).

Additional support comes from the 'neutral' orientation of the endites. Again, the different fate of the 2nd maxilla in Malacostraca, with an enhancement of the 'proximal' endite and a further sub-division of both endite and basipod, might point in this direction. As a whole, this would bring Cephalocarida closer to Branchiopoda and Maxillopoda rather than their remaining in a 'central' position among Crustacea.

Again, it has to be remembered that the possession of a proximal endite characterizes the ground plan of Crustacea s. str. Its diminution in all postmaxillulary limbs of Cephalocarida but its retention in all branchiopod limbs indicates that the condition found in Recent Cephalocarida is derived. Accordingly, a derivation of branchiopod limb types, (as in Sanders 1963b, Fig. 75) cannot be accepted. It also conflicts with the proposed relationships between the different taxa: in Cannon's scheme, the Notostraca with their lobate endites are set apart, while the lobate type of the *Lepidocaris* limb gave rise to a primordial cladoceran, and then in turn to Cladocera and Euanostraca, and the Conchostraca.

In Recent Branchiopoda only the proximal endites of the maxillae are retained, but *Lepidocaris* as well as *Rehbachiella* demonstrate that this status must have been achieved independently in the different lineages. Their prominence in *Rehbachiella* fulfills the branchiopod plan but is clearly different from the condition of Cephalocarida.

It must also be remembered that *Lepidocaris* was most likely not a filter feeder in the strict sense. Eriksson (1934) and Fryer (1985) have clearly shown the derived state of this fossil, which evidently scraped together food with its anterior limbs while swimming in morphological orientation. The deeply invaginated food groove, the length of the setae on the proximal endites, and the number of endites along the limbs, however, demand a reexamination, since neither author had examined the material personally. In

any case, the three sets of armature – anterior row of setae, median spine(s) and posterior row – as described for *Rehbachiella*, are clearly visible in Fryer's (1985) reconstruction (his Fig. 3; see also Fig. 50A herein).

Cannon (1933), probably influenced by phyllocarid limb morphology, hypothesized a long endopod for the ancestral branchiopod in his evolutionary scheme (see above). This is remarkably well borne out by the limb morphology of *Rehbachiella*. His model would give a better fit if he had included (1) the median set of spines between the two marginal rows of enditic setae and (2) the large proximal endite with its characteristic spine, also known from *Rehbachiella* (Figs. 27H, 33). Remarkably, he had illustrated such a spine for *Lepidocaris* (his Figs. 6, 7) and Conchostraca (his Fig. 8). Considering such additions and recognizing the notostracan posterior limb as a reflection of the ancestral plan (immature state), Cannon's scheme is fully supported by *Rehbachiella*.

In summary, *Rehbachiella* possesses characters in the postnaupliar locomotory and feeding apparatuses that, in morphological as well as functional respects, clearly preclude either a malacostracan or a cephalocaridan relationship. Again, in several respects the cephalocaridan limb apparatus is closer to that of Branchiopoda than to Malacostraca. Further studies will have to evaluate how the relationships of the Maxillopoda as the proposed sister group of Branchiopoda conform with the characters recognized in the limb system.

The branchiopod model of the limb apparatus of REHBACHIELLA

The resemblance of the basic filter-feeding apparatus of the Branchiopoda to that of *Rehbachiella* is firstly evident in their structural similarities, such as the deep sternal invagination to form a true thoracic food channel, the limb shape, the prominence of the proximal endites, the posterior direction of the endites, the setal armature (3 sets), its position (closure setae anteriorly, filter setae posteriorly) and orientation. Functional aspects also accord with this, such as the posteriorly open system, with a water current entering all along the series of limbs, the capture area between the limbs, and inter-limb sucking chambers essential for filtration.

It is not surprising that the apparatus of *Rehbachiella*, besides having many similarities with that of Euanostraca, bears also a structural resemblance with that of Phyllopoda, even when no longer used for filter feeding, such as in Notostraca. This is particularly true for the shape of the mandible, which is much coarser in *Rehbachiella* than in the Recent Euanostraca but more similar to those of Notostraca (see below), and the segmented endopod, which seems to be indicated at least in certain Spinicaudata (Fig. 46I, J). In this respect, the euanostracan model is not strictly applicable for reconstructing the mode of locomotion and feeding of *Rehbachiella*, but various of the major differences, such as the large shield, the well-developed maxillae, and the higher number of enditic lobes may simply accord with the ancestry of the fossil.

Again, the fact that the naupliar limbs were better developed in the fossil must be considered with much caution, since the largest stage was most likely still larval, i.e. at a state when the same limbs are also still functional in all Recent Euanostraca. Progressive improvement of the effectiveness of the filter-feeding apparatus of Recent Euanostraca may have led to an applanation of the endites on the limb corms in an axial direction. In this respect, the rows of setae of the posterior set (3) form a closely spaced and even grid, while sets (1) and (2) are close together.

In the origination of modern Euanostraca, concentration processes may have led to fusion of the endites in a different manner. Its evolutionary steps are still recognizable particularly in the proximal ones of the set, which are distinctively separate (e.g., Fig. 46A), by an indentation of the blade-like proximal endite, or by the position of larger spines of the median set (see particularly Fryer 1983). The three sets of enditic setae of *Rehbachiella* are found basally in all Branchiopoda. Other setal types also correspond to those of Branchiopoda in form, function, and position: comb spine (of median set 2; Fig. 35A), brush setae or spines (Fig. 35B–D), and the anterior 'Sperrborsten' (Figs. 37, 38).

Rehbachiella shows no signs of scrapers on the distal parts of the postmandibular limbs, such as are present in *Lepidocaris* or certain Cladocera (Fryer 1985) and non-filtering Euanostraca such as *Branchinecta gigas* or *B. ferox* (Fryer 1966, 1983). With this, a grazing or raptorial habit of *Rehbachiella* is less likely, although the cutting edge of the mandible is less developed than in Recent filter-feeding forms. This is in clear contrast to *Lepidocaris*, which is interpreted as a mobile saprophytic grazer type (Eriksson 1934; Fryer 1985), similar to *Tanymastix* among Euanostraca or certain Cladocera.

Assuming that the *Rehbachiella* type of filtratory appendages represents a rather primordial state among Anostraca, the Lipostraca thus have clearly modified their appendages. *Lepidocaris* grazed off particles with specialized endopodal scraping spines (Scourfield 1926, Fig. 15; Fryer 1985, particularly his Fig. 4). Interestingly, the proximal endites are much as in *Rehbachiella*, having the posterior row of pectinate setae (set 3) and also the rigid spine of the median set (2; Fig. 47F, G; see also Cannon 1933, Figs. 6, 7 for *Lepidocaris* and Fig. 8 for an estherid spinicaudate). With regard to Fryer's illustration (1985, Fig. 3), the anterior set (1) seems to be more important in the anterior feeding limbs than does the posterior set (Fig. 51A). Their loose contact precludes filter activities.

Again, while the anterior limbs with their shortened exopods probably contributed little to locomotion in *Lepidocaris*, its posterior thoracopods are more or less exclu-

sively natatory. The gap between the locomotory 2nd antenna and the set of locomotory thoracopods (Fig. 42G) might explain the retention of the former as a swimming device. By contrast, shortening of the head in Euanostraca (Fig. 42H) and the anterior shifting of the postcephalic locomotory apparatus might have resulted in the possibility of modifying the 2nd antenna. A special life style of *Lepidocaris* is also indicated by the possible lack of compound eyes, the missing shield and the unique furca.

Furcal rami are not only stabilizers of the trunk but can also act as rudders. Thus, the appearance as well as retention of this organ are correlated with these functions. With the possession of a large shield in *Rehbachiella* limiting the appendage manoeuvres, the appearance of articulate furcal rami earlier than in the Recent Euanostraca might, thus, be partly historical and partly due to functional requirements.

When compared with Recent Euanostraca – filter-feeding and carnivorous ones – and *Lepidocaris*, the median edge of the mandibular grinding plate appears much simpler and more primordial in *Rehbachiella*, lacking the complex tooth structures of the pars molaris (e.g., Fryer 1983, Pls. 8–11, for Euanostraca, Scourfield 1926, Pl. 23:3–5, for *Lepidocaris*). It is thus interesting that while similarities in the trunk limbs between *Rehbachiella* and Notostraca are less obvious due to the non-filtering life style of the latter, their mouthparts are rather similarly designed. This is particularly true for the mandible. The few rigid teeth in the posterior part of the cutting edge ('pars incisivus') resemble those of, e.g., *Triops cancriformis* to a remarkable degree, particularly in ventral aspect (even bifid; see Fryer 1988, Figs. 104, 107).

Differences are, however, obvious when viewing from the inner side. Here the notostracan mandible reveals huge vertically oriented complex teeth, while the pars molaris seems to be greatly shortened (Fryer 1988, Fig. 118). The presumed anostracan relationship of *Rehbachiella* suggests that this similarity reflects an ancestral shape in both forms. Again, the fine spinules and setules in the shallow depression of the anterior part, the 'pars molaris', suggest that the feeding habits of *Rehbachiella* differed in detail from those of Notostraca.

Also different from the feeding apparatus of the latter are the paragnaths, which in *Rehbachiella* are located much closer together. Again, the median food groove is deeper and the 1st maxillae function as pushers, while those of Notostraca s. str. are 'toothed' medially (Fryer 1988, Figs. 100, 101 for *Triops cancriformis* and 104 for *T. longicaudatus*), in line with their non-filtratory habit. With regard to the question of whether this mode of feeding is ancestral or derived, the closer similarity between the feeding apparatus of *Rehbachiella* and larval Notostraca gives further support for the assumption of a secondary loss of the filter-feeding habit in this branchiopod group.

At first sight, the lobate or knobbed shape of the endites of *Rehbachiella* differs from that of the flattened endites of filtering Euanostraca. Such projecting endites are, however, present in *Lepidocaris* (Fig. 46F–H) as well as in transitional or certain trunk limbs of other branchiopod taxa, including the non-filtrating Notostraca s. str. (Figs. 46 and 47A, F) and even in the 'no longer filtratory Euanostraca' (e.g., *Branchinecta gigas*, see Fryer 1966). Notostraca s. str. retain this shape in their posterior limbs; these have bulging endites and also all three sets of setae and spines (Fryer 1988, Figs. 5–15).

As a speciality of this group, the endites of the anterior trunk limbs, which are used differently, are not only applanate in anterior–posterior aspect but also slightly deformed posteriorly and elongated medially, extremely drawn out in the anterior limbs (Fig. 47D). The relative shortness and rigidity of the notostracan setation demonstrates again their non-filtratory function. Another difference of Notostraca s. str. is in the proximal endites which, relative to those of *Rehbachiella*, appear to have rotated almost 90° rearwards (Fryer 1988, Fig. 118). In consequence, the posterior row (3) is proximal while the anterior row (1), well-developed in this branchiopodan group, comes to lie on the distal side (cf. also Cannon 1933, pp. 326–327).

Little is known of the thoracopodal endopods of *Rehbachiella*, since they are nowhere completely preserved in specimens representing late stages. From the preliminary photographs of the destroyed UB W54 and the complete endopods of early stages, it is concluded that there are no scraping devices in *Rehbachiella* but simply a cluster of 4–5 setae. Again, from the few photos left showing the specimen prior to breakage, it can be seen that the endopod was much elongated and most likely four-segmented as in the early instars, its podomeres having a similar armature to that of the distal endites of the corm. Thus, there is little evidence that it would eventually become a uniform paddle with marginal setation, as in recent Branchiopoda (e.g., Figs. 14, 37). On the other hand, the branchiopod endopod is by no means always leaf-shaped and uniform, but shows various modifications, ranging from being larger than the exopod to completely absent (examples in Fig. 46A–D).

The importance of the endopods of the anterior trunk limbs of *Lepidocaris* has been mentioned already, but signs of segmentation are missing. In this respect, the *Rehbachiella* endopod apparently stands 'alone' at present, showing neither affinities to the stumpy, five- or six-segmented cephalocaridan endopod nor to maxillopod or even malacostracan ones. The continuation of the enditic armature of the corm onto the endopod might indicate its close affiliation with the filtering process, which would not conflict with a basic design for Branchiopoda. Interestingly, also in spinicaudate Conchostraca the endopods may be of considerable length and, moreover, are feebly segmented (Fig. 46I, J).

Branchiopod exopods may be rather flat and leaf-shaped, particularly in Notostraca and Conchostraca (ex-

amples in Figs. 46 and 47). Yet others are much more like those of *Rehbachiella* and even slightly 'stalked', such as in Kazacharthra (Fig. 47B), or certain Euanostraca (Fig. 46D for *Tanymastix*; for *Branchinella* species see Geddes 1981). The design of *Rehbachiella* would thus not conflict with the suggested affinities, but is also no positive indicator, since such rami occur in the same fashion elsewhere (e.g., 2nd maxilla of nebaliaceans, see Fig. 48D). Since, as in *Rehbachiella*, the plate-like exopods have a basal joint in various Euanostraca, it has been assumed that they closed the apparatus distally to enhance the sucking effect. This presumed function, which Storch (1924, 1925) and Cannon (1927b) were the first to discover, has been extensively discussed and rejected by Eriksson (1934).

Significance of morphological details

Arthropod morphology and life strategies in general are greatly influenced by the possession of an external cuticular skeleton with its many structural possibilities. There can be little doubt that key steps in the evolution of this phylum – and its different branches – are mirrored in the external morphology and, hence, can be traced through it. Since these are the features that can be recognized also in fossil material, special attention is drawn here to such data, including ontogeny and morphogenetic changes. Likewise they are examined for their potential for phylogenetic reconstructions, which is considerable provided that the morphology is adequately known, as is the case with the *orsten* arthropods. The selection of characters of *Rehbachiella* considered in the following text is presumed to be of significance for Branchiopoda and Crustacea in general. As well as the neck organ this includes aspects of head and appendage morphology, and tagmosis.

The major aspects of the crustacean 'labrum', which developed as a special glandular structure at the rear of the primordial hypostome and represents an evolutionary novelty of Crustacea s. str., have been noted by Walossek & Müller (1990) in detail. Their interpretation makes any discussion of the nature of this organ as an appendage unnecessary. If at all, this would refer instead to the phylogenetically older hypostome in front of it (which, in fact, is also uniform in all hitherto discovered arthropods from the Lower Cambrian on). Remnants of the original hypostome are present in various Recent Crustacea, for example in the Cephalocarida, where it is set off from the labral part by a transverse furrow and termed 'clypeus' (e.g., Elofsson & Hessler 1990, Figs. 4, 5).

The 'naupliar eye', considered as one of the two autapomorphies of Crustacea by Lauterbach (1983) cannot be considered in great detail either. As mentioned earlier, the quadripartite naupliar eye cannot have characterized the ground plan of Crustacea s. str., but represents an autapomorphy of Phyllopoda (see also Eberhard 1981; Huvard

1990). Hence the tripartite state is plesiomorphic for Branchiopoda, and for Crustacea in general, but *orsten* arthropods do not contribute to our understanding of the origin of this internal structure due to their preservation of exclusively external details.

Neck organ

A conspicuous structure of the early stages of *Rehbachiella* is the plate-like area on the apex of the head shield with two pairs of pits or pores, one on the surface of the plate, and another at the posterior margin (Fig. 6; Pls. 1:1, 3, 6; 2:7, 8; 3:5; 5:5). Small fillings in some specimens may indicate that the pits demarcated the former position of sensory hairs. This structure becomes more and more poorly developed throughout further development. As it retains its size (about 50 µm in the earliest larvae), it becomes smaller relative to the shield, and, due to its correlation with the anterior head segments, shifts relatively more anteriorly. The whole structure vanishes after a number of instars.

From its position and design it is homologized with the 'neck organ' or 'nuchal organ' of virtually all Recent Branchiopoda (e.g., Claus 1873 for Euanostraca; Gurney 1926 and Martin & Belk 1988 for laevicaudate Conchostraca; Dejdar 1931 for Cladocera and a review of previous literature; Rieder *et al.* 1984 for spinicaudate Conchostraca; e.g., Fryer 1988 for Notostraca). Although there is some variation within the different groups, it is considered as homologous in all of them.

The function of the organ in the fossil is, of course, unknown. In Recent branchiopod larvae it serves as an osmoregulatory organ and seems to be uniform in structure and function (sometimes termed 'salt gland'), though Rieder *et al.* (1984, pp. 437–438) note that this function has not always been demonstrated but only extrapolated from morphological similarity. Older assumptions of, e.g., a respiratory function (e.g., Dejdar 1931) have been clearly disproved for various Branchiopoda from either lineage by ultrastructural studies, which have revealed the typical epithelium of active ionic transport functions (e.g., Croghan 1958a, b; Copeland 1966; Hootman *et al.* 1972; Hootman & Conte 1975; Potts & Durning 1980; Meurice & Goffinet 1982, 1983; Halcrow 1982, 1985; Goffinet & Meurice 1983; Rieder *et al.* 1984). This subject has been reviewed by Potts & Durning (1980), Rieder *et al.* (1984) and, recently, Criel (1991).

The generalized neck organ of Recent Branchiopoda is an oval, watch-glass shaped area on the apex of the head, situated approximately between the 2nd antenna and mandible, mainly encircled by a cuticular ring (e.g., Dejdar 1931, Fig. 5; Botnariuc 1947, Fig. 11; Hootman *et al.* 1972, Figs. 1–3; Dumont & Van de Velde 1976, Figs. 1, 6, 7; Halcrow 1982, Fig. 2; Meurice & Goffinet 1983, Figs. 1, 2, 8, 10;). As described from Cladocera and spinicaudate Conchostraca (Dejdar 1931; Rieder *et al.* 1984), a marginal

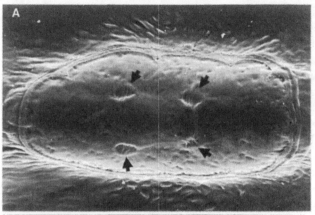

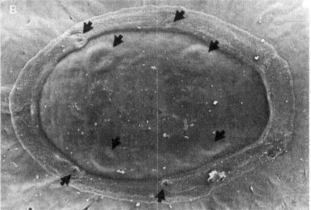

Fig. 52. Neck organs of laevicaudate Conchostraca (Branchiopoda); arrows point to pits and pores associated with the organ. ☐A. Organ of female *Paralimnetis papimi* (length about 80–85 µm). ☐B. Organ of female *Lynceiopsis gracilicornis* (length about 60–65 µm). From Martin & Belk 1988, Fig. 2g, f, by kind permission.

row of cuticular cells, which also forms the cuticular wall ring of the structure, encloses a varying number of inner cells (especially Meurice & Goffinet 1983, also their Figs. 1–4).

Interestingly, Rieder *et al.* describe four nerve cells which reach to the apex of the structure. Two sets of distinctive pits are present in neck organs of laevicaudate Conchostraca (Martin & Belk 1988; see also Fig. 52); a fifth, central one, which they describe, is not unequivocal. If the folded areas in Pls. 1:3; 2:7, 8; 3:5 for *Rehbachiella* can be accepted as outlines of the cells of the organ below the apical membrane, and if at least the inner pits correspond to the conchostracan pits and nerve cells, this would indeed make structural identity at least quite likely. Rieder *et al.* (1984) conclude that the nerve cells may be added secondarily. However, if the neck organ of the marine fossil *Rehbachiella* reflects the ancient state of development of this organ, a lack of nerve cells may rather be the advanced condition.

In Conchostraca it may either reduce eventually (example: *Cyzicus tetracercus*) or alternatively become raised, dome-shaped and even stalked during ontogeny (example: *Limnadia lenticularis*; e.g., Botnariuc 1947, Fig. 39; Rieder *et al.* 1984, Figs. 8, 11, 13). Persistence and even enhance-

Table 5. Occurrence of the neck organ (n.o.), as interpreted herein, among Crustacea (after various authors).

	in larvae	in adults	assoc. structures	funct. organs of adults
Branchiopoda				
Euanostraca	well-developed to huge	vestigial		epipods unknown
Lepidocaris	not described as yet			
Rehbachiella	well-developed vanishing soon	?	2 pits on plate + 2 on posterior margin	?
Spinicaudata	well-developed	stalked in *Limnadia*	4 nerve cells	epipods + neck organ
Laevicaudata	well-developed	present	4 pits + several marginal pits	?epipods
Cladocera	well-developed to missing	huge to missing		epipods alone, together with n.o., or n.o. alone
Notostraca	well-developed	small	?	?epipods + n.o. unknown
Maxillopoda				
Bredocaris	well-developed	present	4 pits on plate	unknown, no epipods
Copepoda	possibly present in larval stages (details and function not clear)			
Facetotecta	'window', but details and function unknown			
other groups	different structures present, but not described in detail as yet			

ment is also known from marine members of the Cladocera (Meurice & Goffinet 1983). In most Euanostraca the neck organ is present only in early larvae, in the hatched nauplius filling almost the entire space of the dorsal shield (Fig. 53B). The organ atrophies in *Artemia salina* as the distal epipods on the thoracopods become increasingly functional. According to Dejdar (1931), this organ still persists after its external effacement in some species, such as *Branchipus stagnalis* and *Siphonophanes (Chirocephalus) grubei*, and it is illustrated as a tiny node by Claus (1873, Pl. 5:16). The function of this structure in the adult, however, has never been examined. In any case it is retained for much longer than in *Rehbachiella*, where it is effaced at least at TS4–5 while the specimen of *Branchipus torvicornis* illustrated by Claus (Fig. 44B) is roughly at TS8.

In larval Notostraca the neck organ is prominent (e.g., Claus 1873, Pl. 6:1B, 2B, 2C for *Triops cancriformis*; Pai 1958, Figs. 2, 3, 7; Fig. 45D–F herein). It becomes smaller subsequently, but persists into the adult, where it is positioned in close contact to the dorsally shifted compound eye (e.g., Claus 1873, Pl. 8:5; Longhurst 1955, especially Fig. 4; Alonso 1985, Fig. 6a–c, m; Fryer 1988, Fig. 3). The function of the adult organ has never been studied.

Remarkably similar to the neck organ of *Rehbachiella* is that of laevicaudate Conchostraca (e.g., Martin & Belk 1988, Fig. 2f, g) and Cladocera. This is even more evident, since in Laevicaudata there are also pits in the bordering ring wall (arrows in Fig. 52). Its persistence into the adult in these two groups, with or without the appearance of epipods, is still not completely understood. Other terms also applied to this structure, such as 'frontal organ' (Martin 1989, Fig. 1) or even 'naupliar eye' (Siewing 1985, Fig. 965), must be rejected since these are in general use for different organs.

The common possession of such a ring-shaped neck organ in Branchiopoda and *Rehbachiella* seems to be a synapomorphy, as other Crustacea lack it. The exception is *Bredocaris*, where it occurs in all developmental stages. This organ (Müller & Walossek 1988b, Pl. 10:1 for a larva and Pl. 3:2 for the adult) is also surrounded by a weakly developed ring and possesses on its surface four pits, in which hairs may have been located. The relationships of *Bredocaris* are clearly with Maxillopoda, and additional support comes form the comparative ontogenetic study herein. Hence, the neck organ may have been present already in the ground plan of Maxillopoda, and, accordingly, may represent a synapomorphy of these and Branchiopoda.

As in *Rehbachiella*, the nearest forms with which it can be compared are, however, only the extant ones. These seem to lack this structure, although there are some indications of it among Copepoda: recently Dibbern & Arlt (1989), for example, have described an oval to triangular 'nuchal organ' for the naupliar stages of the harpacticoid *Mesochra aestuarii* (their Figs. 2B–7B). However, they did not study the nature of this organ in detail to substantiate such a terminology. Another organ of interest was described for calanoid copepods by Nishida (1989) as a 'cephalic dorsal hump'. Further comparative morphological and ultrastructural studies, also for the other maxillopodan taxa, are still scarce, but increasingly under way (Høeg, personal communication, 1991).

Table 5 shows the occurrence of the neck organ, as interpreted above, among Crustacea. A neck organ in this sense is unknown from Cephalocarida, Malacostraca, and Remipedia. So-called 'head pores' on shields have been described from various crustaceans (examples: head pores of cladocerans: Frey 1959; Dumont & Van de Velde 1976 [who regarded the neck organ as one exceptionally large pore]; cirripeds: Walker & Lee 1976), but their relation-ships are unclear as yet.

Among these, a 'dorsal organ' is briefly reported as a glandular–sensory complex from larval decapod Eumala-costraca (Laverack & Barrientos 1985; Barrientos & Lave-rack 1986). This often dome-shaped structure consists of a central pore with a glandular cell underneath (sometimes missing) surrounded by four innervated pits, and the correlation with head segmentation is unclear. These au-thors assume a combination of glandular and sensory functions (chemo- or baro-perception) for this structure, but detailed information is still lacking (as is the case with various other crustacean structures; see also the 'gills' below). Barrientos & Laverack (1986) suggest closer affini-ties between this eumalacostracan organ and the median tubercle of trilobites (e.g., Hanström 1934; see also Fortey & Clarkson 1976 for *Nileus* and other trilobites). A similar structure – a humped area on the head shield with a cluster of pits which seemingly bore sensory bristles – has also been recognized in the Upper Cambrian *Agnostus pisi-formis*, and relationships between these structures have been expressed by Müller & Walossek (1987).

It is difficult to compare this eumalacostracan 'dorsal organ' with the branchiopodan 'neck organ', particularly in regard to its rather vaguely described position and seemingly different functions. There may, however, be a basic connection between the two distinct types, when considering the presumed sensilla of the neck organ of *Rehbachiella* and *Bredocaris*, the pits in laevicaudate Conchostraca, and the nerve cells in the Spinicaudata. It is also worth noting that Rieder *et al.* (1984) found a central cell in the latter group. Hence, it is assumed that the neck organ of Branchiopoda and Maxillopoda represents a composite of two elements: a phylogenetically older organ in the sense of the 'dorsal organ' and a newer organ structure, the 'neck organ', which evolved around it. In terms of position and origin, it would thus be homologous to the ancestral organ, but as a compound structure it has achieved a new function and is regarded as an evolutionary novelty restricted to Maxillopoda and Branchiopoda. Its morphological stasis would indeed be remarkable, in the light of a time lapse of 500 million years and a 'move' of the Branchiopoda into freshwater.

Mauchline (1977) noted integumental sensilla and glands in some Crustacea, but considered non-malacostra-cans briefly (only cladoceran Branchiopoda). In agree-ment with Barrientos & Laverack (1986) he recognized a group of four pores with a central area, but without any discussion of function. Remarkably, these structures may be located either anteriorly (the supposed light sense was challenged by Barrientos & Laverack) or posteriorly on the shield (Isopoda) or on both sides (the leptostracan phyllo-carid *Nebaliopsis typica*). Since advanced larvae of *Reh-bachiella* possess a set of pores in a very similar posterior position (Pl. 11:7), it would be interesting to investigate whether these dorsal structures at the posterior edge of the shields can be related to one another, thus representing an additional 'ancient crustacean character'.

Cephalic shields and carapaces

From their earliest appearance in the fossil record, the head of arthropods has been covered by a shield-like plate, generally considered as product of fusion of the tergites of the anterior segments. Various attempts have been made to evaluate the status of this character for the Crustacea, often involving the presence or absence of a 'carapace' versus a simple head shield. Because the important literature on this subject has been compiled by Newman & Knight (1984), discussion herein focuses on the central problem of the history of dorsal shields and the existence of a 'cara-pace', including its status among Crustacea.

In accordance with Dahl (1983), one of the key problems seems to lie in the traditional misunderstanding of the mode of growth of such shield-like head covers and the postulate of a 'carapace' fold at the rear of the head from which a 'carapace' should grow out laterally and posteri-orly (for different definitions see, e.g., Calman 1909. p. 6; Kaestner 1967, pp. 883–886; Moore & McCormick 1969, pp. R91–93; McLaughlin 1980, p. 2). Neither the morpho-genesis of Recent crustacean shields nor the fully docu-mented ontogeny of *Bredocaris* (Müller & Walossek 1988b) and *Rehbachiella*, described here, give support for such structures. With one single exception – the Onychura (see below) – the shields of Crustacea are products of progressive growth of the naupliar shield, regardless of their eventual size and segmentary equipment. If the term 'carapace' is retained to include those shields that incorpo-rate one or more postcephalic segments, it would be syn-onymous to a 'cephalo-thoracic shield'. Such a shield has been variously achieved by parallel development, and re-cent investigations show that incorporation of thoraco-meres is more widespread than hitherto assumed (Dahl, personal communication, 1990). This may also have a further impact on the discussion of the primary segmenta-tion of the 'head' (see below).

Shields and their segmentary equipment. – The earliest ar-ticulate fossils identifiable as arthropods, such as the re-cently discovered Lower Cambrian *Cassubia* and *Liwia* of NE Poland (cf. Dzik & Lendzion 1988) or of Chengjiang, China (cf. Hou 1987a, b; Chen *et al.* 1989), already have flat head shields of varying size. For example, in *Fuxianhuia* (Hou 1987b), the shield extends freely backwards to cover at least the anterior 4–5 trunk segments (personal observa-tions), but the number of head limbs below is unknown. No one would term such a shield a 'carapace'. In various arthropods of the well-known Burgess Shale-type faunas (cf. Conway Morris *et al.* 1982 and Whittington 1985 for further references; Collins 1987 and Conway Morris 1989a, b for additional sites bearing Burgess Shale-type faunas) the frontal body region bears a shield of varying size. The number of appendiferous segments below and their caudal extension onto the trunk is, in most cases, however, unclear.

Recent detailed studies of multi-segmented trilobites (cf. particularly Cisne 1975, 1981; Whittington 1977, 1980; Whittington & Almond 1987) and the ventral details of *Agnostus pisiformis* (see Müller & Walossek 1987) have shown that trilobites and related forms had no more than the pair of 1st antennae and three pairs of biramous limbs below their head shields, which apparently challenges all earlier hypothetical approaches. Bergström & Brassel (1984) noted five for the Lower Devonian phacopid *Rhenops*. Since the latter is a fairly late trilobite, it is not unlikely that this resulted from the fusion of the subsequent segment to the head in convergence to cephalo-thoracic shields of various arthropods.

Among chelicerae-bearing members of Arachnata (*sensu* Lauterbach 1980, more recently changed into 'Pan-Chelicerata' by Lauterbach 1989), pantopod protonymph larvae have a small dorsal shield which reaches backwards to the first walking leg (2nd post-chelicerate one; cf. Behrens 1984). The same is true for the Upper Cambrian larva D of Müller & Walossek (1986b, 1988a). If these structures lateral to the frontal mouth represent the vestigial 1st antennae, the shield would terminate behind the 4th limb-bearing head segment.

The crustacean head shield. – It is generally accepted that the ground plan of Crustacea should include a shield covering the anterior five appendiferous segments (e.g., Newman & Knight 1984). The description of stem-group crustaceans by Walossek & Müller (1990) confirms this but does not bear out this model for the ground plan of Crustacea s. str., since at least two of the forms have a head that includes only four appendiferous segments. Accordingly, a shield of such kind cannot have characterized the early phase of crustacean evolution.

It is generally accepted that the ground plan of Crustacea should include a shield covering the anterior five appendiferous segments (e.g., Newman & Knight 1984). The description of stem-group crustaceans by Walossek & Müller (1990) confirms this but does not bear out the model for the ground plan of Crustacea s.str., since at least two of the forms have a head that includes only four appendiferous segments.

The process of stepwise coalescence of postnaupliar body segments to form a 'head shield' that embraces at least five limb-bearing segments is reflected in the early life history of two more of the stem-group crustaceans (cf. Müller & Walossek 1990), in *Rehbachiella*, *Bredocaris* and even in extant crustaceans s. str. For example, a shield coalesced only from the anterior four limb-bearing head

segments occurs in the first larval phase of Maxillopoda of the thecostracan lineage (e.g., Ascothoracida: Boxshall & Böttger-Schnack 1988; Cirripedia: Anderson *et al.* 1988; Branchiura: Fryer 1961; *Bredocaris*: Müller & Walossek 1988b). Again, in the protozoëal phase of Eumalacostraca the maxillary segment is still free from the larval head (e.g., Kaestner 1967, Fig. 697; also Newman & Knight 1984, Fig. 1D). In consequence, a shield made of all five limb-bearing head segments back to the 2nd maxilla cannot have characterized the early phase of crustacean evolution.

This does not imply that a larval shield which is coalesced from 3–4 segments cannot be large and extended in any direction. Such shields can be seen, e.g., in various thecostracan Maxillopoda (examples in Müller & Walossek 1988b, Figs. 9, 10; see also Dahms 1987b, Fig. 1 for the large naupliar shield of the harpacticoid copepod *Bryocamptus pygmaeus*), in larvae of laevicaudate Conchostraca (Fig. 44F), or, e.g., in the mentioned protozoëans of Eumalacostraca.

In *Bredocaris* as well as in *Rehbachiella* the posterolateral corners start to extend posteriorly after a few instars, while the midline of the posterior shield grows more slowly. This results in an excavation of the posterior margin. While the shield of *Bredocaris* stops clearly at the rear (Müller & Walossek 1988b, Pl. 3:7, 8) with only its corners wing-like extended, in *Rehbachiella* the shield continues its simple growth to either side very gradually after the final segmentary equipment has developed and thus by far exceeds the *Bredocaris* level. Neither is there a fold appearing at any stage of growth, nor does the shield belong exclusively to one of the 'head segments'.

Thus it seems inconvenient to distinguish terminologically between shields that comprise only four or less appendiferous segments – as in the stem-group crustaceans and other early euarthropods – or five and more – as in the different crustacean taxa. Splitting would also imply that it is necessary to differentiate between transient shields of the ontogenetic stages in *Rehbachiella*, and Crustacea in general.

The naupliar shield of crustaceans may already grow out allometrically by elongation of one or all of its margins. As an extreme case, in the Ostracoda the shield encloses the larva completely from the beginning. The primordial state has now been clarified with reference to the recently discovered punciid ostracodes. Here the shield starts as a little arched univalved shield (Swanson 1989a, b), no larger than in the members of the thecostracan core of Maxillopoda, to which Ostracoda may belong.

Hence enlargement or conservation of size at any developmental state seem to be the strategies that led to the plasticity of shields among Crustacea s. str. Simple, i.e. univalve but not necessarily small, cephalic shields have been described from phyllocarid (see below) and certain eumalacostracan Malacostraca, Cephalocarida, *Rehbachiella* and notostracan Branchiopoda (see below), *Bredocaris* and most of the Recent members of the thecostracan lineage of Maxillopoda, and *Skara* and Mystacocarida of the copepodan lineage of Maxillopoda. Incorporation of postmandibular segments stops behind the 2nd maxillae, while growth of the shield continues until the final shape is achieved. The 'free carapace' of Newman & Knight (1984) is, thus, more or less synonymous with such a 'cephalic shield'.

Incorporation of subsequent, 'postcephalic' segments leads to cephalo-thoracic shields. These had obviously evolved in separate stocks by parallel development: inclusion of one segment occurs for example in both lineages of the Maxillopoda, in Remipedia (see Schram *et al.* 1986; Schram & Lewis 1989), and in the Notostraca, as new studies have shown (Dahl, personal communication, 1990) – which gives additional support to a derived state of Notostraca among Branchiopoda.

The evolutionary path toward the cephalo-thoracic shield of Copepoda may be deducible from the Upper Cambrian Skaracarida, where the free tergite of the 1st thoracomere (maxilliped segment) fits nicely with the posterior shield margin but is 'not yet incorporated' (Müller & Walossek 1985b, Figs. 10, 11; see also Boxshall *et al.* 1984 for the reconstruction of the 'ancestral copepod').

A simple dorsal shield ought to represent the plesiomorphic condition also for Malacostraca, as stated by Newman & Knight (1984). While it seems to be retained in primordial Eumalacostraca, such as in Thermosbaenacea (Cals & Boutin 1985, Fig. A) where it covers freely the segments of the 2nd maxilla and maxilliped, or Spelaeogriphacea (Gordon 1957, Fig. 1), new investigations (Dahl, personal communication, 1990) have revealed that leptostracan Phyllocarida have a cephalo-thoracic shield with a specifically variable number of coalesced thoracomeres. This will surely shed new light on the taxonomic position of the various fossil taxa currently included within the Phyllocarida.

The question of whether or not a 'carapace' is the ancestral state for Eumalacostraca may be answered thus: the shield may reduce in size, but still retain its plesiomorphic state as being cephalic only. Any posterior enlargement and incorporation of one to all thoracomeres (dorsally) represents the apomorphic state. Within the Eumalacostraca – and parallel to Phyllocarida – a range from one to all eight thoracomeres may be incorporated. In accordance with Kaestner (1967, pp. 882–886), all such shields are cephalothoracic.

Hence, the term 'carapace' may at best be restricted to its extreme case, where all thoracomeres are included, as claimed by Newman & Knight (1984). The latter type occurs only in the Eucarida. With regard to the evaluation of relationships between crustacean taxa, it thus seems necessary to consider size and segmentary equipment, but it is not a question of whether a 'carapace' has developed at a certain stage or became reduced again. Enhancement of

rigidity and even mineralization are further secondary processes that may modify a shield considerably, processes most likely derived by convergence in numerous groups.

The extreme is the 'missing' shield of Euanostraca. Yet, in their nauplius it is still present. Since the neck organ fills most of its space, it is most clearly recognizable only in the phase immediately after emergence from the hatching membrane (e.g., Rafiee *et al.* 1986, Fig. 4E; Fig. 53B herein). This may be the reason that it has never been noticed, although it is little different from the feebly developed shield of the *Rehbachiella* nauplius or the hatching metanauplius of Notostraca s. str. (Fig. 45D).

In conclusion, it is recommended simply to drop the term 'carapace' in order to avoid complications. If retained, it should be restricted to the special shield of certain Eumalacostraca, as an autapomorphic character, i.e. only when it is fused with the complete thorax, covering the thoracic gills. In any case, such a shield is nothing more than the extreme of a 'cephalo-thoracic shield', clearly evolved in parallel at least among Maxillopoda, Eumalacostraca, Remipedia, and, according to new evidence, also in leptostracan Phyllocarida and Notostraca. Application of the term 'carapace' to all 'large head shields' in general and inclusion of crustaceans with such shields into a taxon 'Palliata' (e.g., Lauterbach 1974), thus disregarding their well-founded distinctive taxonomic positions, is rejected.

Bivalved shields. – Large bivalved shields are already present among the first shields in the fossil record, found in the Lower Cambrian (Hou 1987c) and the Middle Cambrian (e.g., Brooks & Caster 1956 [shield length 13 cm]; Briggs 1977). This indicates that the bivalved condition is the most likely advanced state of a shield, whatever segmentary equipment it had, and developed convergently several times over. Since the simple cephalic shield must be considered as the plesiomorphic condition among the members of the Crustacea s. str., the bivalved state in the Upper Cambrian Phosphatocopina, Ostracoda, Ascothoracida and Onychura, is merely a homoplasy. This can also be deduced from their different morphogenesis, different formation of a 'hinge', and different degree of fusion with the body.

The shield of Ostracoda is said to be free from the anterior head region and fixed exclusively to the maxillary segment, and to cover the thorax freely. Illustrations by Schulz (1976, Fig. 1, 2) and the development of the recently discovered punciids (Swanson 1989a, b) do not accord with this but merely indicate some compression of the dorsal area of the head. Punciid ontogeny starts with a nauplius bearing a simple shield, while the true hinge progressively develops with the subsequent stages, as in all other Recent Ostracoda.

In consequence, the punciids yield further support for the inclusion of ostracodes into the Maxillopoda, as presumed by, e.g., Schulz (1976) or Grygier (1984), and prob-

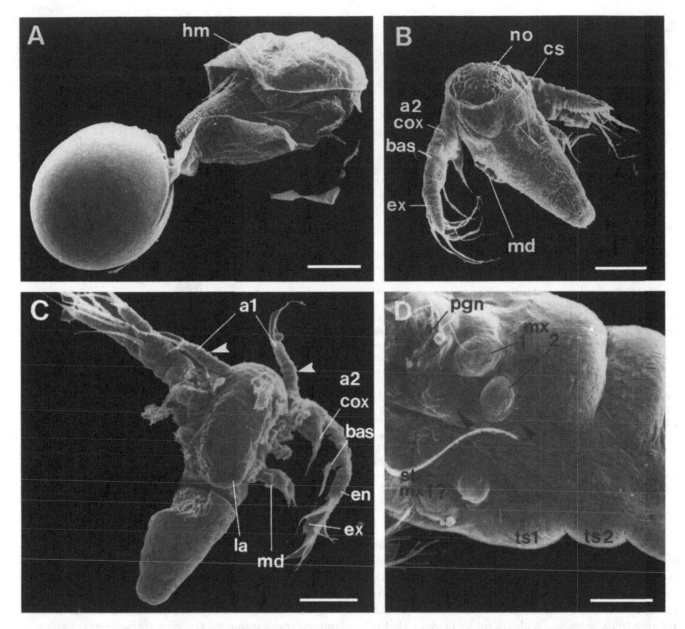

Fig. 53. SEM micrographs of *Artemia franciscana* (by kind permission of the Springer Verlag and T.H. MacRea). □A. Nauplius, just emerged from the cyst but still within hatching membrane (hm; scale bar = 100 μm). □B. Slightly stretched nauplius; cuticle of neck organ rubbed off, exposing the cell layer underneath; shield visible, with its posterior margin reaching back to mandibles. Scale bar 100 μm. □C. Partly distorted nauplius with segmented 1st antenna (arrows). Scale bar = 100 μm. □D. Detail of postoral ventral region, with anlagen of the two pairs of maxillae; arrows point to indistinct segment boundaries. Scale bar 30 μm.

ably to the thecostracan lineage of Maxillopoda (Boxshall & Huys 1989). Segment coalescence stops after the 2nd maxillae. According to Grygier (1984), the shield of Ascothoracida should be fundamentally bivalved, the valves being joined by a simple hinge (his Fig. 2a). Their ontogeny indicates, however, that their shield develops from a simple cephalic shield (process in Brattström 1948, Fig. 25).

The growth of the shields of laevicaudate Conchostraca is entirely different. Its earliest known larva – which is not a nauplius – has a very large shield, obviously fused with the body all along its length. Prior to metamorphosis to an advanced larval type with many limbs, the 'heilophora', a new and completely different shield can be seen below the old cuticle (Fig. 44F). This secondary shield is at least post-mandibular, since the neck organ remains anterior to it. It is wing-like and elongated laterally, posteriorly, and also anteriorly around the labrum. Subsequently this new shield becomes larger and progressively bivalved. Finally a hinge structure is formed posterior to the attachment area, i.e. dorsal to the free thorax (Linder 1945).

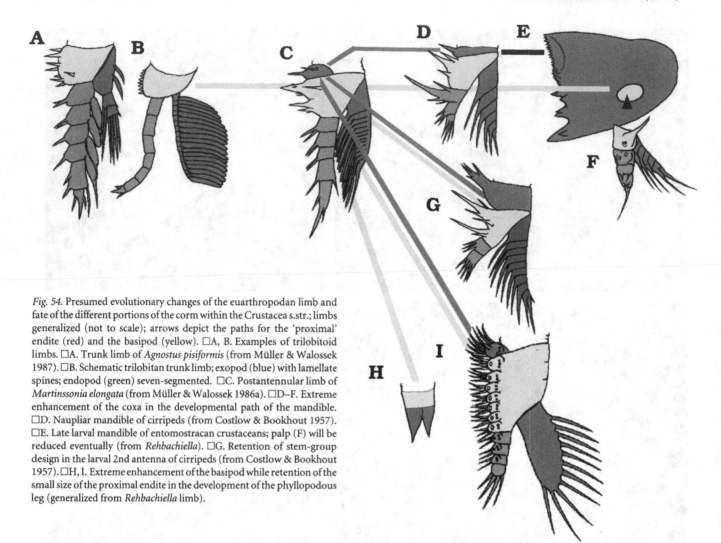

Fig. 54. Presumed evolutionary changes of the euarthropodan limb and fate of the different portions of the corm within the Crustacea s.str.; limbs generalized (not to scale); arrows depict the paths for the 'proximal' endite (red) and the basipod (yellow). □A, B. Examples of trilobitoid limbs. □A. Trunk limb of *Agnostus pisiformis* (from Müller & Walossek 1987). □B. Schematic trilobitan trunk limb; exopod (blue) with lamellate spines; endopod (green) seven-segmented. □C. Postantennular limb of *Martinssonia elongata* (from Müller & Walossek 1986a). □D–F. Extreme enhancement of the coxa in the developmental path of the mandible. □D. Naupliar mandible of cirripeds (from Costlow & Bookhout 1957). □E. Late larval mandible of entomostracan crustaceans; palp (F) will be reduced eventually (from *Rehbachiella*). □G. Retention of stem-group design in the larval 2nd antenna of cirripeds (from Costlow & Bookhout 1957). □H, I. Extreme enhancement of the basipod while retention of the small size of the proximal endite in the development of the phyllopodous leg (generalized from *Rehbachiella* limb).

A similar mode occurs in Spinicaudata, but here the secondary shield starts with much less of an extension and grows out more gradually (e.g., Anderson 1967; Figs. 44D, E, G, 45A–C herein). Strength & Sissom (1975, also their Pls. 2, 3) remarked that the shield grows out from the anterior trunk segment. With regard to Linder's (1945) illustrations, referring to Sars (Linder 1945, Figs. 2b, 6a, b) fusion with body is, however, unclear, and it remains open whether the shield grows out from the maxillary segment or from the anterior trunk region in both the Laevicaudata and the Spinicaudata. Since its derivation and attachment is reported to be from the segment of the 2nd maxilla in Cladocera, it remains unclear whether the sclerotic cover of the front is made up of the anterior four or five appendiferous body segments in the Onychura.

Summary. – Crustacea do not differ from the general arthropod habit of having a dorsal 'head shield', apart from the fact, as assumed, that an additional, fifth appendiferous body somite was already coalesced dorsally (see below) with the shield in the ground plan of Crustacea s. str. All the various crustacean shields represent nothing more than modifications of the basic design in terms of enlargement

and further inclusion of posterior body somites, or alternatively conservation and reduction. The single exception is the secondary shield of Onychura (Conchostraca and Cladocera), as indicated already by Dahl (1983, p. 365). Its origin from either the segment of the 2nd maxilla or the anterior trunk segments, however, still warrants clarification.

It is also difficult to estimate whether a large shield enclosing the whole thorax, such as in *Rehbachiella,* or whether a smaller shield, for example as in *Bredocaris* (Müller & Walossek 1988b, Fig. 4), characterized the ground-plan of Crustacea s. str. The large shield of *Rehbachiella* is remarkably similar to that of certain Notostraca s. str., such as *Lepidurus packardi* (Linder 1952, Pl. 2:1, 2). The sides of the shields, however, do not extend beyond the limb bases in *Rehbachiella,* which is in contrast to the much larger, bottom-dwelling Notostraca. Kazacharthra, on the other hand, have a wide shield, but it is much shorter than that of their possible sister group, the Notostraca (see Fig. 42A), reaching back only to the last, 11th thoracopods. With this, the assumption that a large and long shield represents the derived state, as favoured by Lauterbach (1974, 'carapace'), Hessler & Newman (1975), or Dahl

(1976), remains to be substantiated by detailed examination of segment equipment, shape and size.

Structures in the head region

Eyes. – The different development of the eye region in the two lineages of Branchiopoda has been considered herein as one of the key features for recognizing *Rehbachiella* as a stem-group anostracan. Since only the anterolateral corners enlarge to some extent, the eye area is reconstructed to project beyond the forehead, but its fate during later development remains unknown. Hence, it is difficult at present to estimate the polarity of this feature, and little can be contributed to the question of whether the sessile condition of compound eyes preceded the stalked condition, as favoured by Bowman (1984) or the reverse, as proposed by Hessler & Newman (1975). Taking the stem-group crustaceans, the situation is similar. *Martinssonia* lacks external eyes (Müller & Walossek 1986a), in *Cambropachycope* and *Goticaris* the presumed eye is a huge single uniform facetted forehead structure, and *Henningsmoenicaris* has stalked eyes (Walossek & Müller 1990).

Among the Crustacea s. str., the Malacostraca seem primitively to possess compound eyes. Remipedia have neither naupliar nor compound eyes, similar to the Cephalocarida (recent investigations by Elofsson & Hessler 1990 have led to the identification of a special nuchal organ in this structure, invalidating Burnett's 1981 description of rudimentary compound eyes). Since atrophy of the eye and/or a new structure is clearly apomorphic, the original condition cannot be established, though it is fully agreed that compound eyes belong to the ground plan of Crustacea s. str. Within Maxillopoda compound eyes occur only in the thecostracan lineage (cf. Müller & Walossek 1988b for references). If it is true that Maxillopoda have evolved by paedogenesis (cf. Newman 1983), their sessile eyes might simply reflect the larval state rather than indicating a primordial design.

Sternum and paragnaths. – The sternal region of *Rehbachiella* undergoes a number of changes during ontogeny (e.g., Fig. 25). Of significance is the possession of a separate sternite of the antennal somite in the nauplius (Pl. 1:4), a feature not recognized in extant crustaceans. This portion, obviously related to the position of the 2nd antenna and its prominence in the locomotory and feeding apparatus at this stage, does not seem to atrophy but rather merges with the mandibular sternite, forming its slope into the atrium oris.

If this reflects the evolutionary path in the formation of the anterior part of the sternum, it gives further support for the assumptions of Walossek & Müller (1990) that the whole set of naupliar feeding structures characterize the ground plan of the crown group. Also according to Dahl (1976, p. 164) 'such a double feeding mechanism' should have been present in the common ancestor of the Crustacea and (the naupliar apparatus) 'is a prerequisite for the existence of autonomous larvae'.

At the rear of the mandibular sternite of *Rehbachiella*, a pair of humps grows out gradually, eventually to form the bulging 'paragnaths'. In consequence, the morphogenesis clearly demonstrates that these structures are referable exclusively to this particular sternite. This challenges various postulates about their derivation, such as their being part of the segment of the 2nd maxillae (Sanders 1963b) or belonging to the entire cephalic sternum (Lauterbach 1980, 1986). It also challenges all speculations about their nature as modified appendages (e.g., Claus 1873), already rejected in detail by Eriksson (1934, p. 50).

Appendages

First antenna. – While the nature of the euarthropod 1st antenna as an appendage has been frequently questioned (e.g., Siewing 1963), that of Crustacea clearly is an appendage. This is evident from its musculature (Hessler 1964, in particular) as well as from its pattern of motion integrated within the metachronal beat of the naupliar limb apparatus (e.g., Barlow & Sleigh 1980; Fryer 1983; Moyse 1987).

In its subdivision into a finely annulated shaft and a few cylindrical distal podomeres, equipment with feeding and locomotory setation, and the apical set of setae, the 1st antenna of *Rehbachiella* resembles not only that of *Bredocaris* or *Skara*, but also that of the early larval stages of Recent Eumalacostraca (e.g., Fielder *et al.* 1975, Figs. 1–5 for *Penaeus esculentus*) as well as various larval thecostracan Maxillopoda (particularly Ascothoracida and Cirripedia), and Cephalocarida. A remarkable similarity exists also to the 2nd antenna along its endopod (and to the larval mandible as well).

The similarity in outline and function of the 1st antenna has led, besides the evolutionary novelty 'proximal endite', to the recognition of a group of *orsten* arthropods as stem-group crustaceans and to the suggestion that the modification of this appendage was one of the key steps in crustacean evolution (cf. Walossek & Müller 1990). Within Crustacea, the uniramous state of the 1st antenna represents the plesiomorphic state, while all larger numbers of rami must have evolved secondarily (e.g., the biramous state in adult Malacostraca). Likewise, the number of podomeres seems to have been limited primarily, possible no more than 10–15.

In extant branchiopods the 1st antenna appears much reduced, which has generally led to this feature being regarded as a conspicuous character of this group. In fact, among Euanostraca the larval 1st antenna is by no means generally small (e.g., Kaestner 1967, Figs. 725, 726), nor is it necessarily undivided. According to Hsü (1933) the long 1st antenna of *Chirocephalus* is well-segmented (see also Valousek 1950, Figs. 1–4) and of the same length as the 2nd antenna (Oehmichen 1921 for *Chirocephalus grubei*; Heath

1924 for *Branchinecta occidentalis*). For the latter species even a two-segmented 1st antenna is reported (Heath 1924). Segmentation of the 1st antenna is clearly present in the nauplius of *Artemia*, where it is even enhanced by rows of denticles (Fig. 53C), such that are well-known from various Recent and *orsten* crustacean larvae. Again, *Lepidocaris* (Scourfield 1926, Pl. 22) and its larval stages (Scourfield (1940, Fig. 4) have a feebly developed but well-segmented 1st antenna.

Spinicaudate Conchostraca are remarkably similar to *Rehbachiella* in the division of the 1st antenna into many so-called sensory lobes and the fine annulations on the shaft (e.g., Battish 1981, Fig. 16, for a *Leptestheria* sp. indet.). Again, the 1st antenna of Kazacharthra consists of up to 15 annules, while that of Recent notostracans is small but two-segmented and still larger than the highly reduced 2nd antenna. In consequence, each of the two branchiopodan lineages should basically have possessed a segmented 1st antenna, as supported by the fossils *Lepidocaris*, Kazacharthra, and *Rehbachiella*, and the present-day status is merely the result of convergent reduction in size and/or effacement of segmentation (homoplasy).

Second antenna. – The 2nd antenna of *Rehbachiella* agrees in all major aspects of design with that of non-malacostracan crustaceans as well as that of eumalacostracans during the early larval phase (except of its missing feeding equipment in this group), save for the remipedes. This is most obvious in the annulation of the limb corm, occurring for example in spinicaudate Conchostraca (e.g., Battish 1981, Figs. 17, 24, 32, 41) or larvae of ascothoracid Maxillopoda (e.g., Grygier 1985, Fig. 4J, K), also in its division into coxa and basis (not 'praecoxa' and 'coxa', as stated by Schrehardt 1987a), and in the two rami. Both the coxa and basipod are well equipped for feeding with elongate endites and distal spines (also the opposing set pointing to the labrum, as in Maxillopoda),

This stasis in morphology might well result from the continued use and prominence of this appendage among Crustacea s. str. A major branching off from this pattern occurred in the evolution of the Malacostraca. Again, reduction of this limb late in the ontogeny of Notostraca points to a secondary modification achieved in the course of adapting to a bottom-dwelling life style. Kazacharthra demonstrate not only the plesiomorphic state of the 2nd antennae within the Calmanostraca, being small but well-developed and biramous, but also indicate that the 2nd antennae transformed convergently well within the different branchiopodan lines.

Development of the antennal exopod starts with a relatively few ring-shaped divisions, rather uniformly among crustaceans (often 7–9), and *Rehbachiella* is no exception. The number increases to 18–19 in *Rehbachiella*, which is slightly more than in most Recent Euanostraca (about 15; the exception is *Chirocephalus grubei* with more than 20

annules, according to Valousek 1950) and Kazacharthra (10–15), but less than in Cephalocarida. In the latter this state is kept until adulthood, while it seems clear from the size of fragments of later larval stages that the 2nd antenna progressively reduces in size and equipment in *Rehbachiella*.

The number of endopodal podomeres is four in the earliest larvae of *Rehbachiella*, if the socket of the apical seta represents a further podomere. Its subsequent fusion with the penultimate podomere to a three-segmented state is in accord with observations on *Bredocaris* (Müller & Walossek 1988b). Remarkably, this process occurs in all three naupliar appendages, including the 1st antenna, which points again to its appendage nature.

Among most Recent crustaceans, three or less podomeres seem to be the highest number (e.g., larval ascothoracid Maxillopoda: Boxshall & Böttger-Schnack 1989, Figs. 1–4; larval cirripeds: Walley 1969, Fig. 1). However, if the elongate distal element in Mystacocarida can be accepted as a fourth endopodal podomere (see e.g., Delamare Deboutteville 1954, Fig. 4; Hessler & Sanders 1966, Fig. 3C), the Maxillopoda at least should have had a four-segmented endopod in their ground plan. This would then contrast with *Rehbachiella*, which modifies that character during ontogeny. Kazacharthra had three or four endopodal podomeres, according to McKenzie et at. (1991), while the Upper Cambrian stem-group crustacean *Martinssonia* had a five-segmented endopod (Fig. 48K).

Two-segmented states in various crustaceans, for example in Cephalocarida, would then be derived but by no means synapomorphic to Crustacea. The status of the rows of denticles, probably indicating a faint segmentation into eight portions in *Rehbachiella*, remains unclear, but there is here a remarkable resemblance to a faint subdivision in the conchostracan *Eoleptestheria variabilis* (Botnariuc 1947, Fig. 4i) as well as to the larval 2nd antenna of *Artemia* (Schrehardt 1987a, Figs. 4, 5).

It has been claimed that the 2nd antenna of Branchiopoda is different from that of other Crustacea in the prominence of its corm relative to the rami (cf., e.g., Sanders 1963b). While this cannot be claimed for all branchiopods (e.g., kazacharthran 2nd antennae seem to have had very short corms), in other crustacean larvae also the proximal part may in fact exceed 50%. The reason for such a design is seen in functional demands, mainly for reaching around the elongated labrum toward the vicinity of the mouth. There seem to be various strategies among crustaceans to elongate the inner parts of the 2nd antenna, which are concerned with food intake in the larval phase. In various Recent Branchiopoda this may be achieved by elongation of the basipod portion, while the coxa stays the same size. In *Lepidocaris* the coxa and basipod are rather short while the endopod – two-segmented in the female and three-segmented in the male – is elongated (Scourfield 1926, Pl. 23:1, 2).

Rehbachiella has elongated its endopod by splitting the 2nd podomere. Similarly, in the benthic infaunal Mystacocarida also the endopod is elongated, and splitting seems also to be indicated (Dahl 1952, Fig. 2A). In Malacostraca the same function is fulfilled by the newly developing mandibular palp, since the antennae are at no stage involved in the feeding process. Nevertheless, during larval development the corm of the 2nd antenna may also be large, for example in *Euphausia gibboides*, until the calyptopis stages (Knight 1975; see also Mauchline 1971).

According to Hessler & Newman (1975), the similarity of 2nd antenna and mandible is due to the retention of a basic naupliar morphology (see also their Fig. 4), which Calman (1909) referred to as the 'primary head region of Crustacea'. Walossek & Müller (1990), again, argue that their specific design, with enhanced proximal endites, is among the derived ground plan characters of Crustacea s. str. and cannot have developed earlier. Accordingly, the 2nd antenna as an integral part of the naupliar feeding and locomotory system is apomorphic to Crustacea. But being basically not an 'antenna' but rather a 'pre-mandible' in its specific shape, with locomotory, feeding and grooming functions, its name puts emphasis on a state achieved only within a particular crustacean taxon. Its origin from the first postantennular limb is indisputable, but this does not contribute anything to elucidating relationships of Crustacea with any other high-rank arthropod taxon.

The ancestral condition is still better reflected in larval Cirripedia than in, for example, *Rehbachiella* or *Bredocaris*: in these the coxal endite of the 2nd antenna is very prominent and even gnathobase-like, while the mandibular coxa is less developed in all naupliar stages. The size of the mandible becomes about the same as that of the 2nd antenna at or after the 6th instar (e.g., Bassindale 1936, Figs. 4–6; Costlow & Bookhout 1957; Walley 1969, Fig. 1; Fig. 45H herein).

Within the different crustacean taxa, the 2nd antenna may have departed from its original function by further anterior migration. This led to its centering on locomotion or serving for new functions, such as an attachment device in various parasites, or as a sensory organ among Malacostraca. In the bottom-dwelling Cephalocarida, the feeding function of the 2nd antenna is passed through rapidly (Sanders 1963b), and also among the Maxillopoda the coupled feeding and locomotory functions characterize the 2nd antenna only in its early larval phase. Its subsequent fate may be very diverse, and various members of the thecostracan line lose their head appendages more or less completely. A sensory function, for monitoring flow fields, as in cyclopoid copepods (e.g., Kerfoot *et al.* 1980) is assumed to be secondary.

In Branchiopoda the fate of the 2nd antenna also varies considerably. Feeding and locomotory function may be lost very late in the postlarval differentiation phase of Euanostraca, the locomotory function may even be enhanced, as in Onychura, and the whole limb may almost completely atrophy, as in the Notostraca.

Mandible. – The early larval mandible of *Rehbachiella* resembles the posterior appendages, in particular with regard to the massive basipod and the endopod and the small proximal endite (the future coxa). At this early stage it seemingly reflects the primordial shape of the limbs at the stem-group level. On the other hand, the exopod is designed as that of the 2nd antenna, which clearly contrasts with the paddle-shaped exopods of the postmandibular limbs of virtually all Crustacea s. str., at least in their early phase of morphogenesis.

Various authors have described the euanostracan mandible as uniramous (e.g., Heath 1924; Gauld 1959; Anderson 1967; Baid 1967). A biramous state has been noted by Oehmichen (1921, Fig. 15), Hentschel (1968), Benesch (1969, also his Fig. 24) and Schrehardt (1987a, particularly his Fig. 4). Hence, the biramous state of the *Rehbachiella* mandible, as the probable plesiomorphic condition, does not conflict with supposed anostracan affinities. Sanders (1963a) has considered the uniramous mandible as a distinctive feature of Branchiopoda. In the light of the more recent studies on Euanostraca and *Rehbachiella*, a biramous mandible should have characterized the ground plan of Branchiopoda.

During ontogeny the mandibular coxa enlarges considerably but gradually, while the palp reduces in size, its foramen being no larger than in the nauplius at stage TS13. Degeneration of the palp during late larval development is common not only to Branchiopoda, but occurs also in Cephalocarida. A palp is also unknown from Remipedia, while the biramous larval palp of Eumalacostraca (Recent Phyllocarida have no early larvae) is replaced by a peculiar, large and uniramous palp (Figs. 50C, D, 51A). An explanation for the retention of basipod and rami in Maxillopoda has been given above.

While the autapomorphic condition for Malacostraca is clear, the status for all other crustaceans remains to be clarified. With regard to the mandible of *Rehbachiella*, useful comparative details are in the coxal shape, the basipod and its armature, and the rami. The conclusion of Schrehardt (1986a, 1987a) that a transitory 'larval mandible' is replaced by an 'adult mandible' during development of *Artemia salina* simply neglects the morphogenesis of this appendage among Crustacea. According to Walossek & Müller (1990) the splitting of the limb series in the ground plan of Crustacea also implies the possession of a functional mandible, in the larva as well as the adult. The postulated maxilla-like mandible of *orsten* arthropods, which should reflect a more primordial state of development, as postulated by Lauterbach (1988), was based on his misidentification of the 1st maxilla in a late metanauplius of *Bredocaris* as the mandible.

Maxillae. – The early morphogenesis of the first maxilla of *Rehbachiella* resembles that of *Bredocaris admirabilis.* In the latter, this limb becomes a brush, with retention of a primordial pattern of paired setae and reduction of the exopod. Similarly, the 1st maxilla of *Rehbachiella* never develops more than the proximal endite plus three on the corm (basis). The proximal endite becomes the major element of the limb and develops a very specific armature with surrounding pectinate setae, rows and single spines of various lengths and furnishment with setules (see Figs. 10F, 18G). At a late stage, the bulging endite is positioned immediately behind the elevated paragnaths (Figs. 25D, 33), with its long anteriorly curved proximal setae pointing into the deep cleft between the paragnaths ('paragnath channel'). The 2nd endite is elongate and armed with a number of pusher spines to support the oral transport (this characteristic 2nd endite was very helpful in the identification of fragmentary specimens). Thus, and contrasting with all posterior limbs, the 1st maxilla:

- develops only 4 endites rather than 6 or 8–9,

- has a bulging proximal endite used as a brush and a 2nd endite as a pusher, and

- has an exopod that grows more slowly relative to the posterior limbs, and diminishes in size after TS8.

The inner edge of the larval 1st maxilla of Cephalocarida is also subdivided into four endites, before the proximal one of them elongates. The latter, however, does not develop into a setiferous brush, as in *Rehbachiella, Bredocaris* or Mystacocarida (Hessler & Sanders 1966, Fig. 6A; all with 4 endites), but carries a few rigid spines terminally (e.g., Sanders & Hessler 1964, Fig. 5, for *Lightiella incisa*). In contrast with the situation in all other Crustacea, the malacostracan 1st maxilla does not subdivide the inner edge of the basipod portion of its corm at any stage, while the proximal endite is very pronounced (Figs. 48C for Leptostraca, 48F for Euphausiacea; 48I for larval Decapoda [maxillae mismatched in Siewing 1985, Fig. 23b]; Bowman & Iliffe 1986, Fig. 1K, for the thermosbaenacean *Halosbaena fortunata*).

Hence it seems as if the distinctive shape of the 1st maxilla could indeed serve to support the assumption of an early branching off of the Malacostraca. These uniformly have retained the shape of a limb at the ground-plan level, i.e. with undivided proximal endite and basipod. All other crustaceans have modified this limb by subdividing the inner edge of its corm ('pe' plus basipod) into four endites. Further investigations will have to substantiate whether this occurred once only.

If so the the unity of Cephalocarida, Maxillopoda, and Branchiopoda would be supported – and a synapomorphic character found that could aid in the recognition of the 'Entomostraca' as the sister group of Malacostraca. Since it is assumed that further evolution of the 2nd maxilla

occurred after the separation of Malacostraca, it would also be interesting to investigate the median subdivision of this limb and its rami (examples for different Crustacea in Figs. 46–48). Moreover, examination of the subdivision of limb corms might help in the evaluation of the interrelationships of the Remipedia.

Development of brush-shaped proximal endites in both of the maxillae seems to characterize only the branchiopodan lineage, in accordance with development of the filter-feeding mode of life. In all Recent Branchiopoda, these limbs are represented merely by their large and well equipped proximal endites but nevertheless well functional (cf. Fryer 1983 for Euanostraca, 1988 for Notostraca). Reduction of both maxillae in the first instance appears to be a synapomorphy of Branchiopoda, but the fossil record indicates that these limbs were primitively well-developed: in *Rehbachiella* as well as in *Lepidocaris* (cf. Schram 1986) the 2nd maxilla had the shape of a trunk limb. Moreover, the 1st maxillae are claspers in the male of *Lepidocaris* and also not reduced. Since the degree of reduction is rather different, closer examination is needed of the shape and function of branchiopod maxillae to search for detailed differences also in the Recent taxa to substantiate this assumption of a parallel modification. An indication for this presumed homoplasy is in the different degree of prominence of the two maxillae in the Notostraca (cf. Fryer 1988). Again, also in certain thecostracan taxa (e.g., cirripeds), both maxillae become reduced with the moult to the cypris, but from a clearly different developmental state.

Postmaxillary appendages. – Shape and segmentation of crustacean limbs are largely influenced by functional needs. Although a basic pattern may be still inherent in Recent forms, it is largely obscured by uncertain relationships between taxa and contradictory descriptions. For example, the endopod of Cephalocarida is variously used to define the primordial state, probably influenced by its trilobite-like appearance with the distal claw. Terming this a 'multi-articulate state' (e.g., Hessler 1982b) conceals that its number of podomeres may be just the same as in Malacostraca or in *Martinssonia* (Müller & Walossek 1986a), namely five, such as illustrated by Jones (1961), Gooding (1963), McLaughlin (1976), and Knox & Fenwick (1977). The distal 'claw' may, thus, be nothing more than the cluster of setae found on various endopods of other Crustacea, but modified to rigid spines. On the other hand, this number has a bearing for *Rehbachiella* and *Bredocaris*, since they have four endopodal podomeres maximally.

It has been generally stated that six endites – the proximal endite plus five more on the basipod portion of the corm – are characteristic of Branchiopoda (e.g., Eriksson 1934). This may be partly based on an erroneous comparison with a trilobitoid limb and misinterpretation of the endites as outgrowths of its 'telopodite podomeres' accordingly. It is not unlikely that Calman (1909) unwittingly added to such confusion, since he used the same abbreviation 'en.' for both the endopod of non-branchiopod limbs and the endites of branchiopod limbs (his Figs. 3 and 4). The same happened with the exopod and exites.

But while Eriksson (1934) recognized endopod, exopod and epipods as such, confusion arose particularly by Preuss (1951, 1957) and Siewing (1960). Considering the portion of the proximal endite as 'protopod', these authors named the five subsequent endites of the corm and the endopod together as endites of a six-segmented telopod (particularly Siewing's Fig. 19, comparing the cephalocaridan, the *Lepidocaris*, and the euanostracan limbs). In consequence, the exopod turned into the endopod, and the distal epipod into the exopod.

Although there is a striking contrast between the morphogenesis of these portions, the limb morphology of Cephalocarida, and that of *Lepidocaris*, this same confusion has been continued into more recent papers (e.g., Benesch 1969; Schrehardt 1986a, 1987a, b). Regrettable results of such misinterpretations include discrepancies in the limb musculature, as stressed by Preuss (1951, 1957), or the osmoregulatory function of the 'exopod' of *Artemia*, as claimed by Schrehardt (1987a, b). The terminology used by Claus (1873) and continued by Martin & Belk (1988) and Martin (1989) even misses the endopod completely, which has been interpreted as the distal endite of the corm, while the exopod and the epipods are congruous with the terminology used herein.

On the other hand, Fryer (1983) has commented upon the nature of the proximal endite of Euanostraca as representing a composite structure which raises the number of endites for Recent forms (see also Fig. 46A). After 'rearrangement' of the terminology, referring to the limb morphology of *Rehbachiella* (Figs. 27, 47C), the musculature as well as the limb portions of the euanostracan limb become consistent with that of other Branchiopoda and Crustacea. Further support for the structural homology of the phyllopodous limb with other crustacean limbs comes from the shape of the proximal endite, the position of the paddle-shaped exopod and the appearance of the epipods. This recognition of structural homologies is also facilitated by the design known from the Cephalocarida (Fig. 48A) as well as from *Lepidocaris* (Fig. 46F–H).

Remarkably, not only *Rehbachiella* but also *Lepidocaris* and Kazacharthra (cf. McKenzie *et al.* 1991) seem to lack epipodial structures. It must be remembered, however, that postlarval stages are still unknown in *Rehbachiella*. As is the case for several crustacean features, the epipods, as exites of the outer edge of crustacean limbs, still suffer from apparently inconsistent descriptions, but detailed comparative studies of structure and function of these organs are still lacking. The epipods of Malacostraca, which exclusively stem from the proximal endite ('coxa'), are well-known, and their gill-function seems well-established (e.g., Burnett & Hessler 1973).

Among the Branchiopoda, the Euanostraca are described as having up to three epipods. From their specific design it is most likely that the proximal two, arising from the proximal portion of the corm, merely comprise the portions of a single one, in analogy to the eumalacostracan condition (e.g., Alonso 1985, compare his Figs. 1c for *Branchinecta ferox* with a single proximal epipod, 2k for *Branchipus schaefferi* with a faintly divided one, and 4d for *Chirocephalus diaphanus* with two proximal ones; also Thiery & Champeau 1988, Fig. 3A for *Linderiella massalliensis*).

As regards the distal epipod, several studies have clarified its osmoregulatory function (e.g., Croghan 1958a, b; Copeland 1966; Ewing *et al.* 1974; 'exopod' according to Schrehardt 1987a). The more proximal epipod seems to function as a gill (Schrehardt, in press, as cited in Schrehardt 1987a). Phyllopoda have only one fleshy, hose-shaped epipod immediately proximal to the insertion of the exopod, and clear information about its function could not be found. It remains also unclear whether this epipod is homologous to the distal one of Euanostraca, while it seems clear that a proximal epipod is not developed in the Phyllopoda.

Among the other non-malacostracans, the Remipedia, the Maxillopoda, and the Cephalocarida lack corresponding structures. The nature of the pseudepipod of Cephalocarida has never been clarified. According to Sanders (1955, 1963b, p. 7), Jones (1961, particularly her Fig. 14), Gooding (1963), Shiino (1965), McLaughlin (1976), Knox & Fenwick (1977), and Ito (1989a, particularly his Figs. 3, 4 [SEM-picture]) it stems from the proximal part of the exopod or a common portion rather than from the limb corm. Such a position, together with the setation and the locomotory function, precludes homology with epipods. In the light of the overall similarity between the trunk limbs of *Rehbachiella* and Cephalocarida, it is more likely that this portion represents the proximal elongation of a leaf-shaped exopod, as developed in eumalacostracan limbs (larval, also in 2nd maxilla) and other Crustacea (examples in Figs. 46–48), but which became jointed and mobile in the Cephalocarida for specific needs.

According to Manton (1977) and McLaughlin (1982) the epipods are outgrowths of the outer edge of the crustacean limb corm and, thus, are not rami. Since in all Crustacea s. str. at least the naupliar limbs, the maxillae, and (when present) mostly also the maxillipeds, lack epipods, any presumption of their presence at the basis of crustacean evolution (e.g., Siewing 1960) warrants substantiation. Such presumptions are not in accordance with the fossil record and may rather derive from original misinterpretation of the trilobite limbs.

Lauterbach (1983, referring to his earlier papers) claimed that crustacean epipods should have originated from primordial respiratory 'exopods' by development from marginal feathers that subsequently shifted onto the corm. There is virtually no evidence from the fossil record to substantiate the assumption of such a complicated pathway, nor for the purely speculative and even highly unlikely respiratory function of the exopods of early euarthropods (cf. the recently discovered limb of *Naraoia*, illustrated by Chen *et al.* 1991).

On the other hand, the nature of the crustacean exopod as a locomotory organ is evident for both the naupliar appendages and the postmandibular limbs (particularly Fryer 1983 and 1988; also McLaughlin 1982, p. 202). Its development for this function is fully supported by the morphology of stem-group crustaceans (Müller & Walossek 1986a; Walossek & Müller 1990; Walossek & Szaniawski 1991). Its shape is basically identical in all crustaceans (examples in Figs. 46–48) and clearly different between the naupliar set and posterior set. According to Walossek & Müller (1990) there is no discrepancy in the homology of the trilobitoid limb basis and rami with the basipod and rami of crustaceans.

'Stalked' and paddle-shaped exopods similar to those of *Rehbachiella* can be seen among Branchiopoda, for example in Euanostraca, such as in *Tanymastix stagnalis* (Fig. 46C) or *Parartemia zietziana* (Fig. 46D), but also in Conchostraca (Figs. 46I–K), Cladocera (Fig. 46L), Kazacharthra (47B), and Malacostraca (e.g., the 2nd maxillae of extant Phyllocarida, Fig. 48D, and Mysidacea, Fig. 48H). Respiratory exopods may be developed in those forms where it is very foliate (e.g., recent representatives of Phyllocarida). Yet, this function has not been clarified in detail by comparative ultrastructural and physiological studies, and also does not take account of its marginal setae. Such flattened exopods may just serve as vibratory plates to produce currents or to act as valves to regulate these (secondary, according to Fryer 1988).

Ventrocaudal processes

These outgrowths are a characteristic feature of *Rehbachiella*. They appear at about TS4 and develop in a very similar manner to the furcal rami in having also marginal spines and pores associated with these on the ventral side of the margin. From their position they may possibly have borne sensory bristles, but this is of course purely speculative.

In Branchiopoda large transverse muscles serve as dilatators of the anus (examples: Hsü 1933, Figs. 20, 26, for the euanostracan *Chirocephalus*; Claus 1873, Pls. 7:3, 3', 4 and 8:5 for the notostracan *Triops*; Longhurst 1955, Fig. 13C). Their position is generally associated with development of two structures, an axial ventral furrow in the telson and a pair of posteriorly pointing caudal outgrowths ventrally to the insertions of the furcal rami (same figures of Claus for Notostraca). Besides Branchiopoda, similar structures are present in the Recent Phyllocarida (e.g., Barnard 1914, Pl. 39; Hessler 1984, Fig. 4A; Bowman *et al.* 1985, Fig. 2j, k) and certain Maxillopoda (e.g., Tantulocarida, see Huys 1989,

Fig. 2E; Mystacocarida, see Hessler & Sanders 1966, Fig. 3A, 4F). Such outgrowths are not described from Remipedia, euanostracan Branchiopoda, and the *orsten* fossils *Bredocaris* (Müller & Walossek 1988b) and *Dala* (Müller 1983). Further information on the nature of these, however, has never been given in the descriptions.

Relative to the short outgrowths of other crustaceans, the strong prominence of the processes of *Rehbachiella* are probably a speciality of this form. Their armament of spines furnished with spinules in connection with the furcal rami may point to a participation in a grooming function and even association with the steering of the furca.

Segmental organs

In a few specimens, pores were recognized on the sternal plates of the maxillary segment and the thorax (Pls. 13:7; 24:3). Also, a number of grooves and pits, possibly belonging together, were observed in the sternal region of the segments of the two maxillae (Pls. 20:8; 21:7). Neither the definite position nor the possible function of these structures could be clarified. Further investigations will have to show whether they might have been segmental excretory structures. It is noteworthy that segmental organs, or their derivatives, are noted for *Artemia* in all postantennular segments as far back as the apodous abdomen (e.g., Benesch 1969). According to Schram & Lewis (1989) these are also present in the head region of Remipedia. Caution is thus required in using such a character as an apomorphy of Crustacea, as was done by Lauterbach (1983). Maxillary glands, on the other hand, can be identified only histologically in Recent Euanostraca, and they need not necessarily be expressed externally in *Rehbachiella*.

Tagmosis

Head. – Traditionally the crustacean head is described as basically comprising five appendiferous segments, including the mandibles and two additional mouth parts, the maxillae. This must be specified in so far as this number of segments, if at all, characterizes only the crown group, the Crustacea s. str. Moreover, this refers with certainty only to the dorsal side which coalesces with the growing 'head'. Ventrally, the formation of a sternum through fusion of postoral sternites including both maxillae has not yet been clarified, while for the trunk-limb status of the 2nd maxilla, which lies at the basis of the different crustacean taxa, more evidence has been accumulated, not least from *Rehbachiella*.

Bearing this in mind, and taking account also of the recognized division of the crustacean limb set into a naupliar and a postnaupliar portion, a subdivision into a 'procephalon' and a 'gnathocephalon', as proposed by Siewing (1963) is poorly founded and not compatible with the stepwise and secondary inclusion of the maxillae into a 'head portion' and their subsequent specialization. This process, which can be followed in the ontogeny of *Rehbachiella*, is seen as an important tool for the recognition of further trends of specialization within the different crustaceans, possibly even aiding in finding synapomorphies in these in order to reconstruct the interrelationships of the crown-group members with more confidence.

Trunk. – The trunk region of *Rehbachiella*, considered as thoracic, bears 13 segments, 12 of which have limbs. This interpretation follows Benesch (1969), who convincingly argued in favour of the recognition of the two 'genital' segments of *Artemia* as thoracomeres on the basis of various internal features shared with the thoracic segments but lacking in the abdominal ones. This has already been noted by Baqai (1963) who did not, however, make consequent use of it. With the restudy of the Upper Cambrian *Bredocaris* and *Rehbachiella*, as well as the male *Lepidocaris* (see Schram 1986), more evidence is available to confirm this conclusion for the genital segments of other Crustacea as well. For Maxillopoda this has been postulated by Müller & Walossek (1988b), and confirmed by Huys (1991).

In Cephalocarida the ninth trunk segment has a modified limb (Sanders 1963b) together with the corresponding musculature (Hessler 1964). In consequence, it should be considered as the last thoracomere rather than the only limb-bearing one of the abdominal segments. Hence, in the ground plan of Branchiopoda, Maxillopoda, and Cephalocarida, the thorax would consistently be limb-bearing, while the abdomen consistently comprises no limb-bearing segments.

It remains difficult to evaluate the character state of the number of thoracomeres and abdominal segments. Within Branchiopoda, *Rehbachiella* has the same number as the Euanostraca and *Lepidocaris* (if accepting the reinterpretation). Abdominal segments are, however, not delineated until the TS13 stage, as in the Euanostraca. Thus, it is unclear whether or not more segments would appear in later stages. The segment number of laevicaudate Conchostraca approaches the number of Euanostraca, *Lepidocaris* (see above) and *Rehbachiella*. While Euanostraca have six apodous segments, *Lepidocaris* has only four, and Laevicaudata and Spinicaudata both lack abdominal segments. Notostraca have clearly multiplied their trunk segments and appendages, and shed no light on this question.

According to the above interpretation, Maxillopoda have seven thoracopods plus four abdominal segments in their ground plan, while Cephalocarida have nine plus ten. Malacostraca possess a trunk comprising 14 limb-bearing segments and one apodous one in front of the telson in Phyllocarida, which is generally considered as the basic condition. A specific character of this group is the division of its set of limbs into two morphologically and functionally distinctive series, the 'pereiopods' and the 'pleopods' (see also Dahl 1976, p. 164).

The morphology of *Rehbachiella*, or other known *orsten* arthropods, provides no clue to this problem, but to stimulate further discussion, the hypothesis is presented that the complete set of trunk segments of Malacostraca bearing the 14 pairs of limbs should be regarded as the thoracic region. Accordingly, Malacostraca would have one more limb-bearing thoracomere than the maximum number in non-malacostracans, but merely one apodous abdominal segment in all known taxa, while the number of apodous abdominal segments is variable in non-malaocostracan crustaceans. This may point to convergent evolution of abdominal segments within the different non-malacostracan taxa. Furthermore, the number of thoracomeres may have become independently modified, as indicated by the results from the comparisons of ontogeny patterns between the different Crustacea above.

This approach of interpreting trunk tagmosis in Crustacea assumes a much smaller basic number of trunk segments than proposed by Hessler & Newman (1975) for their 'urcrustacean'. Presuming approximately 15 trunk segments maximally may prove useful not only for the reconstruction of relationships wihin Crustacea and the tagmosis at the ground-plan level of Crustacea s.str.; it may also be relevant for comparisons with other arthropod taxa.

Concluding remarks

In recent years fossil invertebrates have been increasingly acknowledged for their contribution to phylogenetic reconstructions (examples: Schlee 1981, especially for fossils from amber; Naumann 1987 for zygaenid moths; Smith 1984 and Mooi 1990 for echinoderms; Willmann 1981, 1983, 1987, 1989a for mecopteran insects; Haas 1989 for coleoidean cephalopods). The present work on *Rehbachiella* may add further support to this trend.

According to Willmann (1989b, p. 282) fossils can contribute to understanding homology in shared similarities, in recognizing character states, and in clarifying affinities, even if there are no apparent synapomorphies between the members of a monophylum in question. An example of such transgressive features may be seen in some recently discovered *orsten* arthropods. By recognition of their 'proximal endite' as an evolutionary novelty of Crustacea sensu lato (Pan-Crustacea, according to Walossek & Müller 1990), its subsequent modifications can be traced in particular limbs and in different directions within the Crustacea s. str., in certain cases even forming a distinctive portion, the 'coxa'. The original limb basis of the euarthropod plan (e.g., the 'trilobitoid' limb), which carries the two rami, is retained as a plesiomorphy in the crustacean plan ('basipod').

The advantage of this hypothesis is that it assumes only a single evolutionary novelty in the limbs of Crustacea, the 'proximal endite', most likely resulting from adaptation to new locomotory and feeding strategies (parsimony principle). Speculations about complicated to-and-fro shifting of rami (e.g., Lauterbach 1979), later subdivisions of an originally undivided corm (Hessler 1982b), or the fusion of ramal articles to form a 'basipod' (Ito 1989a) are now no longer necessary. This conclusion, drawn from the fossils as well as from the morphogenesis of crustacean appendages, predicts a biramous origin for crustacean limbs. In consequence, and undivided corm, as postulated for the 'urcrustacean' by Hessler & Newman (1975) rather refers to an older evolutionary level of arthropod limbs.

The inclusion of these fossils in the phylogenetic concept of Crustacea, as representatives of their stem group, has allowed the reconstruction of the ground plan of Crustacea s. str. and to found its monophyly on a set of constitutive characters (cf. Walossek & Müller 1990). This set includes, for example, the enhancement of the proximal endites of the second and third naupliar appendages to form distinctive coxal bodies (specialization of '2nd antennae' and 'mandibles').

Willmann (1988, p. 158) pointed to the possibility that significant features, which later representatives of a taxon may still have in common, need not have been developed in early stem-lineage representatives (see also Königsmann 1975 and Schlee 1981). In this sense, the early stem-group crustaceans possessed the 'proximal endite', but they did not yet have distinctive coxal portions in their anterior postantennular limbs, as do the crown-group crustaceans.

The splitting of the naupliar and the postnaupliar limb set as one of the key characters is of particular interest for *Rehbachiella*, since this character, which in fact includes a whole set of details, permits its recognition, in the first instance, as a representative of the crown group. Again, it has aided in the reappraisal of the distinctiveness of filter-feeding apparatuses in Crustacea, as has been stressed by Cannon (1927b). Many detailed studies since then have clearly revealed that such apparatuses are indeed not primordial in structure and function and cannot have characterized the last common ancestor of the Crustacea s. str., as has been hypothesized by Lauterbach on several occasions (e.g., 1980).

According to Tyler (1988, p. 344) the component-function analysis can aid as another 'means for strengthening the foundations of phylogenetic systems'. Filtration function is achieved by various structures among crustaceans, which provides in fact a large set of characters of value for detailed analysis. This has permitted the monophyly of the Branchiopoda to be founded on their complex postnaupliar locomotory and feeding apparatus. It also has led to the indubitable inclusion of *Rehbachiella* within this taxon, since there are so many shared structural and functional elements.

With this suggested ascription, an additional indication is given that specialized 2nd maxillae cannot have characterized the ground plan of Crustacea s. str. Its trunk-limb

state is now known from Cephalocarida (Sanders 1955, 1963a, b) as well as from fossils, such as *Bredocaris*, possibly *Lepidocaris*, and *Rehbachiella* (herein). Again, indications of this state still persist in other Recent Crustacea, such as in Euanostraca (e.g., Snodgrass 1956; Benesch 1969), and in certain Eumalacostraca (e.g., Mauchline 1967). At least throughout the first larval phase of thecostracan Maxillopoda, the 2nd maxilla remains on the larval trunk. Accordingly, the separation of the pathways of limb apparatuses in Crustacea s. str. must already have occurred at a level when this limb was still not 'cephalized' (see also Manton & Anderson 1979). The recognized distinctiveness of the two maxillae in their morphology and function suggests that the same procedure may also be applicable to the 1st maxilla.

Hence, these 'mouthparts' were obviously not originally a functional unit. Their distinction facilitates, in my mind, the recognition of detailed structural differences in the limb sets, which, in consequence, permits synapomorphies to be sought in these for the various crustacean taxa. Such possibilities might be seen in the different filter-feeding habits among Maxillopoda, which use either the thoracopods (Cirripedia) or the 'mouthparts' together with the 1st thoracopod as functional units (all representatives of the copepodan lineage).

In this way, terms like 'cephalic' versus 'thoracic' limbs sets are, strictly speaking, not readily applicable to Crustacea. In fact, only the naupliar apparatus, the 'primary head region' of Calman (1909), is set well apart from the postnaupliar set. The latter 'buds' off additional limbs stepwise to form a 'head', but not necessarily terminating at the level of the 2nd maxilla. As various Crustacea demonstrate, this process may continue with the inclusion of a different number of postmaxillary limbs. In various cases the postmandibular limbs do not link with a functional apparatus with the mandible in the sense of mouthparts, but constitute other units or serve for different functions (e.g., claspers in males of *Lepidocaris*).

This not only is at odds with a subdivision of the crustacean 'head' into a pro- and a gnathocephalon, as already stated above, but also sheds new light on the current dogma of the close inclusion of four limb-bearing postantennular segments into the head of Euarthropoda (e.g., Lauterbach 1973). In the last two decades substantial new evidence has accrued on early Euarthropoda (examples: Cisne 1975, 1981; Whittington, e.g., 1975, 1977, 1979, 1980; and Whittington & Almond 1987 for trilobites and other early Palaeozoic arthropods; Müller & Walossek 1987 for the Upper Cambrian *Agnostus*; Müller & Walossek 1986b, 1988a for an Upper Cambrian chelicerate larva). In all these cases the 'head' bears a pair of antennulae and only three more pairs of appendages. This is also true for the Cambropachycopidae among the supposed stem-group crustaceans, while *Martinssonia* and *Henningsmoenicaris* seem to incorporate a further segment into the head in their later stages of larval development (Walossek & Müller 1990; unclear for the recently discovered *Cambrocaris*, according to Walossek & Szaniawski 1991). This issue should be no longer neglected in future phylogenetic approaches.

The recognition of the naupliar apparatus as a key character in the ground plan of Crustacea s. str. implies that the nauplius, as the 'most oligomeric arthropod larval type', is another essential characteristic of the crown-group crustaceans, as has been suggested for example by Snodgrass (1956) or Cisne (1982; for remarks on Lauterbach's [1988] contrasting views, see above). Possession of a nauplius in *Rehbachiella* is thus only a symplesiomorphy when compared against other crown-group members, but is again an indicator that it is a true crustacean.

Within Branchiopoda, *Rehbachiella* is considered to be a representative of the stem-group of Anostraca, an assumption which is founded on the progressive protrusion of the eye area and the reduction of the naupliar neck organ shared with the Recent Euanostraca. It is not unlikely that more details of the complex filter apparatus, which are still unclear in part due to preservational limitations, may provide further evidence to substantiate this assignment more precisely.

As a consequence of this reconstructed relationship of *Rehbachiella*, the two major branchiopodan lineages Anostraca and Phyllopoda should already have been separate in Upper Cambrian times. Accordingly, their isolated evolution for more than 500 million years explains their morphological distinctiveness and paucity of synapomorphic features, as variously noted (see e.g., Fryer 1987c). This also implies a separate radiation of the major lines into the freshwater environment, confirming the presumptions of Preuss (1951, 1957) – but without the necessity of challenging the Branchiopoda as a valid monophylum, which is founded on the synapomorphic filter-feeding system.

It is apparent that neither the symplesiomorphic characters (shared primitive features) of extant members nor their autapomorphies ('differential characters') help to clarify relationships. The large set of morphological and morphogenetic data of *Rehbachiella* may, thus, serve as a useful tool for future detailed comparisons and possibly also for solving the still unclear interrelationships between Conchostraca and Cladocera. In particular the details of the postcephalic limb apparatus may be of value for further phylogenetic analyses.

Accepting *Rehbachiella* as an ancestral anostracan branchiopod, it cannot readily serve as a model for the 'urcrustacean', but its morphology exhibits various primordial features that have a bearing particularly on those crustaceans variously regarded as 'most primitive'. These are the Cephalocarida and Remipedia, but this may also be expanded to the Phyllocarida among Malacostraca. In the light of a possible sister-group relationship between Maxillopoda and Branchiopoda, as favoured herein on the basis

of the common possession of a larval neck organ and details of the postnaupliar limb apparatus, it seems not unlikely that Cephalocarida may be related to these (as the sister taxon of the common ancestor of Branchiopoda and Maxillopoda). They share more details of the postnaupliar limb apparatus with Branchiopoda than with the Malacostraca, and indeed the cephalocaridan type of limb apparatus could well serve as a 'precursor' of the branchiopodan type, including *Rehbachiella*.

Malacostraca form a distinctive unit, as they are generally understood, and their characteristic features seem to have developed rapidly. The fossil record of Phyllocarida can only be traced with confidence back to the Ordovician from where the first undoubted Archaeostraca are reported, and these, according to Dahl (1983), already show all the typical malacostracan features.

Willmann (1989b, p. 277) noted that 'fossils can provide minimum ages for monophyla and contribute to the knowledge of their distribution in space'. It is hoped that well-preserved fossils, such as *Rehbachiella* and the other *orsten* arthropods, may increase the value of fossils even further. The detailed description of their external features, including information on ontogeny and morphogenesis of structures and function, may contribute in the future to a detailed analysis of relationships in different directions by application of phylogenetic systematics.

The study of *Rehbachiella* has permitted morphogenetic changes to be monitored along the larval sequence that show up in their terminal state in Recent Crustacea, or uncover evolutionary pathways no longer recognizable in Recent material. Examples are the transitional appearance of a fourth article in the endopods of 2nd antenna and mandible, and at the tip of the 1st antenna, the coalescence of the terminal two exopodal articles which later carries two setae, the separate sternite of the 2nd antenna, and the differential coalescence of the maxillary segments on dorsal and ventral side.

Early Crustacea of different lines, such as *Rehbachiella* as an ancestral anostracan branchiopod and *Bredocaris* as an ancestral thecostracan maxillopod, are remarkably similar to one another. However, this does not relate to their generally plesiomorphic status, but reflects their closeness to one another in terms of absolute number of diverging steps from their common ancestors (see also Willmann 1988, particularly his Fig. 9). Hence, their design is indeed 'still fairly close to the body plan of the common ancestor of Crustacea s. str'. This can also be deduced from the various shared similarities, even in minute details down to denticles as ornaments on limbs, with so-called 'primitive' representatives of the different crustacean taxa, e.g., leptostracan Phyllocarida or Cephalocarida.

Regardless of this absolute proximity of relationship, such early forms have already diverged in different directions. Hence, among their character sets the apomorphic characters of the particular groups they belong to are also embodied. However these may no longer occur in the character set of Recent descendants (see above), or few may be left. These may be less prominent structures or at an incipient state of development in the fossils. A structure, such as a limb, may be plesiomorphic in its gross design but uncovers a mosaic-like pattern of evolutionary steps in the development of its components, such as endites, rami, exites, and setation. This differentiated hierarchy in terms of development, which also refers to function, hampers any phylogenetic analysis exclusively based on Recent taxa. This is particularly true when such structures are obscured after a long evolutionary pathway – as possibly valid for the Branchiopoda and Crustacea as a whole. Early fossils can, thus, hint to the first steps of modification of the morphology of a group in question, on condition that they can be clearly positioned.

In this respect, taxa such as the extant Cephalocarida are indeed remarkable. Apart from their apparent modifications due to life in the flocculent layer (also apparent in their ontogeny and nervous system), comparisons with the fossils reveal that they underwent very little change in several important aspects of their morphology (stasis) and show up as more ancestral than, e.g., the Upper Cambrian *Rehbachiella*.

The value of comparative ontogenetic studies and evaluation of morphogenetic changes of function for phylogenetic approaches among Crustacea has been repeatedly demonstrated (e.g., Fryer 1983, 1988 for Branchiopoda; Izawa 1987 and Dahms 1989a for Copepoda; Grygier 1984, 1987 for selected Maxillopoda). The comparisons made herein on the basis of segment increase following Weisz (1946, 1947) revealed not only common strategies among Crustacea but also distinctive taxon-dependent ones. This is apparent not only in the Maxillopoda, but also in the Cephalocarida, supporting their distinctive status. It is hoped that this will stimulate subsequent students of the postembryonic development of crustaceans to refer to the segment pattern. In my view, this will improve comparability and may also permit the inclusion of more modified groups in general comparisons.

Such comparisons also serve for the reconstruction of life habits. In this respect, functional analysis has revealed that, as in feeding nauplii of Recent Crustacea, the nauplius of *Rehbachiella* was actively feeding while swimming, as enforced by the physical demands of the surrounding viscous milieu (at low Reynolds numbers). Since this can be applied also to the larvae of *Bredocaris*, the speculations of Lauterbach (1988) concerning a creeping ancestral larval type for Crustacea, must be rejected.

Two major strategies seem to have largely affected the evolution of Crustacea also: (1) strong paedomorphic influences, shown by the analysis of ontogenetic patterns, as demonstrated by Schminke (1981) for Bathynellacea and also assumed for Maxillopoda by Newman (1983); (2) ho-moplasy, of which the Branchiopoda appear to be a good example.

In the light of the *orsten* fauna it seems more and more likely that the major branchings among Crustacea s. str. have already occurred at least in the Upper Cambrian. Relationships within the Crustacea are yet not sufficiently understood, but the mass of new evidence from Recent and fossil material brought up in the last few years is promising. This may also hold true for discussions concerning the 'Mandibulata', which is reserved for future publications.

Acknowledgements

My special thanks go to Klaus J. Müller, Bonn, for kind permission to examine the *Rehbachiella* material and helpful comments. I am much indebted to Euan N.K. Clarkson, Edinburgh, for his immense work of reviewing the manuscript and improvement of the language, and to Geoffrey A. Fryer, Cumbria, who also kindly reviewed the manuscript from a fundamentally different view of interpreting relationships, and to Stefan Bengtson for his immense editorial effort to shape this big paper for publication. Thanks are also due to various colleagues for valuable suggestions and support (those who also gave their kind permission for inclusion of data, figures or papers in press, are marked by an asterix), particularly to: J. Bergström*, Stockholm, A.C. Cockcroft*, Port Elizabeth, E. Dahl*, Uppsala, H.-U. Dahms*, Oldenburg, G. Fryer*, Cumbria, R.R. Hessler, La Jolla, I. Hinz*, Bonn, J.W. Martin*, Los Angeles, T.H. MacRea*, Halifax, K.G. McKenzie*, Wagga Wagga, K.J. Müller*, Bonn, J.R. Strickler, Woods Hole, and R. Willmann, Oldenhütten/Kiel. The Springer Verlag kindly permitted the inclusion of SEM micrographs (Fig. 53A, B) from a paper of Go *et al.* (1990) published in *Marine Biology*.

My sincerest gratitude goes not least to my wife Kriemhild and our children Sonja, Christian and Lena, who helped me so much by their tolerance and indulgence. K. Walossek and G. Walossek also helped with technical support.

The study has been financially supported by the Deutsche Forschungsgemeinschaft. All specimens examined in this study are deposited in the Institut für Paläontologie in Bonn under nos. UB 644 and 645 (type specimens designated by Müller 1983), UB 771 and UB W3–W95 (new material).

References

Achituv, Y. 1986: The larval development of *Chthamalus dentatus* Krauss (Cirripedia) from South Africa. *Crustaceana 51(3)*, 259–269.
Alonso, M. 1985: A survey of the Spanish Euphyllopoda. *Miscellania Zoologica Barcelona 9*, 179–208.

Alonso, M. & Alcaraz, M. 1984: Huevos resistentes de crustáceos eufilópodos no cladóceros de la peninsula Ibérica: Observación de la morfologia externa mediante técnicas de microscopia electrónica de barrido. *Oecologia aquatica 7*, 73–78.

Amat, F. 1980: Diferenciación y distribución de las poblaciones de *Artemia* (Crustáceo, Branchiópodo) de España. I. Análisis morfológico. Estudios alométricos referidos al crecimiento y a la forma. *Inv. Pesq. 44(1)*, 217–240.

Anderson, D.T. 1965: Embryonic and larval development and segment formation in *Ibla quadrivalvis* Cuv. (Cirripedia). *Australian Journal of Zoology 13*, 1–15.

Anderson, D.T. 1967: Larval development and segment formation in the branchiopod crustaceans *Limnadia stanleyana* King (Conchostraca) and *Artemia salina* (L.) (Anostraca). *Australian Journal of Zoology 15*, 47–91.

Anderson, D.T., Anderson, J.T. & Egan, E.A. 1988: Balanoid barnacles of the Genus *Hexaminus* (Archaeobalanidae: Elminiinae) from Mangroves of New South Wales, including a Description of a New Species. *Records of the Australian Museum 40*: 205–223.

Ax, P. 1985: Stem species and the stem lineage concept. *Cladistics 1(3)*: 279–287.

Ax, P. 1988: *Systematik in der Biologie, Darstellung der stammesgeschichtlichen Ordnung in der lebenden Natur*, UTB 1502, 1–181. Fischer, Stuttgart.

Ax, P. 1989: The integration of fossils in the Phylogenetic System of organisms. *Abhandlungen des naturwissenschaftlichen Vereins in Hamburg (NF) 28*, 27–43.

Baid, J.C. 1967: On the development of *Artemia salina* L. (Crustacea: Anostraca). *Journal of the Natural History Society of Bombay 64*, 432–439.

Baqai, I.U. 1963: Studies on the postembryonic development of the fairy shrimp *Streptocephalus seali* Ryder. *Tulane Studies in Zoology 10(2)*, 91–120.

Barlow, D.I. & Sleigh, M.A. 1980: The propulsion and use of water currents for swimming and feeding in larval and adult *Artemia*. *In* Persoone, G., Sorgeloos, P., Roels, O. & Jaspers, E. (eds.): *The Brine Shrimp Artemia.* 1. Morphology, Genetics, Radiobiology, Toxicology, 61–73. Universa Press, Wetteren.

Barnard, K.H. 1914: 17. Contributions to the Crustacean Fauna of South Africa. 4. A new Species of *Nebalia*. *Annals of the South African Museum 10*, 443–448, pl. 39.

Barrientos, Y. & Laverack, M.S. 1986: The larval crustacean dorsal organ and its relationship to the trilobite median tubercle. *Lethaia 19*, 309–313.

Bassindale, R. 1936: 4. The Developmental Stages of Three English Barnacles, *Balanus balanoides* (Linn.), *Chtamalus stellatus* (Poli), and *Verruca stroemi* (O.F. Müller). *Proceedings of the Zoological Society of London 106*, 57–74.

Bate, R.H., Collins, J.S.H., Robinson, J.E. & Rolfe, W.D.I. 1967: Chapter Arthropoda: Crustacea. *In: The Fossil Record*, 525–563. Geological Society of London, London.

Battish, S.K. 1981: On some Conchostracans from Punjab with the Description of three new Species and a new Subspecies. *Crustaceana 40(2)*, 178–196.

Behrens, W. 1984: Larvalentwicklung und Metamorphose von *Pycnogonum litorale* (Chelicerata, Pantopoda). *Zoomorphologie 104*, 266–279.

Belk, D. 1970: Functions of the conchostracan egg shell. *Crustaceana 19(1)*, 105.

Belk, D. 1982: Branchiopoda. *In* Parker, S. (ed. in chief): *Synopsis and Classification of Living Organisms 2*, 174–180. McGraw-Hill, New York.

Belk, D. 1989: Identification of species in the conchostracan genus *Eulimnadia* by egg shell morphology. *Journal of Crustacean Biology 9(1)*, 115–125.

Belk, D. & Pereira, G. 1982: *Thamnocephalus venezuelensis*, new species (Anostraca: Thamnocephalidae), first report of *Thamnocephalus* in South America. *Journal of Crustacean Biology 2(2)*, 223–226.

Benesch, R. 1969: Zur Ontogenie und Morphologie von *Artemia salina* L.. *Zoologische Jahrbücher, Abt. Anatomie 86*, 307–458.

Bergström, J. 1980: Morphology and systematics of early arthropods. *Abhandlungen des naturwissenschaftlichen Vereins in Hamburg (NF) 23*, 7–42.

Bergström, J. & Brassel, G. 1984: Legs in the trilobite *Rhenops* from the Lower Devonian Hunsrück Slate. *Lethaia 17*, 67–72.

Bergström, J., Briggs, D.E.G., Dahl, E., Rolfe, W.D.I. & Stürmer, W. 1987: *Nahecaris stuertzi*, a phyllocarid crustacean from the Lower Devonian Hunsrück Slate. *Paläontologische Zeitschrift 61(3/4)*, 273–298, figs. 1–14.

Bergström, J. & Gee, D.G. 1985: The Cambrian in Scandinavia. *In* Gee, D.G. & Sturt, B.A. (eds.): *The Caledonide Orogen – Scandinavia and related Areas*, 247–271. Wiley & Sons Ltd., London.

Bernice, R. 1972: Hatching and postembryonic development of *Streptocephalus dichotomus* Baird (Crustacea: Anostraca). *Hydrobiologia 40*, 251–278.

Bishop, J.A. 1968: Aspects of the post-larval life history of *Limnadia stanleyana* King (Crustacea: Conchostraca). *Australian Journal of Zoology 16*, 885–895.

Boden, B.P. 1950: The post-naupliar stages of the crustacean *Euphausia pacifica*. *Transactions of the American Microscopical Society 69*, 373–386.

Borradaile, L.A. 1926: Notes upon crustacean limbs. *The Annals and Magazine of Natural History, series IX, 17(98)*, 193–213, pls. 1–4.

Botnariuc, N. 1947: Contributions à la connaissance des phyllopodes Conchostracés de Roumanie. *Notationes Biologicae 5(1–3)*, 68–159.

Botnariuc, N. 1948: Contribution à la connaissance du développement des Phyllopodes Conchostracés. *Bulletin Biologique de la France et de la Belgique 82*, 31–36.

Bowman, T.E. 1957: A new species of *Mysidopsis* (Crustacea: Mysidacea) from the southeastern coast of the United States. *Proceedings of the United States National Museum 107(3378)*, 1–7.

Bowman, T.E. 1971: The case of the nonubiquitous telson and the fraudulent furca. *Crustaceana 21*, 165–175.

Bowman, T.E. 1984: Stalking the wild crustacean: the significance of sessile and stalked eyes in phylogeny. *Journal of Crustacean Biology 4(1)*, 7–11.

Bowman, T.E. & Iliffe, T.M. 1986: *Halosbaena fortunata*, a new thermosbaenacean crustacean from the Jameos del Agua Marine Lava Cave, Lanzarote, Canary Islands. *Stygologia 2(1/2)*, 84–89.

Bowman, T.E., Yager, J. & Iliffe, T.M. 1985: *Speonebalia cannoni* n. gen., n.sp., from the Caicos Islands, the first hypogean leptostracan (Nebaliacea: Nebaliidae). *Proceedings of the Biological Society of Washington 98(2)*, 439–446.

Boxshall, G.A. 1985: The comparative anatomy of two copepods, a predatory calanoid and a particle-feeding mormonilloid. *Philosophical Transactions of the Royal Society of London B 311*, 303–377.

Boxshall, G.A. & Böttger-Schnack, R. 1988: Unusual ascothoracid nauplii from the Red Sea. *Bulletin of the British Museum of Natural History (Zool.) 54(6)*, 275–283.

Boxshall, G.A., Ferrari, F.D. & Tiemann, H. 1984: Studies on Copepoda II: The ancestral copepod: towards a consensus of opinion at the First International Conference on Copepoda. *Crustaceana 7*, 68–84.

Boxshall, G.A. & Huys, P. 1989: A new tantulocarid, *Stygotantulus stocki*, parasitic on harpacticoid copepods, with an analysis of the phylogenetic relationships with the Maxillopoda. *Journal of Crustacean Biology 9(1)*, 126–140.

Boxshall, G.A. & Lincoln, R.J. 1983: Tantulocarida, a new class of Crustacea ectoparasitic on other crustaceans. *Journal of Crustacean Biology 3(1)*, 1–16.

Boxshall, G.A. & Lincoln, R.J. 1987: The life cycle of the Tantulocarida. *Philosophical Transactions of the Royal Society of London B, Biological Sciences 315*, 267–303.

Brattegard, T. 1970: Marine biological investigations in the Bahamas. 13. Leptostraca from the shallow water in the Bahamas and Southern Florida. *Sarsia 44*, 1–7.

Brattström, H. 1948: Undersökningar över Öresund. 33. On the Larval Development of the Ascothoracid *Ulophysema öresundense* Brattström, Studies on *Ulophysema öresundense* 2. *Lunds Universitets Årsskrift N.F. Avd. 2, 44(5)*, 1–69.

Brendonck, Luc 1989: Redescription of the fairy shrimp, *Streptocephalus proboscideus* (Frauenfeld, 1873)(Crustacea; Anostraca). *Bulletin van het Koninklijk Belgisch Intituut voor Natuurwetenschappen, Biologie 59*, 49–57.

Briggs, D.E.G. 1977: Bivalved arthropods from the Cambrian Burgess Shale of British Columbia. *Pa*laeontology 20(3), 595–621, pls. 67–72.

Briggs, D.E.G. 1978: The morphology, mode of life, and affinities of *Canadaspis perfecta* (Crustacea, Phyllocarida), Middle Cambrian, Burgess Shale, British Columbia. *Philosophical Transactions of the Royal Society, London 281(984)*, 439–487.

Briggs, D.E.G. 1983: Affinities and early evolution of the Crustacea: the evidence of the Cambrian fossils. *In* Schram, F.R. (ed.): *Crustacean Issues, 1. Crustacean Phylogeny*, 1–22. Balkema, Rotterdam.

Briggs, D.E.G. & Collins, D. 1988: A Middle Cambrian Chelicerate from Mount Stephen, British Columbia. *Palaeontology 31(3)*, 779–798, pls. 71–73.

Briggs, D.E.G. & Whittington, H.B. 1985: Modes of life of arthropods from the Burgess Shale, British Columbia. *Transactions of the Royal Society of Edinburgh: Earth Sciences 76*, 149–160.

Broili, F. 1928: Crustaceenfunde aus dem rheinischen Unterdevon. *Sitzungsberichte der mathematisch-naturwissenschaftlichen Klasse der bayerischen Akademie der Wissenschaften*, 197–204, 1 fig., pls. 1–2.

Brooks, H.K. 1955: A crustacean from the Tesnus formation (Pennsylvanian) of Texas. *Journal of Paleontology 29(5)*, 852–856.

Brooks, H.K. & Caster, K.E. 1956: *Pseudarctolepis sharpi*, n.gen., n.sp. (Phyllocarida), from the Wheeler Shale (Middle Cambrian) of Utah. *Journal of Paleontology 30(1)*, 9–14, pl. 2, 6 figs..

Brooks, H.K., Glaessner, M.F., Hahn, G., Hessler, R.R., Holthuis, L.B., Manning, R.B., Moore, R.C. & Rolfe, W.D.I. 1969: Malacostraca. *In* Moore, R.C. (ed.): *Treatise on invertebrate paleontology R, Arthropoda 4(1), Crustacea (exclusive of Ostracoda), Myriapoda, Hexapoda*, R295–R398. Geological Society & University of Kansas Press, Lawrence, Kansas.

Burnett, B.R. 1981: Compound eyes in the cephalocarid crustacean *Hutchinsoniella macracantha*. *Journal of Crustacean Biology 1*, 11–15.

Burnett B.R. & Hessler, R.R. 1973: Thoracic epipodites in the Stomatopoda (Crustacea): a phylogenetic consideration. *Journal of Zoology 169*, 381–392.

Bushnell, J.H. & Byron, E.R. 1979: Morphological variability and distribution of aquatic invertebrates (principally Crustacea) from the Cumberland Peninsula and Frobisher Bay regions, Baffin Island, N.W.T., Canada. *Arctic and Alpine Research 11(2)*, 159–177.

Calman, W.T. 1909: Crustacea. *In* Lankester, E.R. (ed.): *A treatise on zoology, part 7(3)*, 1–346. Adam & Charles Black, London.

Cals, P. & Boutin, C. 1985: Découverte au Cambodge, domain ancien de la Tethys orientale, dun nouveau 'fossile vivant' *Theosbaena cambodjiana* n.g., n.sp. (Crustacea, Thermosbaenacea). *Comptes Rendus hebdomadaires des séances de la Académie des Sciences, Paris ser. III, 300(8)*, 337–340.

Cannon, H.G. 1924: On the development of an estherid crustacean. *Philosophical Transactions of the Royal Society of London, B 212*, 395–430.

Cannon, H.G. 1927a: On the Post-Embryonic Development of the Fairy Shrimp (*Chirocephalus diaphanus*). *Journal of the Linnaean Society London, Zoology 36*, 401–416.

Cannon, H.G. 1927b: On the feeding mechanism of *Nebalia*. *Transactions of the Royal Society of Edinburgh 55(2)*, 355–369.

Cannon, H.G. 1928: On the Feeding Mechanism of the Fairy Shrimp, *Chirocephalus diaphanus* Prévost. *Transactions of the Royal Society of Edinburgh 55(3)*, 807–822.

Cannon, H.G. 1931: Nebaliacea. *Discovery Reports 3*, 199–222.

Cannon, H.G. 1933: On the feeding mechanism of the Branchiopoda. *Philosophical Transactions of the Royal Society of London B 222*, 267–352.

Chen Junyuan, Bergström, J., Lindström, M. & Hou Xianguang 1991: Fossilized Soft-bodied Fauna. The Chengjiang Fauna – Oldest Soft-bodied Fauna on Earth. *National Geographic Research & Exploration 7(1)*, 8–19.

Chen Junyuan, Hou Xianguang & Erdtmann, B.D. 1989: New soft-bodied fossil fauna near the base of the Cambrian system at Chengjiang, Eastern Yunnan, China. *Chinese Academy of Sciences, Developments in Geoscience, Contributions to 28th International Geological Congress, 1989, Washington D.C., U.S.A.*, 265–278, 1 pl.. Science Press, Beijing, China.

Chen Junyuan & Erdtmann, B.D. 1991: Lower Cambrian fossil Lagerstätte from Chengjiang, Yunnan, China: Insights for reconstructing early metazoan life. *Proceedings of an International Symposium held at the University of Camerino 27–31 March 1989*, 57–76.

Chen Peiji 1985: Jurassic *Triops* from South China – with a discussion on the distribution of Notostraca. *Acta Palaeontologica Sinica 24(3)*, 285–296.

Chen Peiji & Zhou Hanzhong 1985: A preliminary study on fossil Kazacharthra from Turpan Basin. *Kexue Tongbao 30(7)*, 950–954.

Cisne, J.L. 1975: Anatomy of *Triarthrus* and the relationships of the Trilobita. *Fossils and Strata 4*, 45–63.

Cisne, J.L. 1981: *Triarthrus eatoni* (Trilobita): anatomy of its exoskeletal, skeletomuscular, and digestive systems. *Palaeontographica Americana 9(53)*, 99–142.

Cisne, J.L. 1982: Origin of the Crustacea. *In* Abele L.G.(ed.): *The Biology of Crustacea, Vol. 1, Systematics, the Fossil Record, and Biogeography*, 65–92. Academic Press Inc., New York.

Clark, A.E. 1932: *Nebalia caboti* n. sp., with Observations on other Nebaliacea. *Transactions of the Royal Society of Canada, 3, 26(5)*, 217–235, 6 pls..

Claus, C. 1873: Zur Kenntnis des Baues und der Entwicklung von *Branchipus stagnalis* und *Apus cancriformis*. *Abhandlungen der königlichen Gesellschaft der Wissenschaften in Göttingen 18*, 93–140, pls. 1–8.

Claus, C. 1886: Untersuchungen über die Organisation und Entwicklung von *Branchipus* und *Artemia* nebst vergleichenden Bemerkungen über andere Phyllopoden. *Arbeiten aus dem Wiener Zoologischen Institut 6(3)*, 1–104; pls. 1–12.

Cockcroft, A.C. 1985: The larval development of *Macropetasma africanum* (Balss, 1913)(Decapoda, Penaeoidea) reared in the laboratory. *Crustaceana 49(1)*, 52–74.

Collins, D. 1987: Palaeontology. Life in the Cambrian seas. *Nature 326(6109)*, p. 127.

Conway Morris, S. 1979: The Burgess Shale (Middle Cambrian) Fauna. *Ann. Rev. Ecol. Syst. 10*, 327–349.

Conway Morris, S. 1989a: Burgess Shale Faunas and the Cambrian Explosion. *Science 246*, 339–346.

Conway Morris, S. 1989b: The persistence of Burgess Shale-type faunas: implications for the evolution of deeper-water faunas. *Transactions of the Royal Society of Edinburgh: Earth Sciences 80*, 271–283.

Conway Morris, S., Whittington, H.B., Briggs, D.E.G., Hughes, C.P. & Bruton, D.L. 1982: Atlas of the Burgess Shale. *In* Conway Morris, S. (ed.): *Atlas of the Burgess Shale.* Palaeontological Association, London.

Copeland, D.E. 1966: A study of the Salt secreting cells in the Brine Shrimp (*Artemia salina*). *Protoplasma 62(4)*, 363–382.

Costlow, J.D. Jr. & Bookhout, C.G. 1957: Larval development of *Balanus eburneus* in the laboratory. *The Biological Bulletin 112*, 313–324.

Costlow, J.D. Jr. & Bookhout, C.G. 1958: Larval development of *Balanus amphitrite* var. *denticulata* Broch reared in the laboratory. *The Biological Bulletin 114*, 284–295.

Coull, B.C. 1988: 3. Ecology of the Marine Meiofauna. *In* Higgins, R.P. & Thiel, H. (eds.): *Introduction to the Study of Meiofauna*, 18–38. Smithsonian Institution Press, Washington, D.C., London.

Criel, G.R.J. & Walgraeve, H.R.M.A. 1989: Molt Staging in *Artemia* Adapted to Drachs System. *Journal of Morphology 199*, 41–52.

Criel, G.R.J. 1991: Ontogeny of *Artemia*. *In* Browne, R.A., Sorgeloos, P. & Trotmann, C.N.A. (eds.): *Artemia Biology*, 155–187. CRC Press, Boca Raton, Ann Arbor, Boston.

Crisp, D.J. 1962: The planktonic stages of the Cirripedia *Balanus balanoides* (L.) and *Balanus balanus* (L.) from north temperate waters. *Crustaceana 3*, 207–221.

Crittenden, R.N. 1981: Morphological characteristics and dimensions of the filter structures of three species of *Daphnia* (Cladocera). *Crustaceana 41(3)*, 233–248.

Croghan, P.C. 1958a. The osmotic and ionic regulation of *Artemia salina* (L.). *Journal of experimental Biology 35*, 219–233.

Croghan, P.C. 1958b. The mechanism of osmotic regulation in *Artemia salina* (L.): The physiology of the branchiae. *Journal of experimental Biology 35*, 234–242.

Dahl, E. 1952: Mystacocarida. *Lunds Universitets Årsskrifter, N.F. Avd. 2, 48(6)*, 3–41.

Dahl, E. 1956: Some Crustacean Relationships. *In* Wingstrand, K.G. (ed.): *Bertil Hanström. Zoological Papers in Honour of his sixty-fifth birthday, Nov. 20th, 1956*, 138–147. Zoological Institute, Lund.

Dahl, E. 1963: Main evolutionary lines among recent Crustacea. *In: Phylogeny and evolution of Crustacea*, 1–15. Museum of Comparative Zoology, Cambridge (U.S.A.) Special Publication.

Dahl, E. 1976: Structural plans as functional models exemplified by the Crustacea Malacostraca. *Zoologica Scripta 5*, 163–166.

Dahl, E. 1983: Phylogenetic systematics and the Crustacea Malacostraca. A problem of prerequisites. *Verhandlungen des naturwissenschaftlichen Vereins in Hamburg (NF) 26*, 355–371.

Dahl, E. 1984: The subclass Phyllocarida (Crustacea) and the status of some early fossils: a neontologists view. *Videnskabelige Meddelelser fra Dansk naturhistorik Forening 145*, 61–76.

Dahl, E. 1985: Crustacea Leptostraca, principles of taxonomy and a revision of European shelf species. *Sarsia 70*, 135–165.

Dahl, E. 1987: Malacostraca maltreated – the case of the Phyllocarida. *Journal of Crustacean Biology 7(4)*, 721–726.

Dahms, H.-U. 1987a: Postembryonic development of *Drescheriella glacialis* Dahms & Dueckmann (Copepoda, Harpacticoida) reared in the Laboratory. *Polar Biology 8*, 81–93.

Dahms, H.-U. 1987b: Die Nauplius-Stadien von *Bryocamptus pygmaeus* (Sars, 1862) (Copepoda, Harpacticoida, Canthocamptidae). *Drosera 1*, 47–58.

Dahms, H.-U. 1989a: Antennule development during copepodite phase of some representatives of Harpacticoida (Copepoda, Crustacea). *Bijdragen tot de Dierkunde 59(3)*, 159–189.

Dahms, H.-U. 1989b: Short Note: First Record of a Lecithotrophic Nauplius in Harpacticoida (Crustacea: Copepoda) Collected from the Weddell Sea (Antarctica). *Polar Biology 10*, 221–224.

Dalley, R. 1984: The larval stages of the oceanic barnacle *Conchoderma auritum* (L.) (Cirripedia, Thoracica). *Crustaceana 46(1)*, 40–54.

Dejdar, E. 1931: Bau und Funktion des sog. 'Haftorgans' bei marinen Cladoceren. (Versuch einer Analyse mit Hilfe vitaler Effektivfärbung. *Zeitschrift für Morphologie und Ökologie der Tiere 21*, 617–628.

Delamare Deboutteville, Cl. 1954: Recherches sur les Crustacés souterrains. III. Le développement postembryonaire des mystacocarides. *Archives de Zoologie expérimentale et générale 91(1)*, 25–34.

Dibbern, S. & Arlt, G. 1989: Post-embryonic development of *Mesochra aestuarii* Gurney, 1921 (Copepoda, Harpacticoida). *Crustaceana 57(3)*, 263–287.

Dumont, H.J. & Velde, I. van de 1976: Some types of head-pores in the Cladocera as seen by scanning electron microscopy and their possible functions. *Biologisch Jaarboek Dodonaea 44*, 135–142.

Dzik, J. & Lendzion, K. 1988: The oldest arthropods of the East European Platform. *Lethaia 21*, 29–38.

[Eberhard, C. 1981: Zur Morphologie und Anatomie der vom Naupliusaugenzentrum des Nervensystems der Crustacea innervierten Strukturen. *Thesis, University of Hamburg*, 1–157.]

Egan, E.A. & Anderson, D.T. 1988: Larval development of the coronuloid barnacles *Austrobalanus imperator* (Darwin), *Tetraclita purpurascens* (Wood) and *Tesseropora rosea* (Krauss) (Cirripedia, Tetraclitidae). *Journal of Natural History 22*, 1379–1405.

Egan, E.A. & Anderson, D.T. 1989: Larval development of the chtamaloid barnacles *Catomerus polymerus* Darwin, *Chamaesipho tasmanica* Foster & Anderson and *Chtamalus antennatus* (Darwin) (Crustacea: Cirripedia. *Zoological Journal of the Linnean Society 95*, 1–28.

Elofsson, R. & Hessler, R.R. 1990: Central nervous system of *Hutchinsoniella macracantha* (Cephalocarida). *Journal of Crustacean Biology 10(3)*, 423–439.

Elofsson, R. & Hessler, R.R. 1991: Sensory morphology in the antennae of the cephalocarid *Hutchinsoniella macracantha* (Cephalocarida). *Journal of Crustacean Biology 11(3)*, 345–355.

Eriksson, S. 1934: Studien über die Fangapparate der Branchiopoden nebst einigen phylogenetischen Bemerkungen. *Zoologiska bidrag från Uppsala 15*, 23–287.

Ewing, R.D., Peterson, G.L. & Conte, F.P. 1974: Larval salt gland of *Artemia salina* nauplii. Localization and characterization of the Sodium+ Potassium-activated Adenosin Triphosphate. *Journal of Comparative Physiology 88*, 217–234.

Feldmann, R.M., Boswell, R.M. & Kammer, T.W. 1986: *Tropidocaris salsiusculus*, a new rhinocaridid (Crustacea: Phyllocarida) from the Upper Devonian Hampshire formation of West Virginia. *Journal of Paleontology 60(2)*, 379–383.

Fielder, D.R., Greenwood, J.G. & Ryall, J.C. 1975: Larval Development of the Tiger Prawn, *Penaeus esculentus* Haswell, 1879 (Decapoda, Penaeidae), Reared in the Laboratory. *Australian Journal of marine and Freshwater Research 26*, 155–175.

Fielder, D.R., Greenwood, J.G. & Jones, M.M. 1979: Larval Development of the Crab *Leptodius exaratus* (Decapoda, Xanthidae), Reared in the Laboratory. *Proceedings of the Royal Society of Queensland 90*, 117–127.

Fortey, R.A. & Clarkson, E.N.K. 1976: The function of the glabellar 'tubercle' in *Nileus* and other trilobites. *Lethaia 9*, 101–106.

Fränsemeier, L. 1939: Zur Frage der Herkunft des metanauplialen Mesoderms und die Segmentbildung bei *Artemia salina*. *Zeitschrift für Wissenschaftliche Zoologie 152*, 439–472.

Frey, D.G. 1959: The taxonomic and phylogenetic significance of the head pores of the Chydoridae (Cladocera). *Internationale Revue der gesamten Hydrobiologie und Hydrogeographie 44*, 27–50.

Fryer, G. 1961: Larval development in the genus *Chonopeltis* (Crustacea: Branchiura). *Proceedings of the Zoological Society of London 137(1)*, 61–69.

Fryer, G. 1963: The functional morphology and feeding mechanism of the Chydorid Cladoceran *Eurycercus lamellatus* (O.F. Müller). *Transactions of the Royal Society Edinburgh 65(14)*, 46–381.

Fryer, G. 1966: *Branchinecta gigas* Lynch, a non-filter-feeding raptatory anostracan, with notes on the feeding habits of certain other anostracans. *Proceedings of the Linnaean Society London 177*, 19–34.

Fryer, G. 1968: Evolution and adaptive radiation in the Chydoridae (Crustacea, Cladocera): a study in comparative functional morphology and ecology. *Philosophical Transactions of the Royal Society of London B, Biological Sciences 254(795)*, 221–385.

Fryer, G. 1974: Evolution and adaptive radiation in the Macrothricidae (Crustacea: Cladocera): a study in comparative functional morphology and physiology. *Philosophical Transactions of the Royal Society of London B, Biological Sciences 269*, 137–274.

Fryer, G. 1983: Functional ontogenetic changes in *Branchinecta ferox* (Milne-Edwards) (Crustacea: Anostraca). *Philosophical Transactions of the Royal Society of London B, Biological Sciences 303(1115)*, 229–343.

Fryer, G. 1985: Structure and habits of living branchiopod crustaceans and their bearing on the interpretation of fossil forms. *Transactions of the Royal Society of Edinburgh: Earth Sciences 76*, 103–113.

Fryer, G. 1987a: Morphology and the classification of the so-called Cladocera. *Hydrobiologia 145*, 19–28.

Fryer, G. 1987b: The feeding mechanism of the Daphniidae (Crustacea: Cladocera): recent suggestions and neglected considerations. *Journal of Plankton Research 9(3)*, 419–432.

Fryer, G. 1987c: A new classification of the branchiopod Crustacea. *Zoological Journal of the Linnean Society 91*, 357–383.

Fryer, G. 1988: Studies on the functional morphology and biology of the Notostraca (Crustacea: Branchiopoda). *Philosophical Transactions of the Royal Society of London B, Biological Sciences 321(1203)*, 27–124.

Gallager, S.M. 1988: Visual observations of the particle manipulation during feeding in larvae of a bivalve mollusc. *Bulletin of Marine Science 43(3)*, 344–365, figs. 1–9.

Garcia-Valdecasas, A. 1984: Morlockiidae new family of Remipedia (Crustacea) from Lanzarote (Canary Islands). *Eos 60*, 329–333.

Gauld, D.T. 1959: Swimming and feeding in crustacean larvae: the nauplius larva. *Proceedings of the Zoological Society of London 132*, 31–50.

Geddes, M.C. 1981: Revision of Australian Species of *Branchinella* (Crustacea: Anostraca). *Australian Journal of marine and Freshwater Research 32*, 253–295.

Gilchrist, B.M. 1960: Growth and Form of the Brine Shrimp *Artemia salina* (L.). *Proceedings of the Zoological Society of London 134*, 221–235.

Go, E.C., Pandey, A.S. & MacRae, T.H. 1990: The effect of inorganic mercury on the emergence and hatching of the brine shrimp, *Artemia franciscana*. *Marine Biology 107*, 93–102.

Goffinet, G. & Meurice, J.Cl. 1983: Ultrastructure de lorgane nucal des embryons d Evadne tergestina (Cladocere marin gymnomere). *Rapp. Comm. int. Mer Médit. 28(9)*, 145–146.

Gooding, R.U. 1963: *Lightiella incisa* sp. nov. (Cephalocarida) from the West Indies. *Crustaceana 5*, 293–314.

Gordon, I. 1957: On *Spelaeogriphus*, a new cavernicolous crustacean from South Africa. *Bulletin of the British Museum (Natural History), Zoology 5(2)*, 29–47, 26 figs..

Greenwood, J.G. & Fielder, D.R. 1979: A Comparative Study of the First and Final Zoeal Stages of Four Species of *Thalamita* (Crustacea: Portunidae). *Micronesica 15(1–2)*, 309–314.

Greenwood, J.G. & Fielder, D.R. 1980: The Zoeal Stages and Megalopa of *Charybdis callianassa* (Herbst)(Decapoda: Portunidae), reared in the Laboratory. *Proceedings of the Royal Society of Queensland 91*, 61–76.

Greenwood, J.G. & Fielder, D.R. 1984: The zoeal stages of *Pilumnopeus serratifrons* (Kinahan, 1856)(Brachyura: Xanthidae) reared under laboratory conditions. *Journal of Natural History 18*, 31–40.

Grygier, M.J. 1983: Ascothoracida and the unity of Maxillopoda. *In* Schram, F.R. (ed.): *Crustacean Issues, 1. Crustacean Phylogeny*, 73–104. Balkema, Rotterdam.

[Grygier, M.J. 1984: Comparative morphology and ontogeny of the Ascothoracida, a step towards a phylogeny of the Maxillopoda. *Ph.D. thesis, University of California at San Diego*, 1–417.]

Grygier, M.J. 1985: Lauridae: Taxonomy and morphology of the ascothoracid crustacean parasites of zoanthids. *Bulletin of Marine Science 36(2)*, 278–303.

Grygier, M.J. 1987: Nauplii, antennular ontogeny, and the position of the Ascothoracida within the Maxillopoda. *Journal of Crustacean Biology 7*, 87–104.

Gurney, R. 1926: The Nauplius larva of *Limnetis gouldi*. *Internationale Revue der gesamten Hydrobiologie und Hydrogeographie 16*, 114–117.

Gurney, R. 1938: The larvae of the decapod Crustacea, Palaemonidae and Alpheidae. *Great Barrier Reef Expedition 1928–29 6(1)*, 1–60, 265 figs. London.

Haas, W. 1989: Suckers and arm hooks in Coleoidea (Cephalopoda, Mollusca) and their bearing for Phylogenetic Systematics. *Abhandlungen des naturwissenschaftlichen Vereins in Hamburg (NF) 23*, 165–185.

Halcrow, K. 1982: Some Ultrastructural features of the nuchal organ of *Daphnia magna* (Crustacea: Branchiopoda). *Canadian Journal of Zoology 60(6)*, 1257–1264.

Halcrow, K. 1985: A note on the significance of the neck organ of *Leptodora kindtii* (Focke) (Crustacea, Cladocera). *Canadian Journal of Zoology 63*, 738–740.

Hamner, W.M. 1988: Biomechanics of filter feeding in the Antarctic Krill *Euphausia superba*: Review of past work and new observations. *Journal of Crustacean Biology 8(2)*, 149–163.

Hanström, Bertil 1934: Über das Vorkommen eines Nackenschildes und eines vierzelligen Sinnesorganes bei den Trilobiten. *Lunds Universitets Arsskrift N.F. 30(7)*, 3–12.

Heath, H. 1924: The external development of certain phyllopods. *Journal of Morphology 38(4)*, 453–483.

Hentschel, E. 1967: Experimentelle Untersuchungen zum Häutungsgeschehen geschlechtsreifer Artemien (*Artemia salina* Leach, Anostraca, Crustacea). *Zoologische Jahrbücher, Abt. Physiologie 73*, 336–342.

Hentschel, E. 1968. Die postembryonalen Entwicklungsstadien von *Artemia salina* Leach bei verschiedenen Temperaturen (Anostraca, Crustacea). *Zoologischer Anzeiger Jena 180*, 372–384.

Hessler, A.Y., Hessler, R.R. & Sanders, H.L. 1970: Reproductive system of *Hutchinsoniella macracantha*. *Science 168*, 1464.

Hessler, R.R. 1964: The Cephalocarida. Comparative skeletomusculature. *Memoirs of the Connecticut Academy of Arts & Sciences 16*, 1–97. New Haven, Connecticut.

Hessler, R.R. 1969: Cephalocarida. *In* Moore, R.C. (ed.): *Treatise on invertebrate paleontology R, Arthropoda 4(1), Crustacea (exclusive of Ostracoda), Myriapoda, Hexapoda*, R120–R128. Geological Society & University of Kansas Press, Lawrence, Kansas.

Hessler, R.R. 1982a: Evolution of arthropod locomotion: a crustacean model. *In* Herreid II, C.F. & Fourtner, C.R. (eds.): *Locomotion and Energetics in Arthropods*, 9–30. Plenum Publishing Corporation.

Hessler, R.R. 1982b: 5. Evolution within the Crustacea. Part 1: General: Remipedia, Branchiopoda, and Malacostraca. *In* Bliss, D.E. (ed.): *The Biology of Crustacea, 1. Systematics, the Fossil Record, and Biogeography*, 150–185. Academic Press, New York & London.

Hessler, R.R. 1984: *Dahlella caldariensis* new genus, new species: a leptostracan (Crustacean, Malacostraca) from deep-sea hydrothermal vents. *Journal of Crustacean Biology 4(4)*, 655–664.

Hessler, R.R. & Elofsson, R. 1991: Excretory system of *Hutchinsoniella macracantha* (Cephalocarida). *Journal of Crustacean Biology 11(3)*, 356–367.

Hessler, R.R. & Newman, W.A. 1975: A trilobitomorph origin for the Crustacea. *Fossils and Strata 4*, 437–459.

Hessler, R.R. & Sanders, H.L. 1965: Bathyal Leptostraca from the continental slope of the north-eastern United States. *Crustaceana 9(1)*, 71–74.

Hessler, R.R. & Sanders, H.L. 1966: *Derocheilocaris typicus* Pennak & Zinn (Mystacocarida) revisited. *Crustaceana 11*, 142–155.

Hessler, R.R. & Sanders, H.L. 1973: Two new species of *Sandersiella* (Cephalocarida), including one from the deep sea. *Crustaceana 24*, 181–196.

Hessler, R.R. & Schram, F.R. 1984: Leptostraca as Living Fossils. *In* Eldredge, N. & Stanley, S.M. (eds.): *Living Fossils*, 187–191. Springer, New York.

Hootman, S.R. & Conte, F.P. 1975: Functional morphology of the Neck Organ in *Artemia salina* Nauplii. *Journal of Morphology 145*, 371–386.

Hootman, S.R., Harris, P.J. & Conte, F.P. 1972: Surface Specialization of the Larval Salt Gland in *Artemia salina* Nauplii. *Journal of Comparative Physiology 79*, 97–104.

Hou, Xianguang 1987a: Two New Arthropods from Lower Cambrian, Chengjiang, Eastern Yunnan. *Acta Palaeontologica Sinica 26(3)*, 236–256.

Hou, Xianguang 1987b: Three New Large Arthropods from Lower Cambrian, Chengjiang, Eastern Yunnan. *Acta Palaeontologica Sinica 26(3)*, 272–285.

Hou, Xianguang 1987c: Early Cambrian Large Bivalved Arthropods from Chengjiang, Eastern Yunnan. *Acta Palaeontologica Sinica 26(3)*, 292–303.

Hsü, F. 1933: Studies on the anatomy and development of a fresh water phyllopod, *Chirocephalus nankinensis* (Shen). *Contributions of the Biological Laboratory of the Science Society of China (Shanghai), Zoological Series 9(4)*, 119–163.

Huvard, A.L. 1990: Ultrastructural study of the naupliar eye of the ostracode *Vargula gramicola* (Crustacea, Ostracoda). *Zoomorphology 110*, 47–51.

Huys, R. 1989: *Dicrotrichura tricincta* gen. et spec. nov.: A new tantulocaridan (Crustacea: Maxillopoda) from the Mediterranean deep waters off Corsica. *Bijdragen tot de Dierkunde 59(4)*, 243–249.

Huys, R. 1991: Tantulocarida (Crustacea: Maxillopoda): a new taxon from the temporary meiobenthos. *P.S.Z.N.I.: Marine Ecology 12(1)*, 1–34.

Ito, T. 1989a: Origin of the limb basis in copepod limbs, with reference to Remipedian and Cephalocarid limbs. *Journal of Crustacean Biology 9(1)*, 85–103.

Ito, T. 1989b: A New Species of *Hansenocaris* (Crustacea: Facetotecta) from Tanabe Bay, Japan. *Publications of the Seto Marine Biological Laboratory 34(1/3)*, 55–72.

Ito, T. & Schram, F.R. 1988: Gonopores and the reproductive system of nectiopodan Remipedia. *Journal of Crustacean Biology 8(2)*, 250–253.

Izawa, K. 1975: On the development of parasitic Copepoda. II. *Colobomatus pupa* Izawa (Cyclopoida: Philichthyidae). *Publications of the Seto Marine Biological Laboratory 22(1/4)*, 147–155.

Izawa, K. 1987: Studies on the Phylogenetic Implications of Ontogenetic Features in the Poecilostome Nauplii (Copepoda: Cyclopoida). *Publications of the Seto Marine Biological Laboratory 32(4/6)*, 151–217.

Jones, M.L. 1961: *Lightiella serendipita* gen.nov., sp.nov., a cephalocarid from San Francisco Bay, California. *Crustaceana 3*: 31–46.

Jurasz, W., Kittel, W. & Presler, P. 1983: Life cycle of *Branchinecta gaini* Daday, 1910, (Branchiopoda, Anostraca) from King George Island, South Shetland Islands. *Polish Polar Research 4(1–4)*, 143–154.

Jux, U. 1985: Phyllocariden (*Aristozoe scaphidimorpha* n. sp.) aus dem hessischen Unterdevon (O. Ems, Rheinisches Schiefergebirge). *Paläontologische Zeitschrift 59(3/4)*, 261–267.

Kaestner, A. 1967: *Lehrbuch der Speziellen Zoologie, Bd. I, Wirbellose, 2. Crustacea*, 2nd ed., 847–1242. Fischer, Stuttgart.

Kensley, B. 1976: The genus *Nebalia* in South and Southwest Africa (Crustacea, Leptostraca). *Cimbebasia ser. A, 4(8)*, 155–162, 5 figs..

Kerfoot, W.C., Kellogg, D.L. (Jr.) & Strickler, J.R. 1980: Visual Observations of Life Zooplankters: Evasion, Escape, and Chemical Defenses. *In* Kerfoot, W.C. (ed.): *Evolution and Ecology of Zooplankton Communities*, 10–27. The University Press of New England.

Knight, M.D. 1973: The nauplius II, metanauplius, and calyptopis stages of *Thysanopoda tricuspidata* Milne-Edwards (Euphausiacea). *Fishery Bulletin 71(1)*, 53–67.

Knight, K.D. 1975: The larval development of Pacific *Euphausia gibboides* (Euphausiacea). *Fishery Bulletin 73(1)*, 145–168.

Knox, G.A. & Fenwick, G.D. 1977: *Chiltoniella elongata* n. gen. et sp. (Crustacea: Cephalocarida) from New Zealand. *Journal of the Royal Society of New Zealand 7(4)*, 425–432.

Koehl, M.A.R. & Strickler, J.R. 1981: Copepod feeding currents: food capture at low Reynolds number. *Limnology and Oceanography 26(6)*, 1062–1073.

Königsmann, E. 1975. Termini der phylogenetischen Systematik. *Biologische Rundschau 13*, 99–115, Jena.

Korinek, V., Krepelova-Machackova, B. & Machacek, J. 1986: Filtering structures of Cladocera and their ecological significance. II. Relation between the concentration of the seston and the size of filtering combs in some species of the genera *Daphnia* and *Ceriodaphnia*. *Vestník ceskoslovenské Spolecnosti zoologicke 50*, 244–258.

Koza, V. & Korinek, V. 1985: Adaptation of the filtration screen in *Daphnia*: Another answer to the selective pressure of the environment. *Archiv für Hydrobiologie, Beih. Ergebn. Limnol. 21*, 193–198.

Lauterbach, K.-E., 1973: Schlüsselereignisse in der Evolution der Stammgruppe der Euarthropoda. *Zoologische Beiträge N.F. 19*, 251–299.

Lauterbach, K.-E., 1974: Über die Herkunft des Carapax der Crustaceen. *Zoologische Beiträge N.F. 20(2)*, 273–327.

Lauterbach, K.-E. 1979: Über die mutmabliche Herkunft der Epipodite der Crustacea. *Zoologischer Anzeiger Jena 202(1/2)*, 33–50.

Lauterbach, K.-E. 1980: Schlüsselereignisse in der Evolution des Grundbauplans der Mandibulata (Arthropoda). *Abhandlungen des naturwissenschaftlichen Vereins in Hamburg (NF) 23*, 105–161.

Lauterbach, K.-E. 1983: Zum Problem der Monophylie der Crustacea. *Verhandlungen des naturwissenschaftlichen Vereins in Hamburg (NF) 26*, 293–320.

Lauterbach, K.-E. 1986: Zum Grundplan der Crustacea. *Verhandlungen des naturwissenschaftlichen Vereins in Hamburg (NF) 28*, 27–63.

Lauterbach, K.-E. 1988: Zur Position angeblicher Crustacea aus dem Ober-Kambrium im Phylogenetischen System der Mandibulata (Arthropoda). *Verhandlungen des naturwissenschaftlichen Vereins in Hamburg (NF) 28*, 27–63.

Lauterbach, K.-E. 1989: Das Pan-Monophylum – Ein Hilfsmittel für die Praxis der Phylogenetischen Systematik. *Zoologischer Anzeiger Jena 223(3/4)*, 139–156.

Laverack, M.S. & Barrientos, Y. 1985: Sensory and other superficial structures in living marine crustaceans. *Transactions of the Royal Society of Edinburgh: Earth Sciences 76*, 123–136.

Lehmann, W.M. 1955: *Vachonia rogeri* n.g.n.sp. ein Branchiopod aus dem unterdevonischen Hunsrückschiefer. *Paläontologische Zeitschrift 29(3/4)*, 126–130.

Lereboullet, M. 1866: Observation sur la génération et le développement de la Limnadie de Hermann (*Limnadia hermanni* Ad. Brogn.). *Annls. Sci. nat. Zoologie ser. 5(5)*, 283–308, pl. 12.

Linder, F. 1941: Contributions to the Morphology and the Taxonomy of the Branchiopoda Anostraca. *Zoologiska Bidrag fran Uppsala 20*, 101–302.

Linder, F. 1943: Über *Nebaliopsis typica* G.O. Sars nebst einigen allgemeinen Bemerkungen über die Leptostraken. *Dana-Report 25*, 1–38, figs. 1–17, pl. 1.

Linder, F. 1945: Affinities within the Branchiopoda with notes on some dubious fossils. *Arkiv för Zoologi 37A(4)*, 1–28.

Linder, F. 1952: Contributions to the morphology and taxonomy of the Branchiopoda Notostraca, with special reference to the North American species. *Proceedings of the United States National Museum 102(3291)*, 1–69.

Lombardi, J. & Ruppert, E.E. 1982: Functional Morphology of Locomotion in *Derocheilocaris typica* (Crustacea, Mystacocarida). *Zoomorphology 100*, 1–10.

Longhurst, A.R. 1955: A Review of the Notostraca. *Bulletin of the British Museum of Natural History, Zoology 3*, 1–57.

Manton, S.M. 1977: *The Arthropoda: Habits, Functional Morphology and Evolution*, 1–527. Clarendon, Oxford.

Manton, S.M. & Anderson, D.T. 1979: Polyphyly and the Evolution of Arthropods. *In* House, M.R. (ed.): *The Origin of Major Invertebrate Groups*, 269–321. Systematic Association Special Vol. 12, Academic Press, London.

Martin, J. 1989: *Eulimnadia belki*, a new clam shrimp from Cozumel, Mexico (Conchostraca: Limnadiidae), with a review of Central and South American species of the genus *Eulimnadia*. *Journal of Crustacean Biology 9(1)*, 104–114.

Martin, J. & Belk, D. 1988: Review of the Clam shrimp family Lynceidae Stebbing, 1902 (Branchiopoda: Conchostraca), in the Americas. *Journal of Crustacean Biology 8(3)*, 451–482.

Martin, J.W., Felgenhauer, B.E. & Abele, L.G. 1986: Redescription of the clam shrimp *Lynceus gracilicornis* (Packard) (Branchiopoda, Conchostraca, Lynceidae) from Florida, with notes on its biology. *Zoologica Scripta 15(3)*, 221–232.

Mattox, N.T. & Velardo, J.T. 1950: Effect of temperature on the development of the eggs of a conchostracan phyllopod, *Caenestheriella gynecia*. *Ecology 31(4)*, 497–506.

Mauchline, J. 1967: Feeding appendages of Euphausiacea (Crustacea). *Journal of Zoology, London 53*, 1–43.

Mauchline, J. 1971: Euphausiacea Larvae. *Conseil International pour lExploration de la Mer, Zooplankton Sheet 135/137*, 2–16.

Mauchline, J. 1977: The integumental sensilla and glands of pelagic Crustacea. *Journal of the Marine Biological Association of the United Kingdom 57*, 973–994.

McKenzie, K.G., Chen Pei-Ji & Majoran, S. 1991: *Almatium gusevi* (Chernyshev 1940): redescription, shield-shapes, and speculations on the reproductive mode (Branchiopoda, Kazacharthra). *Paläontologische Zeitschrift 65*.

McLaughlin, P.A. 1976: A new species of *Lightiella* (Crustacea: Cephalocarida) from the west coast of Florida. *Bulletin of Marine Science 26(4)*, 593–599.

McLaughlin, P.A. 1980: *Comparative morphology of recent Crustacea*, 1–177. Freeman & Co., San Francisco.

McLaughlin, P.A. 1982: Comparative morphology of crustacean appendages. *In* Abele L.G.(ed.): *The Biology of Crustacea, Vol. 1, Systematics, the Fossil Record, and Biogeography*, 197–256. Academic Press Inc., New York.

Meurice, J.Cl. & Goffinet, G. 1982: Structure et fonction de lorgane nucal des Cladoceres marins gymnomeres. *Comptes Rendus hebdomadaires des séances de la Académie des Sciences, Paris ser. III 295*, 693–695.

Meurice, J.Cl. & Goffinet, G. 1983: Ultrastructural evidence of the ion-transporting role of the adult and larval neck organ of the marine gymnomeran Cladocera (Crustacea, Branchiopoda). *Cell and Tissue Research 234*, 351–363.

Mikulic, D.G., Briggs, D.E.G. & Kluessendorf, J. 1985: A Silurian Soft-Bodied Biota. *Science 228*, 715–717.

Mooi, Rich 1990: Paedomorphosis, Aristoteles lantern, and the origin of the sand dollars (Echinodermata: Clypeasteroida). *Paleobiology 16(1)*, 25–48; Lawrence.

Moore, R.C. & McCormick, L. 1969: General features of crustaceans. *In* Moore, R.C. (ed.): *Treatise on invertebrate paleontology R, Arthropoda 4(1), Crustacea (exclusive of Ostracoda), Myriapoda, Hexapoda*, R57–R128. Geological Society & University of Kansas Press, Lawrence, Kansas.: Q92–Q99. Geological Society & University of Kansas Press, Lawrence, Kansas.

Moyse, J. 1987: Larvae of lepadomorph barnacles. *In* Southward, A.J. (ed.): *Crustacean Issues 5, Barnacle Biology*, 329–362. Balkema, Rotterdam.

Müller, K.J. 1964: Ostracoda (Bradorina) mit phosphatischen Gehäusen aus dem Oberkambrium von Schweden. *Neues Jahrbuch für Geologie und Paläontologie, Abhandlungen 121(1)*, 1–55, 5 pls..

Müller, K.J. 1979: Phosphatocopine ostracodes with preserved appendages from the Upper Cambrian of Sweden. *Lethaia 12(1)*, 1–27.

Müller, K.J. 1981a: Arthropods with phosphatized soft parts from the Upper Cambrian 'Orsten' of Sweden. *Short papers for the 2nd International Symposium on the Cambrian System. Open-file Report 81-743*, 147–151.

Müller, K.J. 1981b: Soft parts of Fossils from the Paleozoic Era. *Reports of the DFG, German Research 2*, 13–15.

Müller, K.J. 1982: *Hesslandona unisulcata* sp. nov. (Ostracoda) with phosphatized appendages from Upper Cambrian 'Orsten' of Sweden. *In* Bate, R.H., Robinson, E. & Shepard, L. (eds.): *A research manual of fossil and recent ostracodes*, 276–307. Ellis Horwood, Chichester.

Müller, K.J. 1983: Crustacea with preserved soft parts from the Upper Cambrian of Sweden. *Lethaia 16*, 93–109.

Müller, K.J. 1985: Exceptional preservation in calcareous nodules. *Philosophical Transactions of the Royal Society of London B 311*, 67–73.

Müller, K.J. & Walossek, D. 1985a: A remarkable arthropod fauna from the Upper Cambrian 'Orsten' of Sweden. *Transactions of the Royal Society of Edinburgh: Earth Sciences 76*, 161–172.

Müller, K.J. & Walossek, D. 1985b: Skaracarida, a new order of Crustacea from the Upper Cambrian of Västergötland, Sweden. *Fossils and Strata 17*, 1–65, 17 pls..

Müller, K.J. & Walossek, D. 1986a: *Martinssonia elongata* gen. et sp.n., a crustacean-like euarthropod from the Upper Cambrian 'Orsten' of Sweden. *Zoologica Scripta 15(1)*, 73–92.

Müller, K.J. & Walossek, D. 1986b: Arthropodal larval stages from the Upper Cambrian 'Orsten' of Sweden. *Transactions of the Royal Society of Edinburgh: Earth Sciences 77*, 157–179.

Müller, K.J. & Walossek, D. 1986c: Fossils with preserved soft integument as indicators for a flocculent sedimental zone. *12th International Sedimentological Congress, Abstracts*, 221. Canberra, Australia.

Müller, K.J. & Walossek, D. 1987: Morphology, ontogeny, and life habit of *Agnostus pisiformis* from the Upper Cambrian of Sweden. *Fossils and Strata 19*, 1–124, 33 pls..

Müller, K.J. & Walossek, D. 1988a: Eine parasitische Cheliceraten-Larve aus dem Kambrium. *Fossilien 1*, 40–42.

Müller, K.J. & Walossek, D. 1988b: External morphology and larval development of the Upper Cambrian maxillopod *Bredocaris admirabilis. Fossils and Strata 23*, 1–70, 16 pls..

Müller, K.J. & Walossek, D. 1991: A view through the 'Orsten' window into the world of arthropods 500 Million years ago. *Verhandlungen der Deutschen Zoologischen Gesellschaft 84*, 281–294.

Mura, G. 1986: SEM morphological survey on the egg shell in the italian Anostracans (Crustacea, Branchiopoda). *Hydrobiologia 134*, 273–286.

Mura, G. 1991: SEM morphology of the resting eggs in the species of the genus *Branchinecta* from North America. *Journal of Crustacean Biology 11(3)*, 432–436.

Mura, G. & Thiery, A. 1986: Taxonomic significance of scanning electron microscopy of the euphyllopods resting eggs from Morocco. *Vie Milieu 36(2)*, 125–131.

Mura, G., Accordi, F. & Rampini, M. 1978: Studies on the resting eggs of some fresh water fairy shrimps of the genus *Chirocephalus*: biometry and scanning electron microscopic morphology (Branchiopoda, Anostraca). *Crustaceana 35(2)*, 190–194.

Nauman, C.M. 1987: On the phylogenetic significance of two Miocene zygaenid moths (Insecta, Lepidoptera). – *Paläontologische Zeitschrift 61*: 299–308.

Newman, W.A. 1983: Origin of the Maxillopoda; urmalacostracan ontogeny and progenesis. *In* Schram, F.R. (ed.): *Crustacean Issues, 1. Crustacean Phylogeny*, 105–119. Balkema, Rotterdam.

Newman, W.A. & Knight, M.A. 1984: The carapace and crustacean evolution – a rebuttal. *Journal of Crustacean Biology 4(4)*, 682–687.

Nishida, S. 1989: Distribution, structure and importance of the cephalic dorsal hump, a new sensory organ in calanoid copepods. *Marine Biology 101*, 173–185.

Nival, S. & Ravera, S. 1979: Morphological study of the appendages of the marine cladoceran *Evadne spinifera* Muller by means of the scanning electron microscope. *Journal of Plankton Research 1(3)*, 207–213.

Nourisson, M. 1959: Quelques données relatives au développement post-embryonnaire du crustace phyllopode *Chirocephalus stagnalis* Shaw. *La terre et la vie, Revue d' écologie appliquée 106*, 174–182. Paris.

Novojilov, N. 1957: Un nouvel ordre dArthropodes particuliers: Kazacharthra, du Lias des Monts Ketmen (Kazakhstan SE, U.R.S. S.). *Bulletin de la Societe de Géologie de France 7(6)*, 171–185.

Novojilov, N. 1959: Position systématique des Kazacharthra (Arthropodes) daprès de nouveaux matériaux des monts Ketmen et Sajkan (Kazakhstan SE et NE). *Bulletin de la Societe de Géologie de France 7(1)*, 265–269.

Oehmichen, A. 1921: I. Wissenschaftliche Mitteilungen. 1. Die Entwicklung der äuberen Form des *Branchipus grubei* Dyb.. *Zoologischer Anzeiger Jena 53(11/13)*, 241–253.

Onbé, T. 1984: The developmental stages of *Longipedia americana* (Copepoda: Harpacticoida) reared in the laboratory. *Journal of Crustacean Biology 4(4)*, 615–631.

Ott, J.A. & Novak, R. 1989: Living at an interface: Meiofauna at the oxygen/sulfide boundary of marine sediments. *In* Ryland, J.S. & Tyler, P.A. (eds.): *Reproduction, Genetics and Distribution of Marine Organisms. 23rd European Marine Biology Symposium*, 415–422. Olsen & Olsen, Fredensborg.

Pai, P.G. 1958: On post-embryonic stages of phyllopod crustaceans, *Triops* (*Apus*), *Streptocephalus* and *Estheria*. *Proceedings of the Indian Academy of Science B 48*, 229–250.

Palmer, A.R. 1957: Miocene arthropods from the Mojave Desert, California. *U.S. Geological Survey Professional Paper 294-G*, 237–280, pls. 30–34.

Paulus, H.F. 1979: Eye structure and the monophyly of Arthropoda. *In* Gupta, A.P. (ed.): *Arthropoda Phylogeny*, 299–383. Van Nostrand Reinhold Co., New York.

[Perryman, J.C. 1961: The functional morphology of the skeleto-musculature system of the larval and adult stages of the copepod *Calanus*, together with an account of the changes undergone by this system during larval development. *Ph.D. thesis, University of London*, 1–97.]

Pillai, N.K. 1959: Miscellaneous notes 19. On the Occurrence of *Nebalia longicornis* in Indian waters. *Journal of the Natural History Society of Bombay 56*, 351–353.

Potts, W.T.W. & Durning, C.T. 1980: Physiological evolution in the branchiopods. *Comparative Biochemistry and Physiology 67B*, 475–484.

Preuss, G. 1951: Die Verwandtschaft der Anostraca und Phyllopoda. *Zoologischer Anzeiger Jena 147(3/4)*, 50–63.

Preuss, G. 1957: Die Muskulatur der Gliedmaben von Phyllopoden und Anostraken. *Mitteilungen aus dem Zoologischen Museum Berlin 33(1)*, 221–256.

Pross, A. 1977: Diskussionsbeitrag zur Segmentierung des Cheliceraten-Kopfes. *Zoomorphologie 86*, 183–196.

Quddusi B. Kazmi & Nasima M. Tirmizi 1989: A new species of *Nebalia* from Pakistan (Leptostraca). *Crustaceana 56(3)*, 293–298.

Rafiee, P., Matthews, C.O., Bagshaw, J.C. & MacRea, T.H. 1986: Reversible arrest of *Artemia* development by Cadmium. *Canadian Journal of Zoology 64*, 1633–1641.

Rama Devi, C. & Ranga Reddy, Y. 1989: The complete postembryonic development of *Paradiaptomus greeni* (Gurney, 1906)(Copepoda, Calanoida) reared in the laboratory. *Crustaceana 56(2)*, 141–161.

Rieder, N., Abaffi, P., Hauf, A., Lindel, M. & Weishäupl, H. 1984: Funktionsmorphologische Untersuchungen an den Conchostracen *Leptestheria dahalacensis* und *Limnadia lenticularis* (Crustacea, Phyllopoda, Conchostraca). *Zoologische Beiträge N.F. 28(3)*, 417–444.

Robbins, E.I., Porter, K.G. & Haberyan, K.A. 1985: Pellet microfossils: Possible evidence for metazoan life in Early Proterozoic time. *Proceedings of the National Academy of Sciences 82*, 5809–5813.

Rolfe, W.D.I. 1963: Morphology of the telson in *Ceratiocaris? cornwalliensis* (Crustacea: Phyllocarida) from Czechoslovakia. *Journal of Paleontology 37(2)*, 486–488.

Rolfe, W.D.I. 1969: Phyllocarida. *In* Moore, R.C. (ed.): *Treatise on invertebrate paleontology R, Arthropoda 4(1), Crustacea (exclusive of Ostracoda), Myriapoda, Hexapoda*, R296–R331. Geological Society & University of Kansas Press, Lawrence, Kansas.

Rolfe, W.D.I. 1981: Phyllocarida and the origin of the Malacostraca. *Géobios 14(1)*, 17–27, figs. 1–4, pl. 1.

Salman, S.D. 1987: Larval development of *Caridina babaulti basrensis* Al-Adhub & Hamzah (Decapoda, Caridea, Atyidae) reared in the laboratory. *Crustaceana 52(3)*, 229–244.

Sanders, H.L. 1955: The Cephalocarida, a new subclass of Crustacea from Long Island Sound. *Proceedings of the National Academy of Sciences 41(1)*, 61–66.

Sanders, H.L. 1963a: XIII. Significance of the Cephalocarida. *In* Whittington, H.B. & Rolfe, W.D.I. (eds.): *Phylogeny and evolution of Crustacea*, 163–175. Museum of Comparative Zoology Special Publications, Cambridge, Massachusetts.

Sanders, H.L. 1963b: The Cephalocarida. Functional morphology, larval development, comparative external anatomy. *Memoirs of the Connecticut Academy of Arts & Science 15*, 1–80.

Sanders, H.L. & Hessler, R.R. 1964: The larval development of *Lightiella incisa* Gooding. *Crustaceana 7*, 81–97.

Schlee, Dieter 1981: Grundsätze der phylogenetischen Systematik (Eine praxisorientierte Übersicht). *Paläontologische Zeitschrift 55(1)*, 11–30.

Schminke, H.K. 1976: The ubiquitous telson and the deceptive furca. *Crustaceana 30*, 293–300.

Schminke, H.K. 1981: Adaptation of Bathynellacea (Crustacea, Syncarida) to Life in the Interstitial ('Zoea Theory'). *Internationale Revue der gesamten Hydrobiologie und Hydrogeographie 66(4)*, 575–637.

Schram, F.R. 1982: The fossil record and evolution of Crustacea. *In* Abele, L.G. (ed.): *The Biology of Crustacea, 1. Systematics, the Fossil Record, and Biogeography*, 93–107. Academic Press, New York & London.

Schram, F.R. 1986: *Crustacea*, 1–606. Oxford University Press, New York, Oxford.

Schram, F.R. & Emerson, M.J. 1986: The Great Tesnus Fossil Expedition of 1985. *Environment Southwest 515*, 16–21.

Schram, F.R. & Lewis, C.A. 1989: Functional morphology of feeding in the Nectiopoda. *In* Felgenhauer, B.E., Watling, L. & Thistle, A.B. (eds.): *Crustacean Issues, 6. Functional morphology of feeding and grooming in Crustacea*, 15–26. Balkema, Rotterdam, Brookfield.

Schram, F.R. & Malzahn, E. 1984: The fossil Leptostracan *Rhabdouraea bentzi* (Malzahn, 1958). *Transactions of the San Diego Society of Natural History 20(6)*, 95–98. figs. 1–2.

Schram, F.R., Yager, J. & Emerson, M.J. 1986: Remipedia. Part I. Systematics. *San Diego Society of Natural History, Memoir 15*, 1–60.

Schrehardt, A. 1986a: Der Salinenkrebs *Artemia*. 1. Organisation des erwachsenen Tieres. *Mikrokosmos 75(8)*, 230–235.

Schrehardt, A. 1986b: Der Salinenkrebs *Artemia*. 2. Die postembryonale Entwicklung. *Mikrokosmos 75(11)*, 334–340.

Schrehardt, A. 1987a: A scanning electron-microscope study of the post-embryonic development of *Artemia*. *In* Sorgeloos, P. Bengtson, D.A., Decleir, W. & Jasper, E. (eds.): *Artemia Research and its Applications. 1. Morphology, Genetics, Strain characterization, Toxicology*, 5–32. Universa Press, Wetteren, Belgium.

Schrehardt, A. 1987b: Ultrastructural investigations of the filter-feeding apparatus and the alimentary canal of *Artemia*. *In* Sorgeloos, P. Bengtson, D.A., Decleir, W. & Jasper, E. (eds.): *Artemia Research and its Applications. 1. Morphology, Genetics, Strain characterization, Toxicology*, 33–52. Universa Press, Wetteren, Belgium.

[Schulz, K. 1976: Das Chitinskelett der Podocopida (Ostracoda, Crustacea) und die Frage der Metamerie dieser Gruppe. *Doctoral thesis, University of Hamburg*, 1–167.]

Scourfield, D.J. 1926: On a new type of crustacean from the Old Red Sandstone (Rhynie Chert Bed, Aberdeenshire) – *Lepidocaris rhyniensis* gen. et sp. nov.. *Philosophical Transactions of the Royal Society of London B 214*, 153–187.

Scourfield, D.J. 1940: Two new and nearly complete specimens of young stages of the Devonian fossil crustacean *Lepidocaris rhyniensis*. *Proceedings of the Linnaean Society London 152*, 290–298.

Seiple, E. 1983: Miocene insects and arthropods in California. *California Geology 11*, 246–248.

Sergeev, V. 1990. A new species of *Daphniopsis* (Crustacea: Anomopoda: Daphniidae) from Australian salt lakes. *Hydrobiologia 190*, 1–7.

Shiino, S.M. 1965: *Sandersiella acuminata* gen. et sp. nov., a cephalocarid from Japanese waters. *Crustaceana 9*, 181–191.

Siewing, R. 1960: Neuere Ergebnisse der Verwandtschaftsforschung bei den Crustaceen. *Wissenschaftliche Zeitschrift der Universität Rostock, Mathematisch-Naturwissenschaftliche Reihe 9(3)*, 343–358.

Siewing, R. 1963. Zum Problem der Arthropodenkopfsegmentierung. *Zoologischer Anzeiger Jena 170*, 429–468.

Siewing, R. (ed.) 1985: *Lehrbuch der Zoologie, II, Systematik, 3rd edn.*, 1–1107. Fischer, Stuttgart.

Smirnov, N.N. 1970: Cladocera (Crustacea) from Permian deposits in Eastern Kazakhstan (title translated from Russian). *Palaeontological Journal 3*, 95–100.

Smith, A.B. 1984: Classification of the Echinodermata. *Palaeontology 27(3)*, 431–459.

Snodgrass, R.E. 1956. Crustacean Metamorphoses. *Smithsonian Miscellaneous Collections 131(10)*, 1–78.

Storch, O. 1924: Der Phyllopoden-Fangapparat I. *Internationale Revue der gesamten Hydrobiologie und Hydrogeographie 12*, 369–391.

Storch, O. 1925: Der Phyllopoden-Fangapparat II. *Internationale Revue der gesamten Hydrobiologie und Hydrogeographie 13*, 78–93.

Strength, N.E. & Sissom, S.L. 1975. A morphological study of the postembryonic stages of *Eulimnadia texana* Packard (Conchostraca, Crustacea). *The Texas Journal of Science 26(1, 2)*, 137–154.

Strickler, J.R. 1975: Feeding currents in calanoid copepods: two new hypotheses. *In* Laverack, M.S. (ed.): *Physiological adaptations of marine animals, Symp. Soc. Exp. Biol. 39*, 459–485.

Stürmer, W. & Bergström, J. 1976: The arthropods *Mimetaster* and *Vachonisia* from the Devonian Hunsrück Shale. *Paläontologische Zeitschrift 52*, 57–81.

Swanson, K.M. 1989a: Ostracod Phylogeny and Evolution – a manawan perspective. *Courier des Forschungsinstitutes Senckenberg 113*, 11–20.

Swanson, K.M. 1989b: *Manawa staceyi* n.sp. (Punciidae, Ostracoda): soft anatomy and ontogeny. *Courier des Forschungsinstitutes Senckenberg 113*, 235–249.

Tasch, P. 1963: XI. Evolution of the Branchiopoda. Phylogeny and Evolution of Crustacea. Museum of Comparative Zoology, Special Publication, 145–157.

Tasch, P. 1969: *Branchiopoda*. *In* Moore, R.C. (ed.): *Treatise on invertebrate paleontology R, Arthropoda 4(1) Crustacea (exclusive of Ostracoda), Myriapoda, Hexapoda*, R128–R191. Geological Society & University of Kansas Press, Lawrence, Kansas.

Thiele, J. 1904: Die Leptostraken. *Deutsche Tiefsee-Expedition auf dem Dampfer 'Valdivia', 1898, 99, 8(1)*, 3–26.

Thiery, A. & Champeau, A. 1988: *Linderiella massaliensis*, new species (Anostraca: Linderiellidae), a fairy shrimp from southeastern France, its ecology and distribution. *Journal of Crustacean Biology 8(1)*, 70–78.

Tyler, S. 1988: The role of function in determination of homology and convergence – examples from invertebrate adhesive organs. *In* Ax, P., Ehlers, U. & Sopott-Ehlers, B. (eds.): *Free-living and Symbiotic Plathelminthes*, 331–347. Fischer, Stuttgart, New York.

Ulrich, E.O. & Bassler, R.S. 1931: Cambrian bivalved Crustacea of the order Conchostraca. *Proceedings of the U.S. Natural Museum 2847, 78(4)*, 1–130.

Vader, W. 1973. *Nebalia typhlops* in Western Norway (Crustacea, Leptostraca). *Sarsia 53*, 25–28.

Valousek, B. 1950: *Chirocephalopsis grubii* Dybowski nascens et crescens observatus est. *Acta Academiae Scientarium Naturalium Moravo-Silesiacae 22(5), F230*, 159–182.

Vincx, M. & Heip, C. 1979: Larval development and biology of *Canuella perplexa* T. and A. Scott, 1893 (Copepoda, Harpacticoida). *Cahiers de Biologie Marine 20*, 281–299.

Wägele, J.W. 1983: *Nebalia marerubri*, sp. nov. aus dem Roten Meer (Crustacea: Phyllocarida: Leptostraca). *Journal of Natural History 17*, 127–138.

Wakabara, Y. 1965: On *Nebalia* sp. from Brazil (Leptostraca). *Crustaceana 9(3)*, 245–248.

Wakabara, Y. 1976. *Paranebalia fortunata* n.sp. from New Zealand (Crustacea, Leptostraca, Nebaliacea). *Journal of the Royal Society of New Zealand 6(3)*, 297–300.

Wakabara, Y. & Mizoguchi, S.M. 1976. Notes and News. Record of *Sandersiella bathyalis* Hessler & Sanders, 1973 (Cephalocarida) from Brazil. *Crustaceana 30(2)*, 220–221.

Walker, G. & Lee, V.E. 1976: Surface structures and sense organs of the cypris larva of *Balanus balanoides* as seen by scanning and transmission electron microscopy. *Journal of Zoology, London 178*, 161–172.

Walley, L.J. 1969: Studies on the larval structure and metamorphosis of *Balanus balanoides* (L.). *Philosophical Transactions of the Royal Society of London B 256*, 237–280.

Walossek, D. & Müller, K.J. 1989: A second type A-nauplius from the Upper Cambrian 'Orsten' of Sweden. *Lethaia 22*, 301–306.

Walossek, D. & Müller, K.J. 1990: Upper Cambrian stem-lineage crustaceans and their bearing upon the monophyly of Crustacea and the position of *Agnostus*. *Lethaia 23(4)*, 409–427.

Walossek, D. & Müller, K.J. 1991: *Henningsmoenicaris* n.g. for *Henningsmoenia* Walossek & Müller, 1990 – correction of name. *Lethaia 24(2)*, 138.

Walossek, D. & Müller, K.J. 1992: The 'Alum Shale Window' – contributions of the 'Orsten' arthropods to the phylogeny of Crustacea. *Acta Zoologica 73*, 305–312.

Walossek, D. & Szaniawski, H. 1991: *Cambrocaris baltica* n.gen. n. sp., a possible stem-lineage crustacean from the Upper Cambrian of Sweden. *Lethaia 24(4)*, 363–378.

Warren, H.S. 1938: The segmental excretory glands of *Artemia salina* Linn. var. *principalis* Simon. (The Brine Shrimp). *Journal of Morphology 62*, 263–297.

Watling, L. 1981: An alternative phylogeny of peracarid crustaceans. *Journal of Crustacean Biology 1(2)*, 201–210.

Watling, L. 1983: Peracaridan disunity and its bearing on eumalacostracan phylogeny with a redefinition of eumalacostracan superorders. *In* Schram, F.R. (ed.): *Crustacean Issues, 1. Crustacean Phylogeny*, 213–228. Balkema, Rotterdam.

Watling, L. 1988: Small-scale features of marine sediments and their importance to the study of deposit-feeding. *Marine Ecology – Progress Series 47*, 135–144.

Watling, L. 1989: A classification system for crustacean setae based on the homology concept. *In* Felgenhauer, B.E., Watling, L. & Thistle, A.B. (eds.): *Crustacean Issues, 6. Functional morphology of feeding and grooming in Crustacea*, 15–26. Balkema, Rotterdam, Brookfield.

Weisz, P.B. 1946: The space–time pattern of segment formation in *Artemia salina*. *The Biological Bulletin 91(2)*, 119–140.

Weisz, P.B. 1947: The histological pattern of metameric development in *Artemia salina*. *Journal of Morphology 81*, 45–95.

Westheide, Wilfried 1987: Progenesis as a principle in meiofauna evolution. *Journal of Natural History 21*, 843–854.

Whittington, H.B. 1974: *Yohoia* Walcott and *Plenocaris* n.gen., arthropods from the Burgess Shale, middle Cambrian, British Columbia. *Bulletin of the geological Survey of Canada 231*, 1–21.

Whittington, H.B. 1975: Trilobites with appendages from the Middle Cambrian, Burgess Shale, British Columbia. *Fossils and Strata 4*, 97–136.

Whittington, H.B. 1977: The Middle Cambrian trilobite *Naraoia*, Burgess Shale, British Columbia. *Philosophical Transactions of the Royal Society of London B, Biological Sciences 280(974)*, 409–443.

Whittington, H.B. 1979: Early Arthropods, their Appendages and Relationships. *In* House, M.R. (ed.): The Origin of Major Invertebrate Groups. *Systematic Association Special Vol. 12.*, 253–268. Academic Press, London and New York.

Whittington, H.B. 1980: Exoskeleton, moult stage, appendage morphology, and habits of the Middle Cambrian trilobite *Olenoides serratus*. *Palaeontology 23(1)*, 171–204.

Whittington, H.B. 1985: *The Burgess Shale*. Yale University Press, New Haven. 151 pp.

Whittington, H.B. & Almond, J.E 1987: Appendages and habits of the Upper Ordovician trilobite *Triarthrus eatoni*. *Philosophical Transactions of the Royal Society of London B, Biological Sciences 317*, 1–46.

Williamson, D.E. 1982: *Larval morphology and diversity. In* Abele, L.G. (ed.): *The Biology of Crustacea, 2. Embryology, morphology and genetics*, 43–110. Academic Press, New York & London.

Willmann, R. 1981: Phylogenie und Verbreitungsgeschichte der Eomemeropidae (Insecta; Mecoptera). Ein Beispiel für die Anwendung der phylogenetischen Systematik in der Paläontologie. *Paläontologische Zeitschrift 55(1)*, 31–49.

Willmann, R. 1983: Widersprüchliche Rekonstruktionen der Phylogenese am Beispiel der Ordnung Mecoptera (Schnabelfliegen; Insecta; Holometabola). *Paläontologische Zeitschrift 57(3/4)*, 285–308.

Willmann, R. 1987 : The phylogenetic system of the Mecoptera. *Systematic Entomology 12*, 519–524.

Willmann, R. 1988: Makroevolution aus paläontologischer Sicht. *Sitzungsberichte der Gesellschaft Naturforschender Freunde zu Berlin (N.F.) 28*, 137–162.

Willmann, R. 1989a: Evolution und Phylogenetisches System der Mecoptera. *Abhandlungen der senckenbergianischen naturforschenden Gesellschaft 544*, 1–122.

Willmann, R. 1989b: Palaeontology and the systematization of natural taxa. *Abhandlungen des naturwissenschaftlichen Vereins in Hamburg (NF) 28*, 267–291.

Yager, J. 1981: Remipedia, a new Class of Crustacea from a marine Cave in the Bahamas. *Journal of Crustacean Biology 1*, 328–333.

Yager, J. 1987a: *Cryptocorynetes haptodiscus*, new genus , new species, and *Speleonectes benjamini*, new species, of remipede crustaceans from anchialine caves in the Bahamas, with remarks on distribution and ecology. *Proceedings of the Biological Society of Washington 100(2)*, 302–320.

Yager, J. 1987b: *Speleonectes tulumensis* n. sp. (Crustacea: Remipedia) from two anchialine cenotes of the Yucatan Peninsula, Mexico. *Stygologia 3(2)*, 160–166.

Yager, J. 1989a: The male reproductive system, sperm, and spermatophores of the primitive, hermaphroditic, remipede crustacean *Speleo-*

nectes benjamini. Invertebrate Reproduction and Development 15, 75–81.

Yager, J. 1989b: *Pleomothra apletochelis* and *Godzillognomus frondosus*, two new genera and species of remipede crustaceans (Godzilliidae) from anchialine caves of the Bahamas. *Bulletin of Marine Science 44(3)*, 1195–1206.

Yager, J. & Schram, F.R. 1986: *Lasionectes entrichoma*, new genus, new species, (Crustacea: Remipedia) from anchialine caves in the Turks and Caicos, British West Indies. *Proceedings of the Biological Society of Washington 99(1)*, 65–70.

Addendum

Note added in proof. – During a stay at the Palaeontological Institution of the Russian Academy of Sciences in Moscow in September, 1992, I learnt that the displays of the museum contain kazacharthrans and notostracans from the Lower Mesozoic as well as laevicaudate conchostracans from the Upper Jurassic. Moreover, Upper Jurassic euanostracans were kindly shown to me at the Palaeoentomological Department, some of which even had egg sacs preserved and sexually modified antennae. I have been unable to obtain references as yet, but at least some of these remarkable finds have been described already in the 1960's in Russian journals.

After submission of the manuscript, one additional published record of type-A larvae has come to my knowledge, and another one has just appeared:

Roy, K. & Fåhræus, L.1989: Tremadocian (Early Ordovician) nauplius-like larvae from the Middle Arm Point Formation, Bay of Islands, western Newfoundland. *Canadian Journal of Earth Sciences 26*, 1802–1806.

Walossek, D., Hinz-Schallreuter, I., Shergold, J.H. & Müller, K.J. 1993: Three-dimensional preservation of arthropod integument from the Middle Cambrian of Australia. *Lethaia 26*, 7–15.

Plates

Plate 1

1–4: Stage L1A, UB 3; 5–7: Stage L2A

1 Lateral view (anterior to the right) of almost complete but slightly crumpled specimen with appendages broken off distally; cephalic shield (cs) feebly demarcated (an = protruded anus; a1 = 1st antenna; a2 = 2nd antenna; dcsp = dorsocaudal spine; fsp = spine of incipient furca; i tr = incipient trunk; la = labrum; md = mandible; no = 'neck organ'). Scale bar 30 μm.

2 Dorsal view; left appendages more complete than right ones but sunken into glue (a1 = 1st antenna; cs = cephalic shield; dcsp = dorsocaudal spine; ex md = mandibular exopod). Scale bar 30 μm.

3 Dorsal view of shield; centre with smooth, watch glass-shaped area of neck organ; two sets of pores are associated with this organ: one in about the middle of the plate and another at the posterior margin (arrows). Scale bar 10 μm.

4 Ventral view; elevated sternum (stn) consisting of the sternal bars of the two postantennulary segments; sternite of 2nd antenna larger than that of mandible (arrow); long enditic spines (esp) of the 2nd antenna reaching to the mouth below the labrum (la); mandibular basipod (bas md) larger than coxa; lower left: exopod of 2nd antenna (cox; fsp = furcal spine). Scale bar 30 μm.

5 UB 4. Ventral view; appendages only partly preserved (a1, a2, md), labrum flexed posteriorly onto the sternum, probably due to shrinkage; mandibular coxa (cox md) with one terminal spine and two setae; on the larval trunk a pair of setae (arrows) indicates the maxillulary segment; furcal spine (fsp) accompanied by a small spinule laterally (spl). Scale bar 30 μm.

6 UB 5. Dorsal view of much depressed specimen fixed to the stub on its right appendages and the dorsocaudal spine (dcsp); shield collapsed and wrinkled save for the neck organ; also the blisters of presumed compound eye (ce?) in front of the shield are collapsed (a1 = 1st antenna; en a2 = endopod of 2nd antenna). Scale bar 30 μm.

7 UB 6. Ventrolateral view; eye not preserved; appendages (a1; a2; md) partly preserved, enabling the inside of the postoral chamber and bulging sternum to be seen (stn; fsp = furcal spine; s mx1 = seta of incipient 1st maxilla). Scale bar 30 μm.

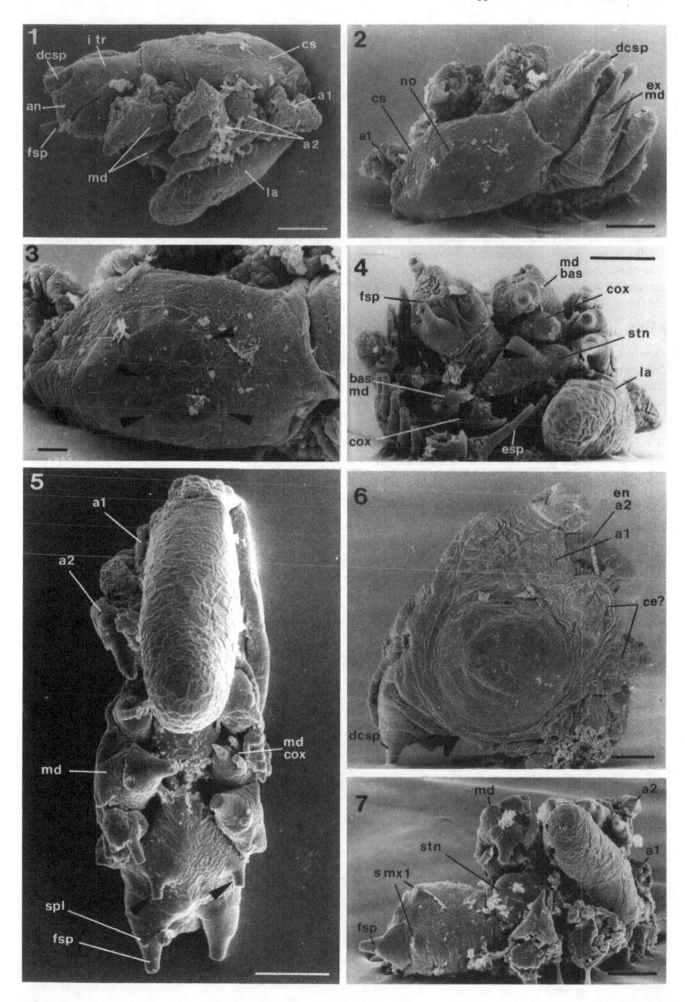

Plate 2

Stage L2A continued

1, 2, 6–8: UB 7

1   Lateral view of slightly compressed specimen; appendages laterally stretched; dorsocaudal spine (dcsp) arising from broad basis (a1 = 1st antenna). Scale bar 30 µm.

2   Ventral view; 1st antennae (a1) posteriorly flexed, some of the setae partly preserved; 2nd antenna (a2) and mandible (md) of right side almost complete, their enditic spines pointing into postlabral feeding chamber (bas = basipod; fsp = furcal spine; s mx1 = seta indicating incipient segment of 1st maxilla). Scale bar 30 µm.

3, 4: UB 8

3   Ventrolateral view, seen slightly from anterior; anterior body portion distorted, but trunk well-inflated; mandibles (md) laterally stretched (a1, 2 = antennae; ce = distorted blisters of compound eye; la = labrum). Scale bar 30 µm.

4   Lateral view; shield poorly demarcated posteriorly (arrow); dorsocaudal spine broken off (dcsp). Scale bar 30 µm.

5   UB 9 (same specimen as in 9, 10; Pl. 30:1). Almost complete 2nd antenna (a2) and mandible (md), lack-

ing most of the enditic spines and setae; arrow points to small terminal endopodal segment (the 4th one!) which had carried a seta originally (bas = basipod; cox = coxa; en = endopod; ex = exopod; la = labrum). Scale bar 30 µm.

6   Same specimen as in 1, 2. View into larval feeding chamber, bordered by labrum (la) anteriorly, trunk (i tr) posteriorly, sternum (stn) dorsally (= proximally), and appendages (a1, a2, md) laterally (gns = two gnathobasic setae of mandibular coxa). Scale bar 10 µm.

7   Part of shield with neck organ; frame shows area of Fig. 8. Scale bar 15 µm.

8   Detail of neck organ with two of the four pores (arrows), one in about the middle and one at the posterior margin. Scale bar 5 µm.

9   Same specimen as in 5. Compound eye collapsed and wrinkled; frame shows area of 10, enclosing a tubercle or pore of unknown nature in the anterior portion of the midventral lobe (mvl) between the eye lobes (lo ce; a1 = insertion area of 1st antenna; a2 = 2nd antenna; la = labrum). Scale bar 10 µm.

10 Detail of midventral lobe with tubercle or pore. Scale bar 1 µm.

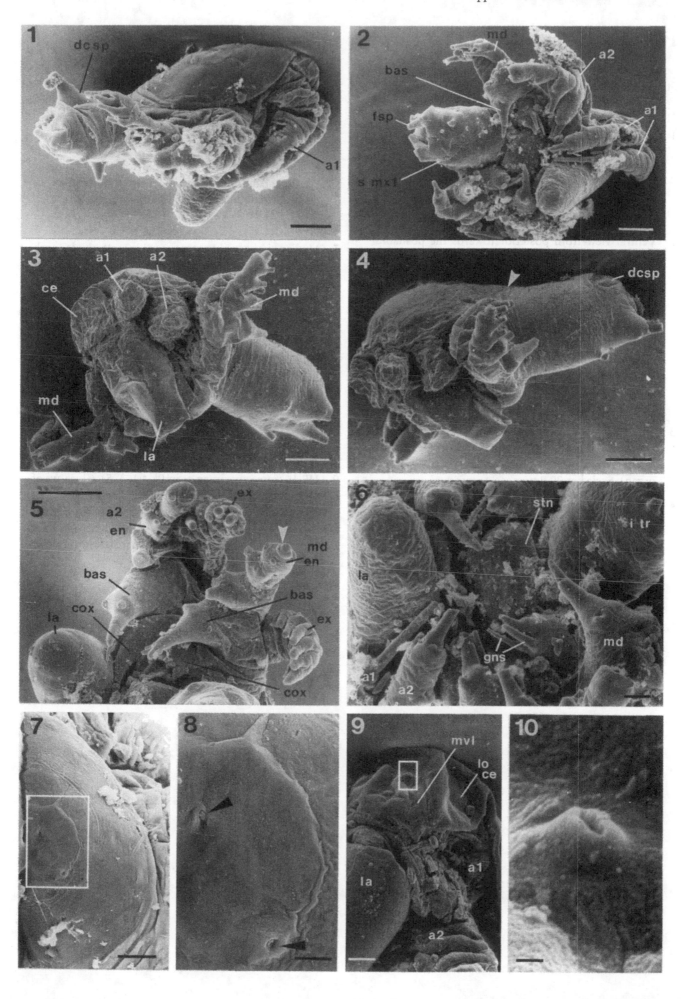

Plate 3

Stage L3A

1, 2: UB 10

1   Ventrolateral view; eye area (ce) collapsed; appendages (a1, a2, md) partly preserved and laterally stretched; 1st maxilla developed as bifid limb bud (mx1 rud); furcal rami (fr) slightly flattened and with three terminal spines (dcsp = dorsocaudal spine; la = labrum). Scale bar 30 μm.

2   Ventral view from posterior towards the postlabral feeding chamber; mandibular coxa forming a distinct segment, its gnathobase (gn) being larger than in preceding stage; inner edge tipped by a few spinules; two gnathobasic setae (gns); basipodal masticatory spine (msp bas) surrounded by setae originally (broken off); sternum (stn) bulging; membrane originally around the anus (an), which is located between dorsocaudal spine (dcsp) and the bases of the furcal rami, is not preserved (ao = atrium oris; a2 = 2nd antennae). Scale bar 30 μm.

3   UB 11. Dorsal view; compound eye (ce) collapsed and attached to the shield; posterior margin of shield indistinct (arrow) (dcsp = dorsocaudal spine; no = neck organ). Scale bar 30 μm.

4   UB 12. Lateral view; body little deformed but eye and appendages only poorly preserved; furcal rami (fr) with setae and dorsocaudal spine almost complete. Scale bar 30 μm.

5, 6: UB 13

5   Dorsal view of cephalic shield with neck organ (no) bordered by faint folds (arrows point to pores); specimen lacking the trunk. Scale bar 30 μm.

6   Close-up of eye region with lateral lobes of compound eye (lo ce) and midventral lobe (mvl); frame encloses pit on latter lobe. Scale bar 15 μm.

7, 8: UB 14

7   Median view of right antennae (a1, a2) attached to each other and mandible (md) behind the slightly crushed labrum (la); (bas = basipod; cox = coxa; en = endopod; mx1 rud = rudimentary 1st maxilla; stn = sternum). Scale bar 30 μm.

8   Close-up view of almost complete but slightly collapsed 1st antenna (a1); shaft subdivided into about 15 incomplete ringlets (posterior surface pliable); distal part composed of three more or less tubular segments and a tiny hump (arrow) carrying the apical seta; setae (s) of the shaft reaching into the space between labrum (la) and endopod of 2nd antenna (en a2; md = mandible). Scale bar 30 μm.

9   UB 15 (same specimen as in Pl. 4:1, 3–5). Close-up of mandibular gnathobase with marginal spinules (spl); gnathobasic seta (gns) with a spiral row of fine setules (arrow); prickles on the surface of the gnathobase indicate setules originally located there (on lower right: basipod). Scale bar 10 μm.

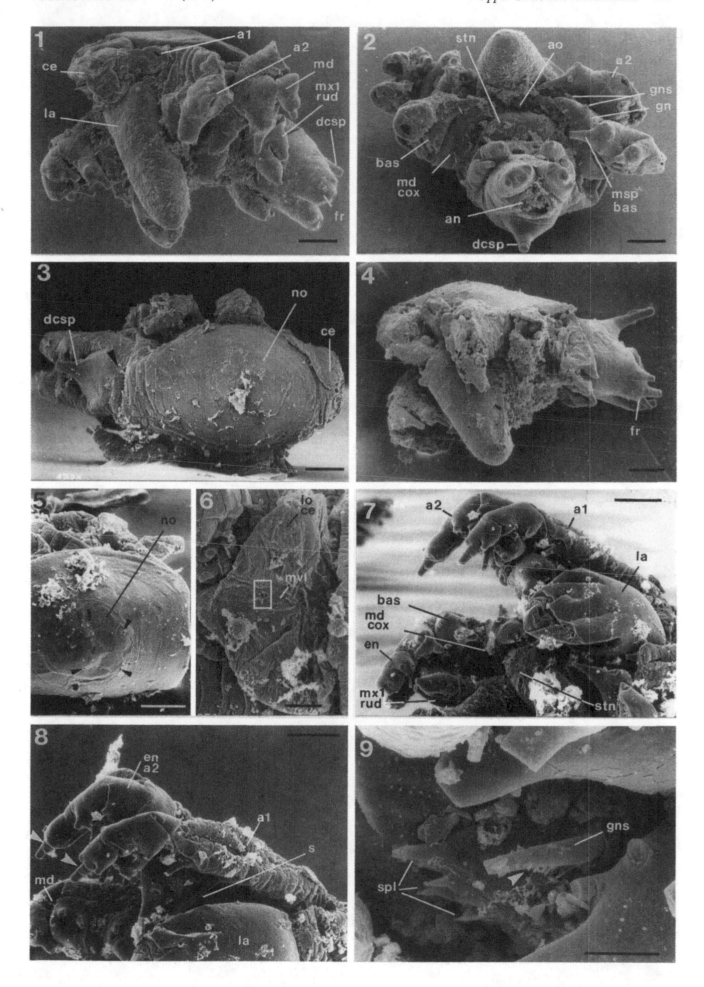

Plate 4

Stage L3A continued

1   Same specimen as in 3–5; Pl. 3:9. Ventrolateral view of crumpled specimen, which still shows details in beautiful preservation; arrow points to bifid tip of basipodal masticatory spine of mandible (an = anus; a1, 2 = antennae; dcsp = dorsocaudal spine; ex = exopod; fr = furcal ramus; la = labrum; md = mandible; mx1 rud = rudimentary 1st maxilla). Scale bar 30 μm.

2   UB 16. Ventral view, seen somewhat from posterior, of collapsed specimen; appendages posteriorly flexed; striae on posterior edge of elevated sternum indicate the future development of the paragnaths in that area (arrows; see also Pl. 4:7; ce = collapsed compound eye; la = labrum). Scale bar 30 μm.

3–5: Same specimen as in 1

3   Exopod of 2nd antenna with posterodistally oriented setal sockets that preform the orientation of the setae; proximal and distal setae thinner than those in the middle of the ramus; antennal endopod broken off; arrow points to the small terminal 4th segment of the mandibular endopod. Scale bar 10 μm.

4   Anterior view of antennal exopod, showing that there are more segments or ringlets on the ramus than setae (see also Pls. 9:2, 10:8, 11:1 and 19:3, 21:4 for series B; den = small denticles, e.g., on mandibular endopod). Scale bar 10 μm.

5   On the anterior appendages several of the setae are preserved almost to their entire length, giving an impression of the original setation (a1 = 1st antenna). Scale bar 15 μm.

6   UB 17. Ventral view of right furcal ramus with three marginal spines, the median one being the largest; a tubercle or pit is located close to the median spine (arrow); short rows of small denticles (den) are also positioned on the ventral side of the ramus (see also Pl. 7:6). Scale bar 5 μm.

7   UB 18. Ventral view of postlabral area and trunk; two sets of striae indicate the future development of the paragnaths (arrows); between the rudimentary 1st maxillae (mx1 rud) a slight elevation indicates the formation of a future sternal bar on this segment (ao = atrium oris; gn md = mandibular gnathobase; stl = remnants of sternal setules). Scale bar 15 μm.

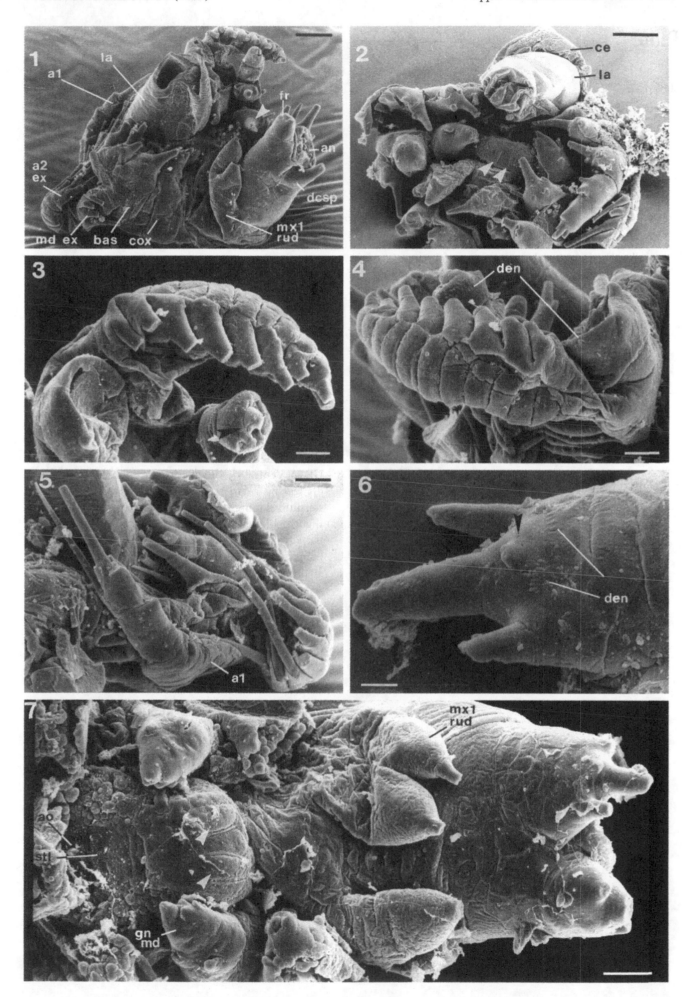

Plate 5

Stage L4A

1   UB 19 (same specimen as in 5). Dorsolateral view of fragmentary specimen lacking anterior head region and appendages anterior to 1st maxilla (mx1); margins of shield (cs) distinct; furrow running down post the 1st maxilla, showing that maxillary segment (arrow) is not incorporated within the larval head; 1st maxilla with paddle-shaped exopod (ex); 2nd maxilla rudimentary (mx2 rud) similar to 1st maxilla of preceding larval stage; dorsocaudal spine (dcsp) seemingly smaller than in L3A and more anterior of the anus (an; fr = furcal ramus; no = neck organ). Scale bar 30 μm.

2, 3: UB 20 (same specimen as in 6; Pl. 6:1, 2)

2   Lateral view of somewhat crumpled specimen; right antennae and maxillae poorly preserved, but large mandibular coxa (md cox) and exopod (ex) still present; posteriorly curvature of the exopod, a common mode of preservation in *orsten* crustaceans, probably resulting from shrinkage of the intersegmental membranes after death (see also Pl. 6:7 and Müller & Walossek 1988b for *Bredocaris admirabilis*; la = labrum). Scale bar 30 μm.

3   Ventral view; left appendages better preserved than right set; note the distinct mandibular coxal body (cox md) with its gnathobase (gn) angled against it; basipod (bas) prominent, with rounded endite carrying a robust masticatory spine and some setae around it; 1st maxilla (mx1) enlarged and now medially subdivided but segmentation unclear; enditic lobes and proximal endopodal segments with paired setae (a2 = 2nd antenna; lo ce = lobes of collapsed compound eye; mvl = midventral lobe). Scale bar 30 μm.

4   UB 21. Anteroventral view of distorted specimen; compound eye (ce) coarsely preserved but inflated showing the original size of this organ. Scale bar 30 μm.

5   Same specimen as in 1. Close-up of pores (arrows) on neck organ; boundaries of the organ almost effaced. Scale bar 10 μm.

6   Same specimen as in 2. View of mandible from posterior; coxal body (cox md) well-sclerotized and with angled, blade-like gnathobase (gn); triturating inner edge with one larger spinule or tooth posteriorly (pt) and some smaller ones anterior to it; basipod and coxa distinctly articulating with one another; basipod sunken into 'palp foramen' due to shrinkage of its joint membrane (plpf); exopod (ex) sharply posteriorly flexed (la = labrum; stn = sternum). Scale bar 30 μm.

7   UB 22. Posterior side of labrum with few tubercles or pits of probably sensory function (see also Pls. 9:7; 23:7; 34:4). Scale bar 10 μm.

8   UB 23. Ventral view of distorted specimen; trunk torn off leaving a large hole behind the sternum (stn); sternum sloping toward the mouth; arrow points to slightly bulging lateral corners of sternum indicative of the advanced development of the paragnaths (md = mandible; mx1 = fragment of 1st maxilla). Scale bar 30 μm.

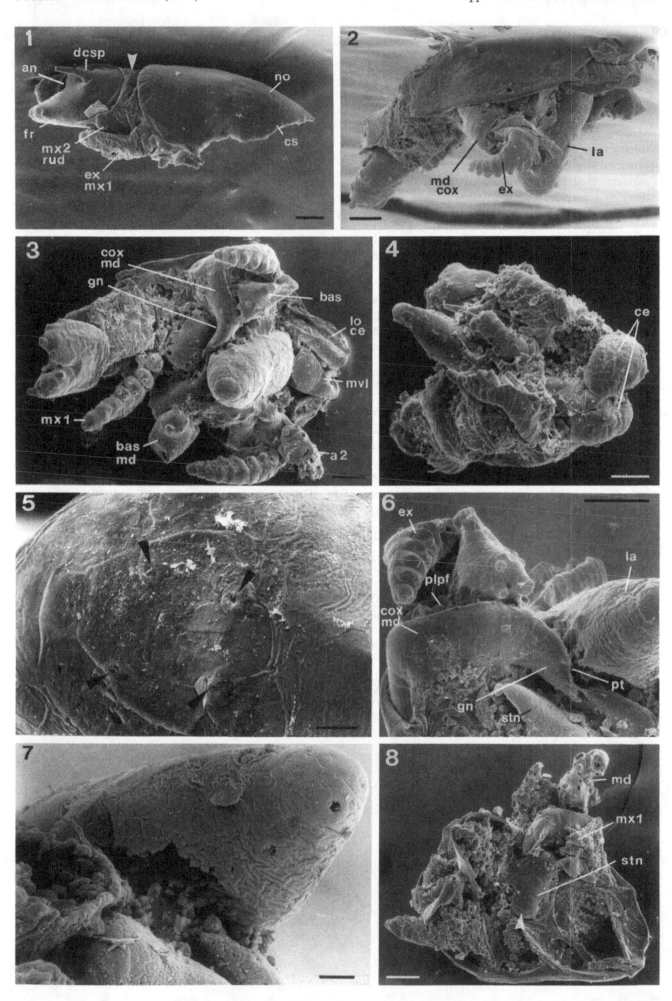

Plate 6

1–3: Stage L4A continued; 4–9: TS1iA

1, 2: Same specimen as in Pl. 5:2, 3, 6)

1 Median view of 1st maxilla. Proximal endite (pe) with three setae or spines, next three endites with two setae; proximal two endopodal (en) segments with one or two setae, rounded distal segment with group of two or three setae apically. Scale bar 10 μm.

2 Anterior view of the same appendage; arrow points to insertion area of exopod, originally arising from the sloping outer margin of the limb corm, not preserved (arrow). Scale bar 10 μm.

3 UB 24. Rear of trunk viewed from posterior; dorso-caudal spine broken off (sharp-edged fracture, dcsp); membrane covering the anus (an) protruded and almost destroyed; few denticles (den) positioned around the furcal spines. Scale bar 10 μm.

4–6: UB 25 (same specimen as in Pl. 7:1)

4 Lateral view of slightly deformed specimen with almost complete right appendages (a1, a2, md, mx1, mx2); shield slightly arched in the anterior third of its length, more roof-like than in preceding stages; anterior margin of shield somewhat recessed and raised behind the eye; eye area (ce) large and projecting from the forehead; trunk pushed into head (arrow). Scale bar 30 μm.

5 Subventral view; labrum (la) deformed distally; subdivision of 1st antenna (a1) into annulated 'shaft' and distal part with three tubular segments well-recognizable; right furcal ramus (fr) preserved in part (a2 = 2nd antenna; en = endopod; ex = exopod; mx1 = 1st maxilla). Scale bar 30 μm.

6 View of the median surfaces of the appendages, enabled by breakage of the distal ends of the left set (compare with 4); outer edges of antennal coxa and basipod with annulations that continue into the annular segmentation of the exopod (arrows). Scale bar 30 μm.

7 UB 26. Lateral view of slightly collapsed specimen; exopod (ex) of rudimentary 2nd maxilla (mx2 rud) with two setae (a2 = 2nd antenna; ce = compound eye; fr = furca ramus; la = labrum; md = mandible; mx1 = 1st maxilla). Scale bar 30 μm.

8 UB 27. Lateral view; eye area, labrum, and appendages fragmentarily preserved; right furcal ramus broken off; on the dorsal surface of the trunk an incomplete furrow ending laterally indicates the appearance of a new segment behind the maxillary one (arrow; latter segment free from the shield); trunk portion behind the maxillary segment named 'abdomen' from now on. Scale bar 30 μm.

9 UB 28 (same specimen as in Pls. 7:2; 30:2). View of postlabral feeding chamber, surrounded by the appendages; some of the enditic spines are still present; labrum broken off distally, inner space filled with coarse particles. Scale bar 30 μm.

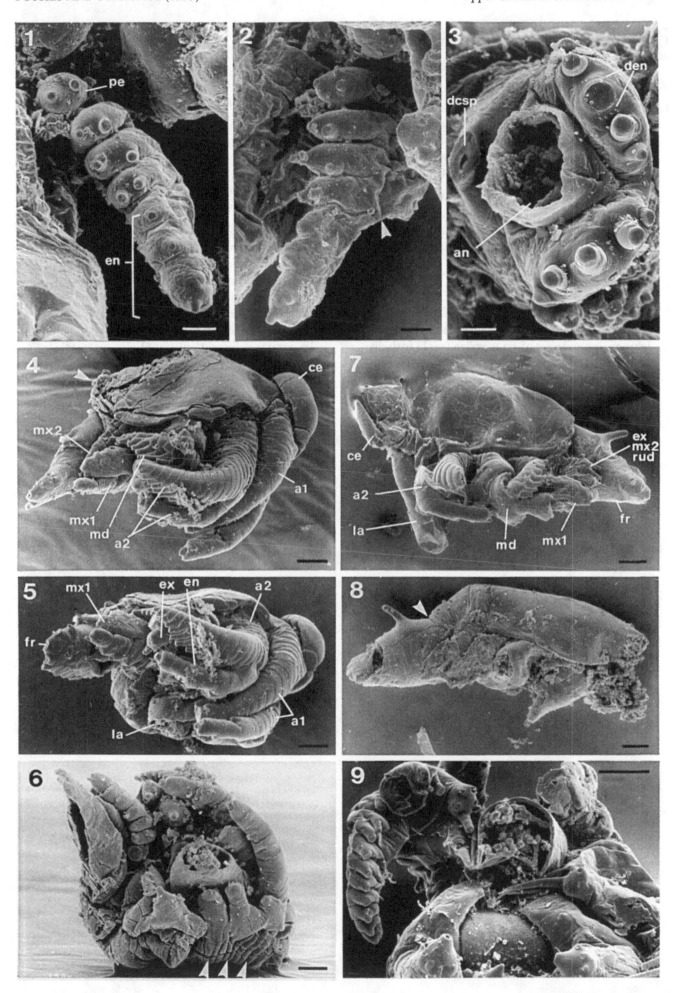

Plate 7

Stage TS1iA continued

1   Same specimen as in Pl. 6:4–6). View of exopods of 2nd antenna (a2) and mandible (md); note the continuation of the outer annulation of the antennal corm into that of the exopod (see also Pl. 6:6) lower margin: shield (cs) with numerous cracks. Scale bar 30 μm.

2   Same specimen as in Pls. 6:9; 33:2. Median view of distal end of 1st maxilla (mx1) and the rudimentary 2nd maxilla (mx2); latter limb with paired setules (stl) on inner edge, probably indicative of future segmentation. Scale bar 10 μm.

3–4: UB 29

3   Ventral view of almost complete postlabral region and appendages; antennal spines (esp) reach around the labrum (la); segment of 1st maxillae (mx1) with distinct sternal bar (st; md = mandible; mx2 rud = rudimentary 2nd maxilla; pe = distinct proximal endite). Scale bar 30 μm.

4   View of one of the enditic spines of the antennal coxa with long setules (gn = mandibular gnathobase; la = labrum). Scale bar 10 μm.

5   UB 30. Close-up of the distal end of the maxillary exopod carrying two spine-like setae; lateral one provided with a number of setules. Scale bar 3 μm.

6   UB 31. Close-up of abdomen with furcal rami; some of the spines are preserved almost to their entire length; arrows point to pores immediately anteroventrally to the spines; in this specimen it seems as if they formed sockets of thin setae, but their nature is unclear. Scale bar 10 μm.

7   UB 32. Lateral view of crumpled and partly preserved specimen, seen slightly from posterior; arrow points to posterior border of incipient 1st thoracomere recognizable as a fold on the dorsal surface; dorsocaudal spine (dcsp) still present but markedly thinner than in early larval stages (an = anus; cox md = large mandibular coxa; fr = furcal ramus; mx1, 2 = maxillae). Scale bar 30 μm.

8   UB 33. Ventrolateral view of incompletely and coarsely preserved specimen, seen slightly from posterior; dorsocaudal spine (dcsp) preserved with its entire length; anal field (anf) protruded probably due to decomposition effects (see also Müller & Walossek 1988b, Pl. 6:4; cox md = coxal body of mandible; fr = furcal ramus; la = labrum). Scale bar 30 μm.

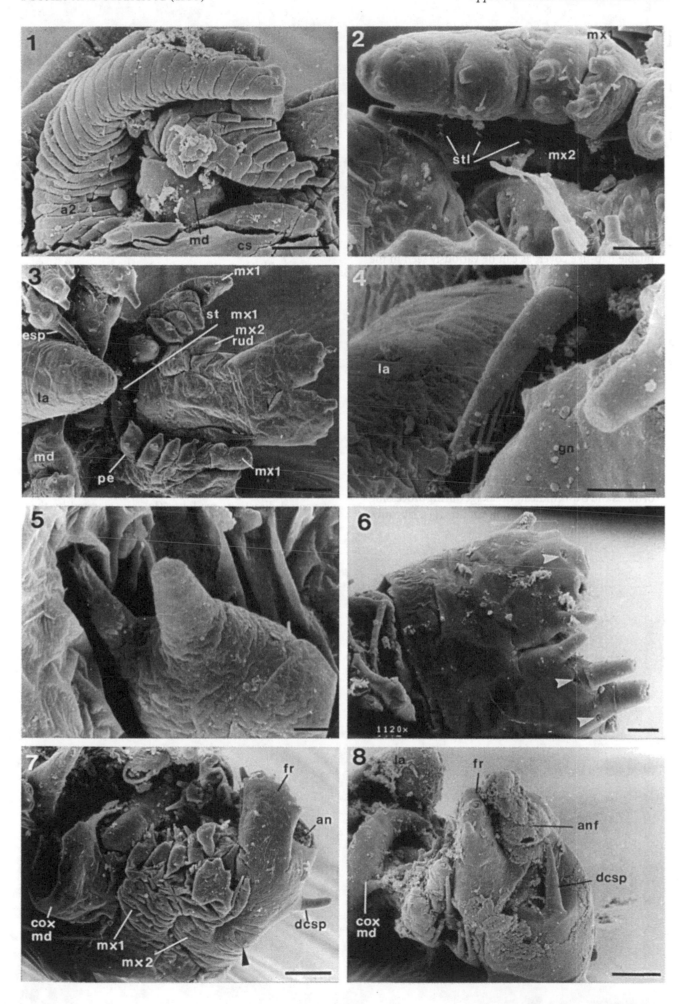

Plate 8

1–4: Stage TS1A; 5, 6: Stage TS2iA; 7, 8: Stage TS2A

1, 2: UB 34

1   Lateral view of incompletely preserved specimen; 2nd maxilla (mx2) enlarged and with small paddle-shaped exopod (ex); segment of 2nd maxilla free from head (arrow); no signs of dorsocaudal spine in this specimen; eye area (ce) protruding from forehead (ths1 = 1st thoracomere). Scale bar 30 µm.

2   Ventrolateral view of right side; anal opening with pliable flap-like cover; as can be variously seen in the material the anal membrane is blown up and distorted (arrow; see also Pls. 7:8). Scale bar 30 µm.

3, 4: UB 35

3   Anterior view of fragmentary specimen; appendages broken off, their bases being covered by foreign particles; shield roof-like in this view, depressions on either side approximately above the 1st antennae may be muscle impressions (see also Müller & Walossek 1985b for Skaracarida, e.g., Pl. 3:3); eye area distorted but seemingly arising from a rather narrow basis (arrow; la = labrum). Scale bar 30 µm.

4   Ventral view; sharp-edged and flat surfaces of fracture of limbs and trunk indicate breakage during processing rather than incomplete phosphatization (la = labrum). Scale bar 30 µm.

5, 6: UB 36

5   Lateral view of stretched, incompletely preserved specimen; behind the 1st thoracomere (ths1) with its rudimentary limb (thp1), the incipient 2nd segment can be seen, incompletely delineated from the abdomen by a furrow which ends laterally (arrow); exopod (ex) of 1st trunk limb with two setae (ce = compound eye). Scale bar 100 µm.

6   Ventral view of postmaxillulary body and trunk with rounded furcal rami; both maxillary and 1st thoracomere with distinct segment borders ventrally, the former probably with a faint sternal bar; anus (an) protruded as in 2. Scale bar 30 µm.

7, 8: UB 37 (same specimen as in Pls. 9:1–3; 30:3)

7   Lateral view of right side of slightly deformed but almost complete specimen; large eye lobes (ce) well-inflated and protruding from the head; 1st antennae (a1) and exopod of 2nd antenna (a2) fragmentary, other limbs beautifully preserved (md, mx1, 2); maxillary segment still free from head but partly covered by the shield (arrow; ths1, 2 = thoracomeres). Scale bar 100 µm.

8   Lateral view of left side, seen from a somewhat antero-ventral direction; eye lobes (lo ce) separated by the bulging midventral lobe (mvl); basis of eye area seemingly constricted (ths1, 2 = thoracomeres). Scale bar 100 µm.

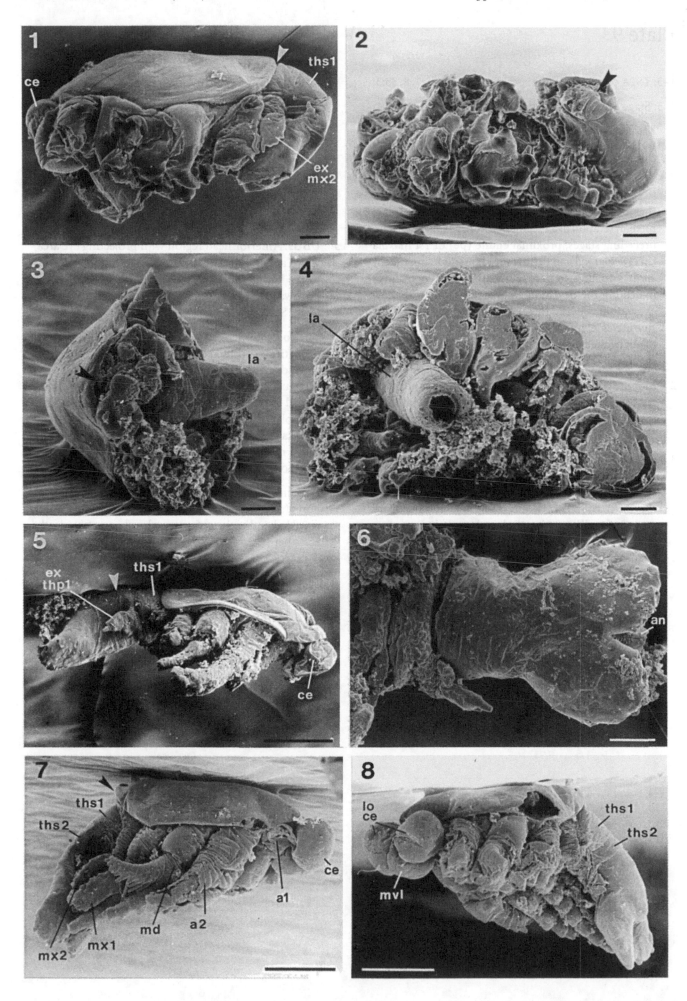

Plate 9

1–3: TS2A continued; same specimen as in Pls. 8:7, 8; 33:3;

4, 5: Stage TS3iA; 6, 7: Stage TS3A, UB 40

1   Median view of right postantennulary appendages, facilitated by partial preservation of left limbs and breakage of the labrum (la); mandible (md) positioned at posterior edge of labrum, while the 2nd antenna (a2) is set slightly more anteriorly; mandibular basipod (bas) with about 8 setae around the median spine; median setation of all segments more advanced than in preceding stages; proximal endite (pe) of 1st maxilla (mx1) enlarged and further differentiated; subsequent endites smaller; sternum (stn) with two distinct humps indicating the developing paragnaths, and a shallow excavation between these; posterior to the humps the sternite is elevated and bears a shallow depression medially; sternites of both maxillae compressed due to shrinkage (mx2 = 2nd maxilla). Scale bar 30 µm.

2   Mandibular exopod; segmentation of ramus and number of setae are not congruent: arrows point to interfering setae (see also Pls. 4:4, 10:8, 11:1 compared to 19:3, 21:4 for series B). Scale bar 10 µm.

3   Posterior end of cephalic shield and anterior trunk region; from the mid-point of the posterior emargination of the shield a furrow runs ventrally (arrows) past the 1st maxilla (mx1), indicating that the maxillary segment is still located on the larval trunk (in *Bredocaris* a similar furrow demarcates the boundary behind the head; Müller & Walossek 1988b, their Pl. 3:7, 8); shafts

of limbs finely folded to enhance flexibility (mx2 = 2nd maxilla; ths1, 2 = 1st and 2nd thoracomeres). Scale bar 15 µm.

4   UB 38. Lateral view of almost complete but much deformed specimen; trunk twisted about 90° counter clockwise; 1st thoracopod (thp1) now with paddle-shaped exopod; 2nd one rudimentary (thp2 rud; md = mandible; ths3i = incipient 3rd thoracomere). Scale bar 30 µm.

5   UB 39. Anterior head region with projecting eye area, which appears to arise from a stalk-like hump anterior to the labrum and below the anterior shield margin (arrow; lo = eye lobes; mvl = crumpled midventral lobe). Scale bar 30 µm.

6   Ventrolateral view, seen somewhat from the posterior, of relatively complete specimen with laterally stretched appendages; posterior side of labrum (la) somewhat triangular, reaching between the medially pointing angled mandibular gnathobases (gn); anal field (anf) crumpled; most likely the wrinkles on the abdomen were caused by ventral flexure after death (arrow; thp2 rud = rudimentary 2nd thoracopod; ths1–3 = 1st to 3rd thoracomeres). Scale bar 30 µm.

7   Posterior surface of labrum with a few tubercles or pores (arrows; see also Pl. 5:7; 23:7; 34:4); cuticle finely wrinkled indicative of some shrinkage after death. Scale bar 10 µm.

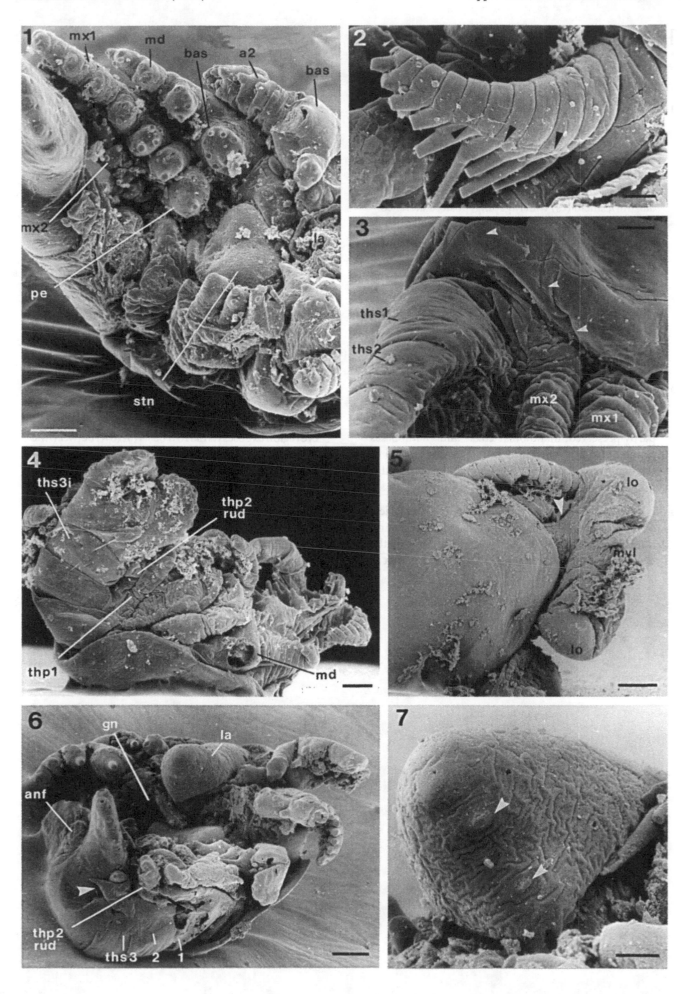

Plate 10

1, 2: Stage TS4A, UB 41; 3–8: Stage TS5A

1   Ventral view of fragmentary specimen; few structures preserved (la, cox md, stn); labrum slightly depressed anteriorly, triangular posterior side (tip broken) declining toward the gnathobases and deflexes anteriorly to form the ceiling of the atrium oris, which provides space for the gnathobases to move underneath the labrum); shield (cs) widely gaping, posterolateral corners wing-like extended posteriorly; posterior edge of shield emarginated; trunk (tr) flexed ventrally originally, but fragmentary (il = soft inner lamella). Scale bar 50 μm.

2   Close-up of postlabral feeding chamber; mandibular gnathobases sharply angled against coxal body, cutting edge with several spinules or denticles of different size; paragnaths (pgn) much elevated and separated by a deep furrow (pt = posterior spine). Scale bar 30 μm.

3   UB 42 (Same specimen as in 8; Pl. 11:1, 8; destroyed). Ventral view of specimen with complete and far anteriorly stretched 2nd antennae; other appendages only partly preserved, mandibular coxae and gnathobases exposed; trunk pushed into anterior body, twisted to the right; furcal rami complete, paddle-shaped (il = inner lamella). Scale bar 100 μm.

4   UB 43 (same specimen as in 7; Pl. 11:6, 7). Lateral view of incomplete specimen; shield inflated (decay?) and inner lamella exposed; maxillary segment free from the head (mx2s), dorsal side apparently softer than that in

succeeding thoracomeres (ths1–5); 2nd trunk limb (thp2) with paddle-shaped exopod (la = labrum). Scale bar 50 μm.

5   UB 44 (same specimen as in Pls. 11:4, 5; 30:4). Ventral view of fragmentary specimen; parts of appendages well preserved; proximal endites of both maxillae (pe mx1, 2) much larger than the other endites; sternum inflated, particularly at its rear (gas production after death?); labrum (la) broken off posteriorly, permitting a view of the shovel-like gnathobases (gn md); sternum (stn) covered with numerous delicate setules (see also Pl. 11:4; a2 = 2nd antenna; thp1, 2 = 1st and 2nd thoracopods). Scale bar 100 μm.

6   UB 45 (same specimen as in Pl. 11:2, 3). Ventral view of incomplete specimen with collapsed ventral structures lying within the gaping shield; left mandible, maxillae and 1st thoracopod (thp1) present, the latter being slightly deformed but almost complete. Scale bar 30 μm.

7   Same specimen as in 4 seen from anterior. Breakage of eye region exposes its narrow basis (arrow) between the insertions of the 1st antennae (see also Pl. 9:5; a1 = 1st antenna; la = labrum; il = inner lamella; tr = distorted trunk). Scale bar 30 μm.

8   Same specimen as in 3. View of the complete 2nd antennae; setal sockets of exopod still present; segmentation not congruent with setation (see also Pls. 4:4, 9:2, 11:1); arrow points to row of setae on proximal edge of proximal maxillulary endite. Scale bar 30 μm.

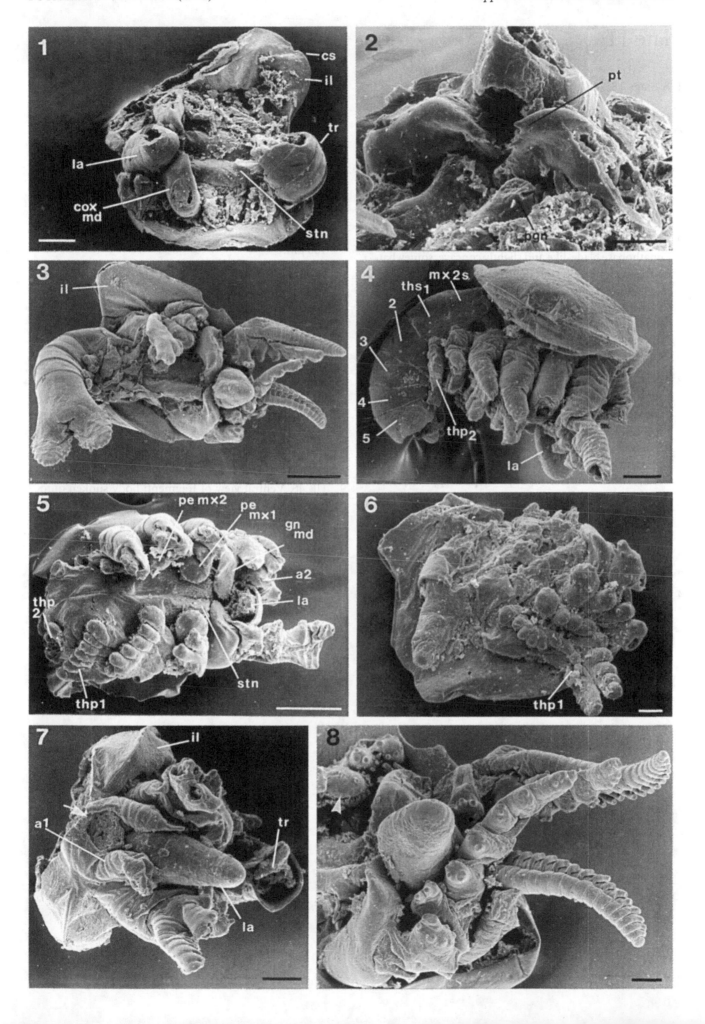

Plate 11

1–8: Stage TS5A continued; 9, 10: Stage TS6iA

1   Same specimen as in 8, Pl. 10:3, 8. Posterior view of antennal exopod; annulation not completely circular but interrupted by a short membranous area posterior to the setal sockets (arrows point to ringlets not congruent with setation; see also Pls. 4:4; 9:2; 10:8). Scale bar 30 μm.

2, 3: Same specimen as in Pl. 10:6.

2   Coxal gnathobase (gn) and basipod (bas) of mandible; gnathobase blade-like and slightly concave, somewhat thickened medially and denticulate; gnathobasic seta broken off (gns); basipodal spine oval in cross-section, with 8 setae around its basis; lower right: endites of 1st maxilla, proximal endite (pe mx1) with marginal row of setae and spines on its median surface originally (arrows). Scale bar 10 μm.

3   Collapsed but almost complete 2nd maxilla (mx2) and 1st thoracopod (thp1); unclear whether maxillary corm has five or six endites, and the endopod (en) four respectively three segments accordingly; endites progressively smaller and more distally oriented from proximal to distal, with frontal and posterior row of setae; median surface slightly humped, tipped by one or few short spines (arrows); exopod (ex) paddle-shaped, with straight inner margin and gently curved outer one carrying a row of setae (pe = proximal endite). Scale bar 30 μm.

4, 5: Same specimen as in Pls. 10:5; 33:4.

4   View of sternum (md, mx1, 2); sternite of 1st maxilla coalesced, also boundary between sternites of maxilla segments feeble; sternal setules arranged in short, curved rows; fewer setules on maxillary sternite; sternum invaginated medially; arrows point to depressions on maxillary segment and between its sternite and that of 1st thoracomere (pe = proximal endite; pgn = paragnaths). Scale bar 30 μm.

5   Lateral view of appendages showing the incipient subdivision of the more firmly sclerotized outer edges of the postmandibular limbs (arrows); 2nd endite of 1st maxilla (end2 mx1) elongated and with three spines medially; note the small size of the mandibular exopod (md ex). Scale bar 30 μm.

6, 7: Same specimen as in Pl. 10:4, 8.

6   Dorsal view of shield; much of the cuticle around the apex is broken off; frame shows area of 7. Scale bar 100 μm.

7   Close-up of small hump with two sets of three pores close to the middle of the posterior emargination of the shield; the nature of this pore-bearing field which could be observed in several other specimens as well, is unknown (e.g., UB 52, TS7A, detail not figured. Scale bar 10 μm.

8   Same specimen as in 1. Anteriorly deformed trunk with abdomen, furca and incipient ventrocaudal processes (i vcp) as small humps which form the sockets of short spines (broken off); arrows point to some of the pores close to the furcal spines; ventral cuticle of posterior thoracomeres collapsed (cs = shield). Scale bar 30 μm.

9   UB 46. Ventral view of fragment; body sunken onto gaping shield; appendages incomplete (except left 2nd antenna ex a2) and coarsely preserved, posterior ones even missing; trunk sharply ventrally flexed, its joint membranes being overstretched; furca broken off (an = T-shaped anal slit; la = labrum). Scale bar 100 μm.

10 UB 47. Fragment without anterior head portion, shield, and furcal rami; limbs also distorted with exception of the mandibular coxae; gnathobases (gn) angled against the coxal body and anteriorly tilted toward the labrum (ths6i = incipient 6th thoracomere; pt = posterior tooth; st = sternum with paragnaths). Scale bar 30 μm.

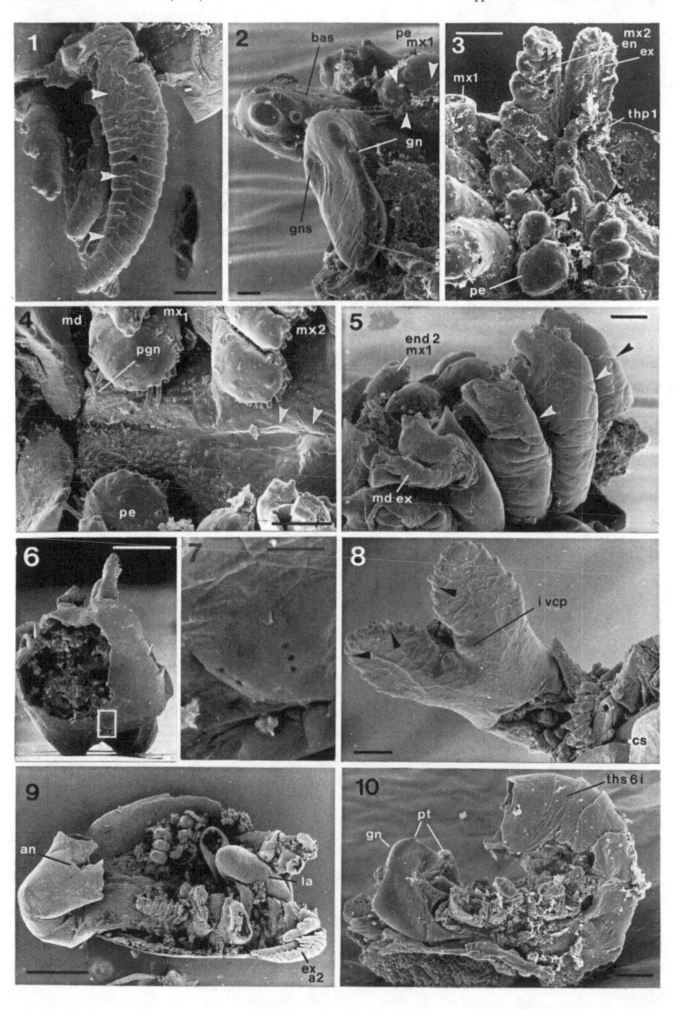

Plate 12

1–5: Stage TS6A; 6, 7: Stage TS7iA, UB 50

1–3: UB 48

1   Ventral view, seen somewhat from anterior; shield deformed due to collapsing; trunk twisted to the left; anterior appendages and eye area not preserved, post-mandibular limbs preserved with their proximal portions being laterally stretched (cs = shield; la = labrum; vcp = ventrocaudal processes). Scale bar 50 μm.

2   Closer view of partly preserved left appendages (a2, md, mx1, 2, thp1, 2); original position of enditic setae indicated by small tubercles (la = labrum; plpf = palp foramen). Scale bar 30 μm.

3   Posterior view of abdomen with ventrocaudal processes (vcp), now with three spines, furcal rami (fr) and T-shaped anus (an; po = pits or pores corresponding to furcal spines; saf = supra-anal flap). Scale bar 30 μm.

4, 5: UB 49.

4   Ventral view of trunk fragment; rudimentary limb of 5th thoracomere (thp5 rud) probably uniramous; ventral side of last segment soft, lacking appendages; some spines of the furcal rami and the ventrocaudal processes are preserved. Scale bar 30 μm.

5   Close-up of right furcal ramus, with some of the spines almost completely preserved; arrows point to some of the pores anteroventrally to the spines; furcal ramus still not fully articulated, but already with signs of the future joint (i j). Scale bar 10 μm.

6   View of fragmentary and twisted specimen, lacking most of the head details and appendages (an = anus; st fgr = sternal food groove). Scale bar 50 μm.

7   Ventral view of posterior end of trunk with articulate furcal rami; ventrocaudal processes (vcp) with two or three spines; furrow on posterior end of abdomen extending between the processes (arrow); structure may be caused by distortion, but can be seen on all larger specimens (see also Pls. 15:2; 26:1); penultimate complete thoracomere with remnants of rudimentary limbs; ventral surface less sclerotized than the dorsal cuticle (j = joint). Scale bar 30 μm.

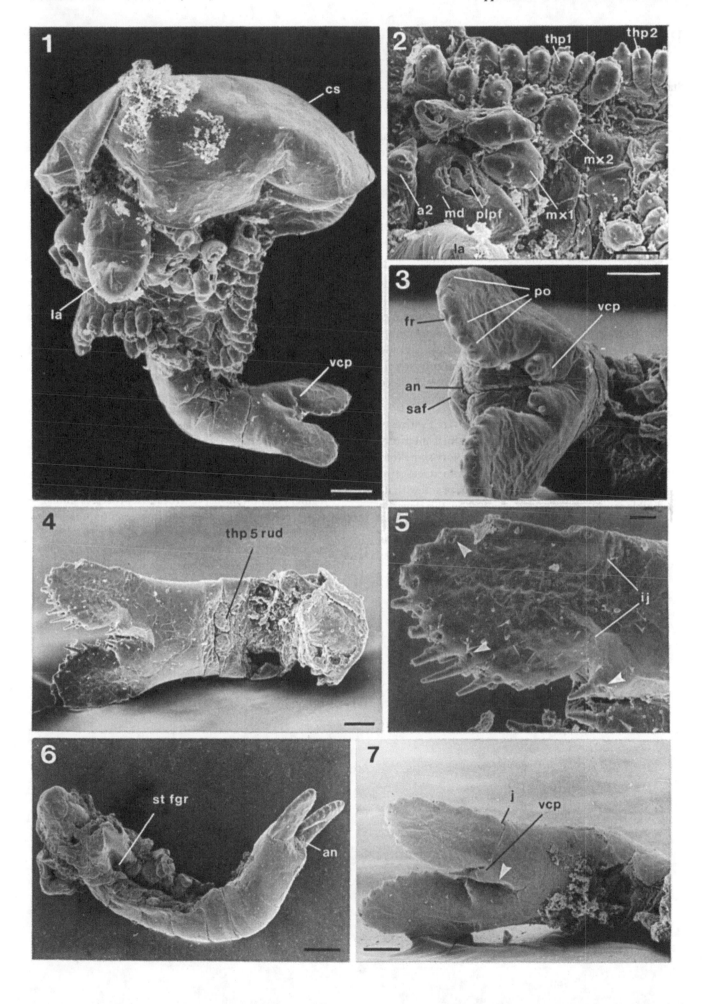

Plate 13

Stage TS7A

1–4, 8: UB 51 (same specimen as in Pl. 31:1); 5–7: UB 52

1 Lateral view. Appendages only partly preserved; lateral margin of shield and furca broken off; anterior surface of labrum (la) with constriction (arrow), distal part distorted. Scale bar 100 μm.

2 Anterior view of shield with ventrolaterally curving anterior margin; compound eye and 1st antennae not preserved; next limbs only represented by their proximal portions (la = labrum). Scale bar 30 μm.

3 Ventral view of postlabral region; most of the appendages recognizable but distorted (md, mx1, 2, thp1–5); 6th thoracopod rudimentary (thp6 rud, see also Fig. 7B); trunk broken within 5th segment during mounting of the specimen; deep food groove (fgr) running from between the paragnaths (pgn) posteriorly; sternites of head region seemingly fused to form a single sternum. Scale bar 50 μm.

4 Close-up of mouth region; sternum covered with tiny setules; paragnaths (pgn) slightly deformed anteriorly due to compression; basipodal masticatory spine (bas msp) of mandible distally branching into at least two spinules; one of the circumstanding setae still with its double row of setules (arrow); from about this stage the posterior edge of the labrum (la) forms a ridge, while the flanks are slightly depressed (with setules); it is possible that the ceiling of the atrium oris (ao) was less sclerotized than the other parts of the labrum (gn = gnathobase). Scale bar 10 μm.

5 Second antenna of fragmentary specimen, with stretched corm; in this particular specimen it is not the basipod (bas) but the proximal endopodal segment (en1) that gives rise to the exopod (ex), probably a defect resulting in an enlargement of the number of external ringlets (arrows; cox = coxa). Scale bar 30 μm.

6 View of median surfaces of maxillae (mx1, 2); proximal endite of 1st one (pe) bulging and with numerous setae around its proximal edge; subsequent endite (end2) elongate, with 3 distal spines and some setae at the sides; 3rd endite (end3) with 2 sets of setae separated by 1 enditic spine medially; next endite (end4, probably the last one of the corm, with axially elongated surface (gn = gnathobase; la = labrum). Scale bar 30 μm.

7 Collapsed sternal region with paired pits on the sternites of the maxillary and two anterior thoracomeres (arrows; see also Pls. 20:8, 21:7; fgr = median food groove; mx1, 2 = maxillae; thp1 = 1st thoracopod). Scale bar 30 μm.

8 Ventral view of posterior end of trunk with last two segments and abdomen; ventral surface of 7th trunk segment membranous; ventrocaudal processes similar as in preceding stage; only right furcal ramus preserved in part (fu = furrow between processes, here clearly enhanced by distortion; po = pore; thp6 rud = collapsed rudimentary 6th thoracopod). Scale bar 30 μm.

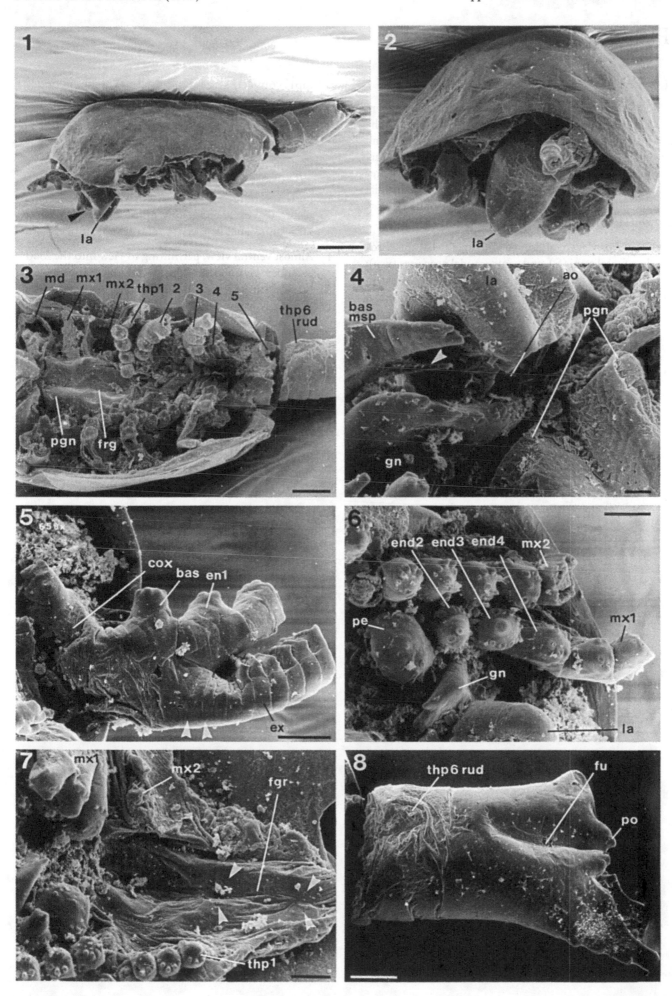

Plate 14

1, 2: Stage TS8iA, UB 53; 3–6: Stage TS8A, UB 54

1   Lateral view of strongly curved fragmentary specimen; anterior thoracomeres broken which exposes the coarsely phosphatized internal filling; cuticle of limb fragments rather coarsely preserved, while dorsal surface of thoracomeres and abdomen (abd) is well-preserved; furcal rami broken off distally (am = arthrodial membrane; app = fragments of thoracopods; cs = shield; ths2–8i = thoracomeres). Scale bar 30 μm.

2   Posterior view of abdomen with anal field (anf), broken furcal rami (fr), and ventrocaudal processes (vcp); thin cuticle of anal field only coarsely preserved in sharp contrast to that of abdomen and ventrocaudal processes indicative of the softness of the former; on left side: enditic setae of thoracopods (saf = supra-anal flap). Scale bar 30 μm.

3   Lateral view of slightly distorted but almost complete specimen, probably of stage TS8, prior to breakage (remains see Figs. 5, 6; size of shield 600 μm); shield not covering the rami of the limbs; labrum with distinct bend on anterior surface (arrow; see also Pl. 13:1); 1st maxilla (mx1) most likely much smaller than subsequent limbs; 2nd maxilla and 1st thoracopod (mx2, thp1) complete save for the distal end of the endopods (en); limb bases long, slender, convex anteriorly and concave posteriorly; exopods (ex) much elongate, paddle-shaped, and with marginal row of setae (only sockets preserved; corms (co) distinctly divided into at least three major portions on outer side; posterior appendages more or less fragmentary; some of the posterior thoracomeres seem to have developed short, pleura-like extensions (an = anus; fr = furcal ramus; saf = supra-anal flap; vcp = ventrocaudal process).

4   Almost ventral view, permitting to view into the food chamber between the postmandibular appendages (la = labrum). Size of shield 600 μm.

5   Close-up of right 1st maxilla (after distortion); setae and spines still with their subordinate setules; more setules on the endites (end2–4); outer edge of corm only bisected (en, ex; j = joint). Scale bar 30 μm.

6   Posterior view of proximal part of left 2nd maxilla; endites with numerous bipectinate setae; lower right: note the orientation of the setules toward the centre of the enditic surface (pe = proximal endite of left 1st maxilla). Scale bar 30 μm.

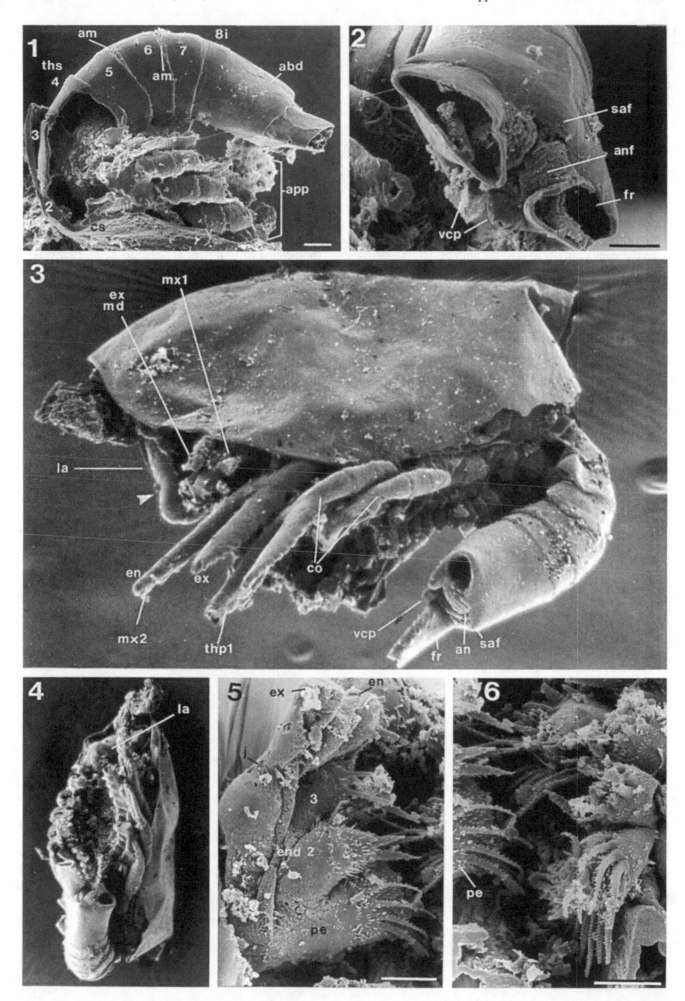

Plate 15

Stage TS10A, UB 55 (same specimen as in Pls. 16:1–7; 31:2, 3)

1   View of left side of stretched and somewhat deformed specimen; most of shield (cs) broken off; eye area, 1st antennae, and furca missing; left appendages fragmentary (a2, md, mx1, 2, thp1–7); labrum (la) with distinct constriction on its anterior surface and raised tip (see also Pls. 13:1; 14:3); left 2nd thoracopod torn off but still attached to the subsequent limb; flanks of anterior thoracomeres with shallow humps from which a groove runs ventrally toward the limb bases (arrows); this structure becomes less apparent in the posterior segments and is absent posterior to the 5th one; on the other hand, the posteroventral margin becomes slightly liberated in the 5th–8th segments, forming feeble pleura-like extensions (abd = abdomen; en = endopod; ths1–10 = thoracomeres). Scale bar 100 μm.

2   Ventral view, showing the lateral compression of the specimen; last, 10th thoracomere (ths10) apodous (abd = abdomen; fr = insertion of furcal ramus; fu = furrow between ventrocaudal processes vcp, see also Pls. 12:7 and 26:1). Scale bar 100 μm.

3   View of right side; right set of limbs broken off behind the 3rd or 4th thoracopod, which renders visible the inner surface of the left series with their enditic setation; note the distinctive subdivision on the outer edge of the limbs (cs = shield; la = labrum; mx1, 2 = maxillae; pe = proximal endite; thp2–9 = thoracopods). Scale bar 100 μm.

4   Median view of postlabral region with distorted left paragnath (pgn), proximal endite of 1st maxilla (pe mx1), proximal portion of 2nd maxilla (mx2), and 1st thoracopod (thp1); right appendages completely distorted and covered with foreign particles (la = labrum; pe mx2r = proximal endite of right 2nd maxilla). Scale bar 30 μm.

5   Close-up of proximal endite of 1st maxilla; median surface with rigid brush spines of varying size, a small scale-like structure of unknown nature (arrow), and tiny setules; posterior row of setulate setae running around proximal margin and reaching far anteriorly (ps; s mx2 = seta of 2nd maxilla lying on the endite; la = labrum; pgn = collapsed paragnaths). Scale bar 10 μm.

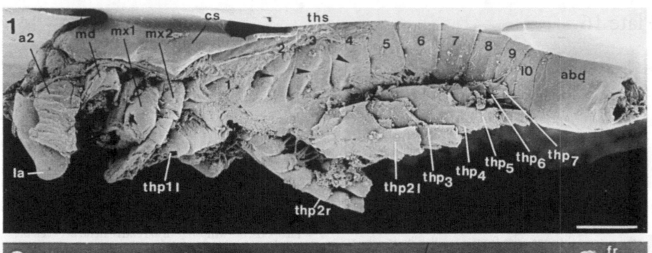

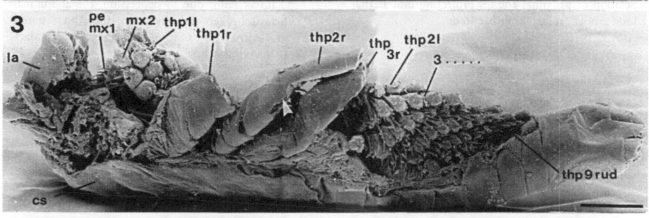

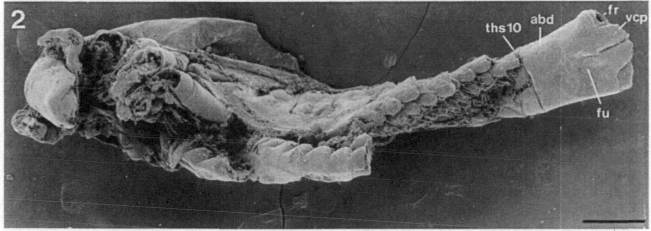

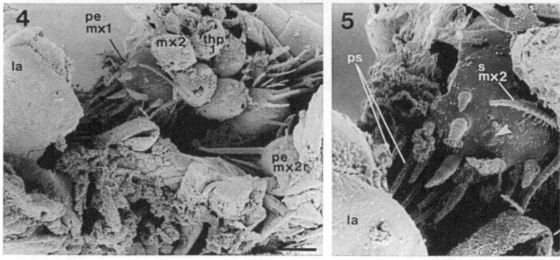

Plate 16

Stage TS10A continued

1–7: Same specimen as in Pls. 15; 34:2, 3

1   Detailed view of proximal maxillary endite with one of the plumose setae (s) and a smaller, brush-like seta (bs) covered with setules in a more irregular pattern; enditic surface furnished with tiny setules, in particular on anterior side. Scale bar 10 μm.

2   Close-up of bipectinate setae of varying lengths and thickness of the posterior edge of the endites; those of the marginal fringe arise from a swollen basis, while others have a small socket which may indicate an articulation of these; note the different orientation of the opposing rows of setules on the setae in such a way that the gap between them always opens towards the centre of the enditic surface. Scale bar 10 μm.

3   Closer view of endites of 2nd to 4th thoracopods; not only the orientation of the rows of setules changes on the setae around the endites but also the distance between the setules becomes larger distally; setae and/ or spines contacted those of the following legs; arrow points to a large enditic spine which is enlarged in Fig. 5. Scale bar 30 μm.

4   Setation of proximal endites projecting into sternal food groove (st fgr). Scale bar 10 μm.

5   Enditic comb spine (csp) of median surface with fringe of setules or denticles distally; proximal portion of spine with few widely spaced setules only. Scale bar 10 μm.

6   Close-up of bipectinate setae with rows of setules; average distance between the setules about 2 μm, slightly shorter towards the basis of the seta. Scale bar 3 μm.

7   Complete slender, pectinate setae of the posterior thoracopods; the whip-like distal ends extend between the medially pointed endites of at least the subsequent legs; concomitant with the tapering of the setae the rows of setules approach each other, while the distance between the setules increases. Scale bar 10 μm.

8, 9: UB 56

8   Posterior end of abdomen (abd) with paddle-shaped ventrocaudal processes (vcp) and large, well-articulated (j) and almost oval furcal rami; as the furcal rami the ventrocaudal processes have pores (po) corresponding to the marginal spines (pfsp = primary row of furcal spines). Scale bar 30 μm.

9   View of abdomen (abd) with ventrocaudal processes and furcal rami; due to depression, the abdominal cuticle is broken dorsally (an = anus; po = pores; pfsp = primary row of furcal spines; sfsp = secondary row; ths = thoracic segments). Scale bar 30 μm.

Plate 17

1: Stage TS13iA; 2–4: TS13A

1   UB 57. Dorsal view of fragmentary specimen, lacking entire head, appendages, and furcal rami; posterior thoracomere incipient (arrow); boundaries between the anterior thoracomeres deeply incised laterally; same segments bipartite immediately dorsal to insertions of limbs (see also Pl. 15:1); no distinct pleural extensions developed (abd = abdomen; fr = fragmentary furcal rami; ths1–13i = thoracomeres). Scale bar 100 µm.

2   UB 58 (same specimen as in Pl. 31:4). Ventral view of largest specimen; anterior head region with eyes, antennae (a1, a2), and distal part of labrum (la) distorted, trunk torn off; sternites sunken onto the shield, appendages partly preserved and laterally stretched; anterior surface of labrum with swelling anterior to constriction (arrow); mandibular coxae (md cox) large, their broad gnathobases pointing underneath the labrum; posterior part of cutting edge with large, flattened teeth; palp foramen (plpf) much smaller relative to preceding stages; 1st maxilla (mx1) preserved with prominent proximal endite (pe), and 3 more endites;

limb broken off distally but apparently shorter than subsequent limbs; limbs compressed in anterior–posterior direction, endites posteriorly oriented (en = endopod; end = endite; ex = exopod; gn = gnathobase; mx2 = 2nd maxilla; thp1–6 = thoracopods). Scale bar 30 µm.

3   Lateral view; breakage of shield renders visible the insertions of the anterior appendages (a2–md); exopod (ex) of 1st maxilla (mx1) much thinner than those of subsequent limbs (only thp4 marked = 4th thoracopod); corms of thoracopods with distinct segmentation on outer edge (arrows; cs = shield). Scale bar 100 µm.

4   Close-up of mouth area; labrum broken off distally, exposing coarse internal filling (fi la) within subtriangular cavity; coxal endite of 2nd antenna preserved, seemingly much smaller than in preceding stages (a2); on lower middle: left lobe of paragnaths (pgn) with striation on anterior surface (pe = proximal endite of 1st maxilla; plpf = insertion area of basipod; pt = posterior tooth). Scale bar 30 µm.

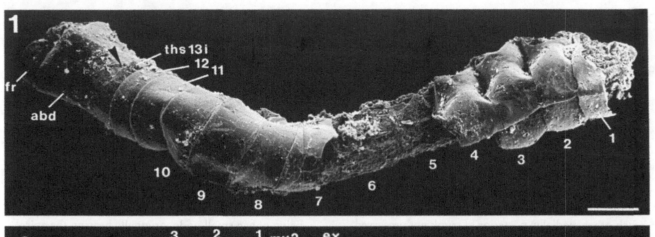

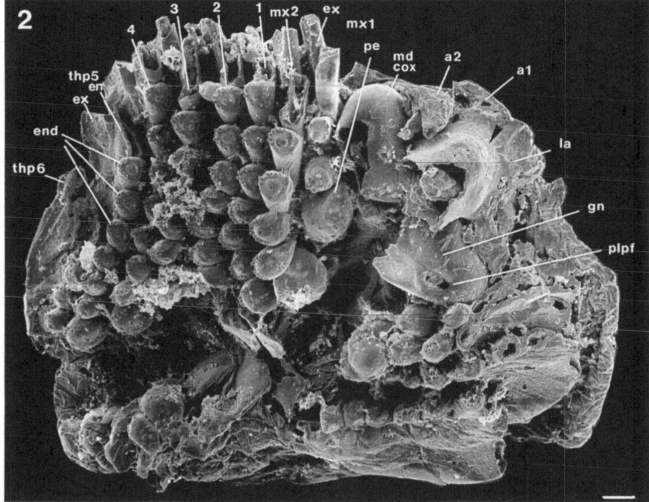

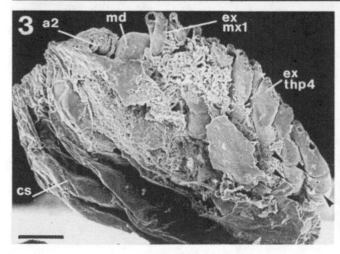

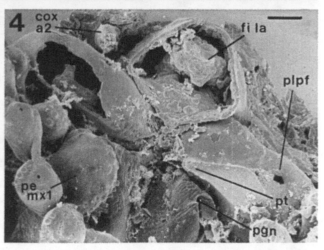

Plate 18

1–6: Stage L4B; 7, 8: Stage TS2B

1   UB 59. Ventral view of somewhat collapsed specimen; eye lobes (lo ce) and midventral lobe (mvl) distorted; 1st antennae broken off distally (a1); 2nd antenna (a2) and mandible (md) rather well-preserved, stretched laterally; mandibular coxa (cox) with flattened gnathobase; basipod (bas) with central spine and 6–7 setae around it; 2nd maxillae visible as a pair of bilobate rudiments (mx2 rud) on larval trunk, approaching each other medially; posterior of trunk not preserved (la = labrum; mx1 = 1st maxilla). Scale bar 30 μm.

2–6: UB 60

2   Lateral view of almost complete specimen; body collapsed and seemingly pulled somewhat out of the shield; proximal parts of antennae (a1, a2) directed anteriorly; arrow points to remains of antennal sternite; mandible and 1st maxilla broken off; segment of 2nd maxilla (mx2s) not coalesced with the head; dorsocaudal spine is missing (see L4A, Pls. 5; 6:1–3), while the furcal ramus (fr) is paddle-shaped and with about 7 marginal spines originally (en = endopod; ex = exopod; ce = compound eye; la = labrum). Scale bar 30 μm.

3   Almost anterior view of head with eye area (lo ce, mvl) projecting beyond the anterior shield margin; 1st antennae (a1) inserting behind the eye lobes, curving inward and anteriorly in front of the labrum in this specimen; 2nd antennae (a2) inserting behind the former, but slightly more laterally (en = endopod; ex = exopod; fr = furcal ramus). Scale bar 30 μm.

4   Ventral view; coxal and basipodal endites of 2nd antenna (a2) drawn out into spines (esp, broken off) accompanied by thinner setae; inner edge of 1st antenna (a1) also carrying setae (s) to aid in food intake (an = anus; bas = basipod; md = mandible; mx1, 2 = maxillae; rud = rudimentary). Scale bar 30 μm.

5   Close-up of endites and armament of left mandible; blade-like gnathobase (gn) with several spinules marginally, posterior one slightly set off (pt); masticatory spine of basipod (bas) broken off, circumstanding setae (s) are recognizable by their sockets; endopodal articles (en md) decreasing in size, distal one (4th) being only a small node, originally bearing the apical seta (arrow; also Pls. 21:3, 22:5; a2 = 2nd antenna; la = labrum). Scale bar 10 μm.

6   Posterior view of collapsed trunk; dorsal spine not recognizable; maxillary segment (mx2s) not coalesced; anal field (anf) collapsed; marginal row of spines of furcal rami (fr) with row of spines recognizable only by their sockets (la = labrum; mx2 rud = rudimentary 2nd maxilla). Scale bar 30 μm.

7   UB 61 (same specimen as in Pl. 19:1). Body twisted between head and trunk; mandibular coxae (cox md) prominent; gnathobases directed towards the mouth; basipod of left mandible somewhat pulled out of its joint (am = joint membrane); right mandible still with short eight-segmented exopod (ex); trunk with maxillary segment, 2 thoracomeres (ths1, 2) and abdomen (abd) with furcal rami (an = anus; cox a2 = coxa of left 2nd antenna; fsp = furcal spines; mx1 = 1st maxilla). Scale bar 30 μm.

8   UB 62 (same specimen as in Pl. 19:2, 3). Ventral view of distorted specimen with laterally stretched appendages (a1, a2, md, mx1); 2nd maxilla broken off; rudimentary 1st thoracopod preserved but collapsed (thp1 rud); trunk end not preserved. Scale bar 30 μm.

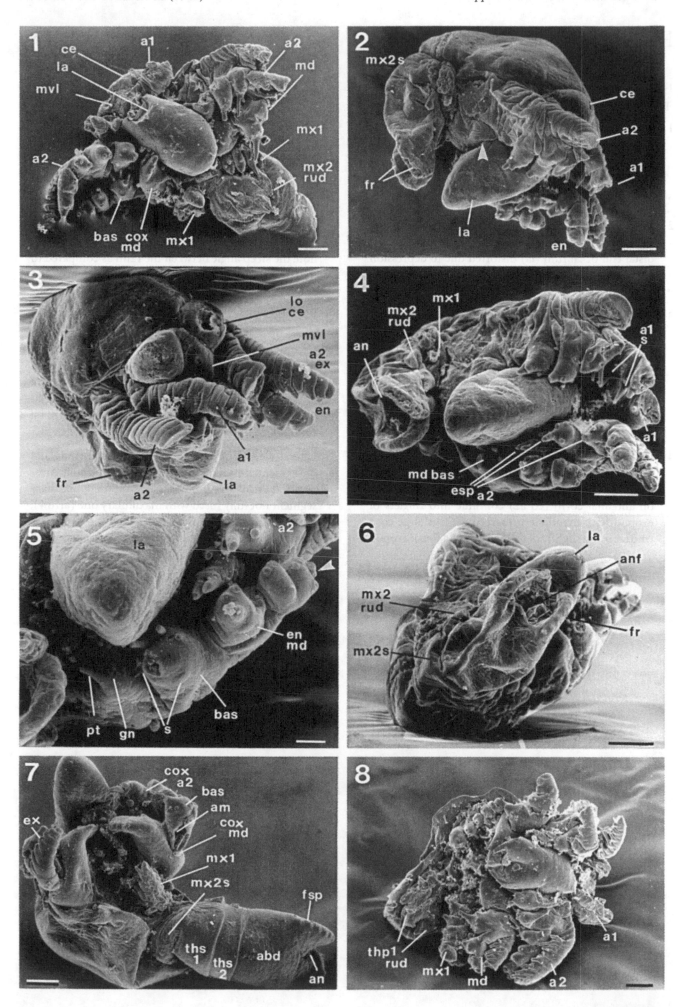

Plate 19

1–3: Stage TS2B continued; 4–7: Stage TS3B, UB 63 (same specimen as in Pl. 32:1)

1   Same specimen as in Pl. 18:7. Posterodorsal view of trunk with abdomen (abd), anal field (anf), T- or Y-shaped anus (an), and furcal rami; angle between rami almost 90°. Scale bar 15 μm.

2, 3: Same specimen as in Pl. 18:8

2   Ventral view of right side appendages: 2nd antenna (en, ex a2), mandible (md), and 1st maxilla (mx1); distal end of labrum (la) distorted due to depression (cs = shield); proximal endite (pe) of 1st maxilla much larger than subsequent endites. Scale bar 30 μm.

3   Close-up of almost complete left 2nd antenna; ringlets of exopod increasing in length progressively; each seta corresponds to a ramal ringlet (see also Pl. 21:4); proximal endopodal article (en1) drawn out medially into elongate endite similar to basipod (bas) and coxa (cox); division of 2nd article (2) incipient (arrow); 3rd one (3) distorted distally (esp = enditic spines; la = labrum). Scale bar 10 μm.

4   Lateral view of appendages, which seem to be pulled out of the body, exposing inner lamella (il) and joint membranes; 1st antenna (a1) incomplete; annulations on antennal corm (co a2) continue into ringlets of exopod (ex); mandible preserved with its large coxa (md cox) and slightly deformed basipod (bas); exopod broken off distally; corm of 1st maxilla (mx1) slightly shrunken; 2nd maxilla (mx2) also slightly collapsed, with folded shaft (see also Pls. 7:7; 8:7 and 9:3); paddle-shaped exopod with 4–5 setae on distal margin (en = endopod). Scale bar 30 μm.

5   Almost anterior view of slightly laterally compressed specimen; prominent eye area (lo ce, mvl) incompletely preserved; 1st antennae (a1) broken off distally; appendages ventrolaterally oriented; labrum tapered distally; furcal rami with partly preserved spines (fsp); note the steep angle between the rami as compared to the corresponding stage of series A (see Pl. 9:6; a2 = 2nd antenna; cs = shield; en md = endopod of mandible; il = inner lamella; mx1 = 1st maxilla). Scale bar 30 μm.

6   Close-up of postoral feeding chamber; various enditic setae and spines preserved, even with their subordinate setules; arrow points to one of the more rarely preserved shorter enditic spines with irregular pattern of setules (a2 = 2nd antenna; fr = furcal rami; la = labrum; md = mandible; pe mx1 = proximal endite of 1st maxilla). Scale bar 10 μm.

7   Lateral view of region between mandible (md) and unsegmented abdomen (abd); due to the peculiar preservation the pliable limb bases are exposed; maxillary segment not fully fused with maxillulary one (arrow); posterior segments (ths1–3) only lightly sclerotized; exopod of rudimentary 1st thoracopod (thp1 rud) still carrying its terminal spine. Scale bar 30 μm.

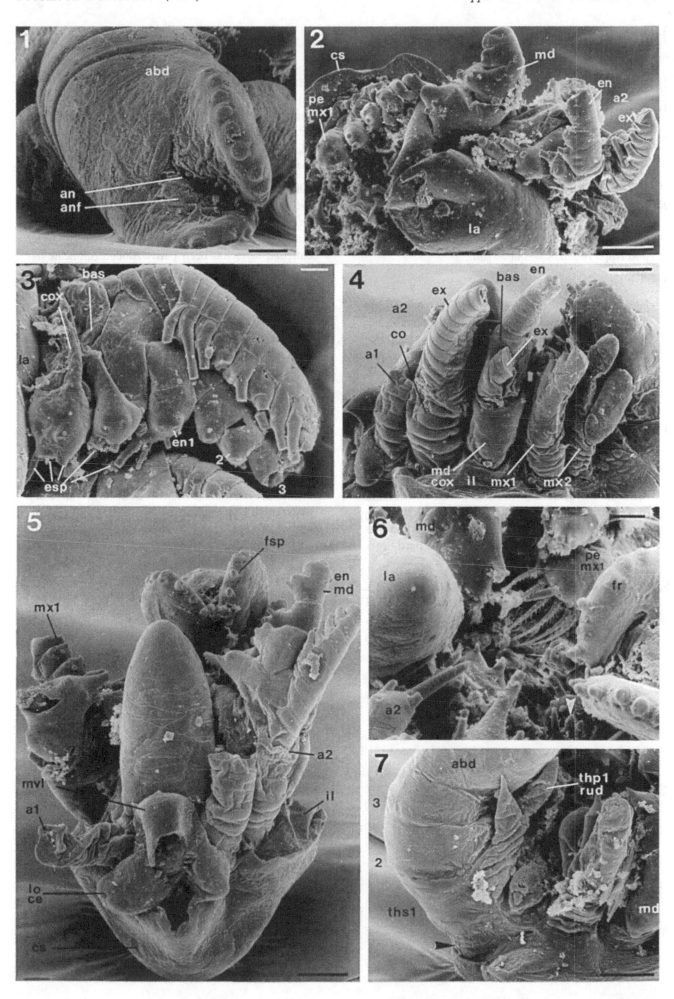

Plate 20

1–4: Stage TS4iB; 5–8: Stage TS4B

1, 2: UB 64

1   Dorsal view of fragmentary specimen with 3 thoracomeres and a 4th incipient one (ths4i); membrane of anal field protruding (see also Pls. 7:7; 8:5; 9:4; 11:1; 17:1). Scale bar 30 μm.

2   Lateral view; arrow points to dorsal furrow demarcating the new segment; lobate structure below the 3rd segment may represent the rudimentary 3rd thoracopod (thp3 rud?). Scale bar 30 μm.

3, 4: UB 65

3   Lateral view of distorted specimen; segment of 2nd maxilla (mx2s) not integrated within the larval head; appendages distorted (mx1, 2, thp1); abdomen (abd) incompletely phosphatized, visible by coin-like crystallites and void spaces (arrows; see also Müller & Walossek 1985b, Pl. 16:2, for Skaracarida). Scale bar 30 μm.

4   Ventral view of anterior head region; eye area collapsed, labrum distorted distally; coxae of 2nd antenna (cox a2) well sclerotized and with distinct joint to basipod (bas); note the orientation of the endites and their mode of setation with smaller setae anterior and more robust spines terminally; mandibular gnathobases (md) angled against the coxal body (both palps missing); one of the gnathobasic setae is partly preserved (gns). Scale bar 30 μm.

5   UB 66 (same specimen as in Pls. 21:2, 3; 22:2, 5). Ventral view; much of the ventral surface is concealed by a huge mass of phosphatic matter; labrum with distinctive constriction on its anterior surface (arrow);

2nd endopodal segment of 2nd antenna (en a2) divided into two (2a, b); furcal rami broadly rounded and oval-shaped, with about 9–10 marginal spines; rami not yet articulated (abd = abdomen; a1 = 1st antenna, see also Pl. 21:2; a2 = 2nd antenna; md = mandible; mx1 = 1st maxilla). Scale bar 50 μm.

6   UB 67 (same specimen as in Pl. 21:7). Anterior view of incomplete specimen with only slightly deformed anterior of shield; eye lobes project from the forehead (compare with next figure); it seems as if the outer cuticular layer has been partly rubbed off. Scale bar 30 μm.

7, 8: UB 68

7   Lateral view of rather coarsely preserved specimen without trunk; appendages not preserved save for the proximal parts of 2nd antenna (a2) and mandible (md cox); eye area complete (lo ce; mvl) and protruding well beyond the anterior shield margin; specimen seems to be inflated, probably by gas production due to decay prior to fossilization; inner lamella (il) exposed between limbs and shield margin (a1 = 1st antenna). Scale bar 30 μm.

8   Ventral view; midventral lobe (mvl) set off from labrum (la); sternum (stn) with deep furrow medially ('paragnath channel'); arrows point to groove of unknown nature at its slightly narrower posterior end (see also Pls. 13:7, 21:7); maxillary sternite probably coalesced with sternum, but it is unclear whether this segment is still free from the head dorsally; body torn off behind the 2nd maxillae (cox md = mandibular coxa; il = inner lamella; mx1, 2 = insertions of maxillae). Scale bar 30 μm.

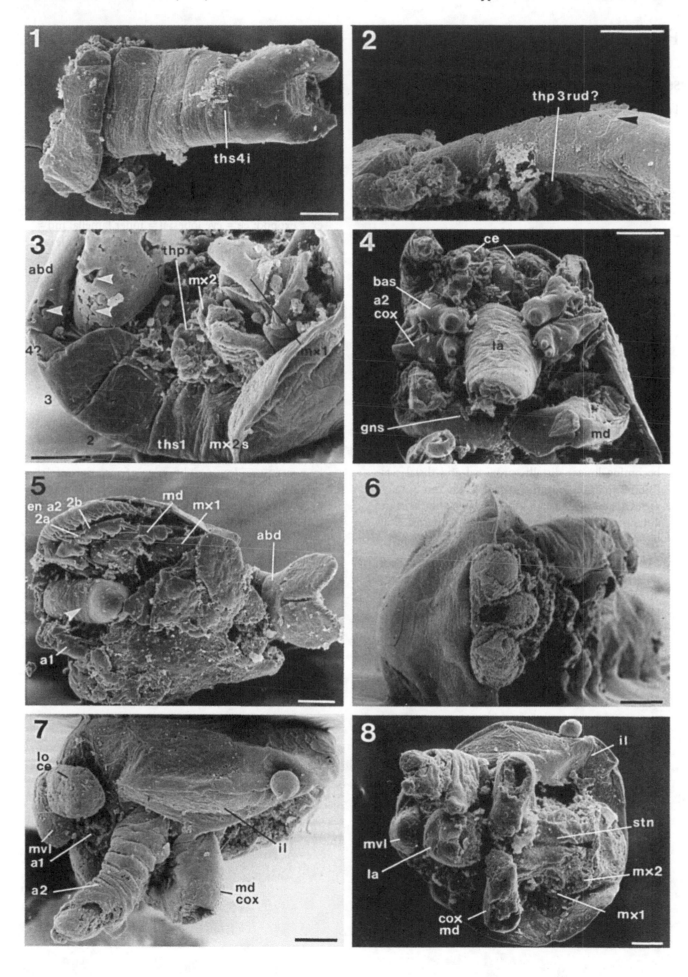

Plate 21

Stage TS4B continued

1   UB 69 (same specimen as in 5). Close-up of slightly collapsed eye area (lo ce, mvl) bordering the insertions of 1st antennae (a1); crack running through mid-ventral lobe and labrum caused by drying out of the adhesive tape; arrows point to thin ridges on the labral surface recognized in various specimens (epicuticular structures?; la = labrum). Scale bar 10 μm.

2   Same specimen as in 3, Pls. 20:5; 21:2, 5). Close-up of partly covered and crumpled 1st antenna with some of the setae from the distal portion; distal part divided into elongate segments which still show a faint incomplete annulation of the same size as those of the proximal 'shaft' (see also Pl. 34:5). Scale bar 15 μm.

3   View of 2nd antenna (a2) and mandible (md); proximal endopodal article (en1) of antenna nesting deeply in basipod (bas); 2nd one divided into two (en2a, b); distal article (en3) with a robust seta apically; its socket represents the rudimentary 4th article (see also on mandibular endopod and Pls. 18:5, 22:5); annules of exopods accord with median setation (compare with series A); proximal two annules lacking setae; ringlets becoming longer distally, proximal and distal setae thinner than those in the middle of the ramus (as in series A!); mandibular endopod (md en) shorter in 2nd antenna, its exopod is concealed by the antenna. Scale bar 30 μm.

4   UB 70 (same specimen as in Pl. 22:1, 6). Eight-segmented mandibular exopod arising from a narrow joint at the sloping other edge of the basipod (bas); proximal two ringlets sharing a seta; setae changing in size from proximal to distal but accord with annulation

(see previous Fig.); ridges on the ringlets demarcate the boundary between sclerotic half-ring on outer side and setal socket on inner edge (en = endopod). Scale bar 10 μm.

5   Same specimen as in 1. Ventrolateral view; partial breakage of limbs permits a view of the postoral food chamber (a1, 2 = insertions of antennae; en = endopod; esp = enditic spine of proximal endopodal article; ex = exopod; ce = eye lobes; il = inner lamella; md = mandible; mx1, 2 = maxillae; stn = sternum). Scale bar 30 μm.

6   UB 71. Close-up of proximal endites of 2nd antenna (a2), mandible (md), and 1st maxilla (mx1); masticatory spine (msp) of mandibular basipodal endite (bas) now oval in cross-section, circumstanding setae preserved only as their insertions; fringe of pectinate setae (ps) and some spines (sp) preserved on proximal endite of 1st maxilla (pe). Scale bar 30 μm.

7   Same specimen as in Pl. 20:6. Detailed view of partly preserved sternum; cuticle of paragnaths (pgn) rubbed off; deep furrow between them reaches posteriorly toward a depression with a pair of pores (po) of unknown nature; lower right: edge of insertion of 1st maxilla, indicating that the groove lies within maxillulary sternite (compare with Pls. 13:7 and 20:8). Scale bar 10 μm.

8   UB 72. Posterior view of fragmentary specimen with sharply ventrally flexed trunk; due to this, the arthrodial membranes between the thoracomeres are widely stretched and exposed (am); note the steep angle between mandibular gnathobase (gn) and coxal body (md cox; abd = abdomen; cs = shield; fr = furcal ramus; plp = remains of palp). Scale bar 30 μm.

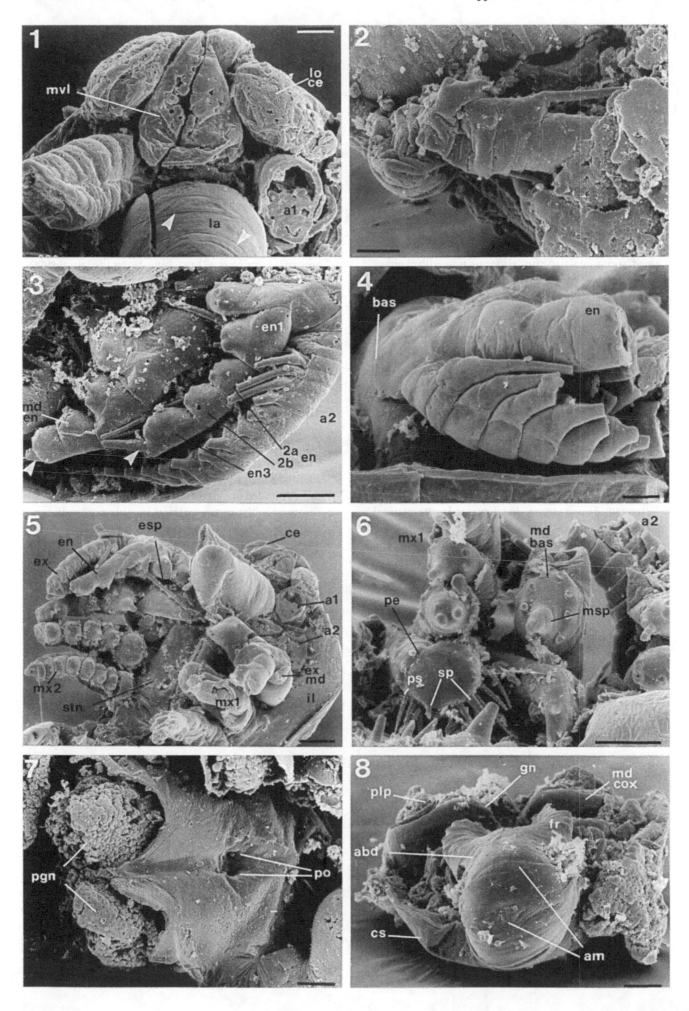

Plate 22

Stage TS4B continued

1   Same specimen as in Pl. 21:4. Ventral view of trunk with incipient ventrocaudal processes (i vcp) and furcal rami; thoracomeres dorsoventrally depressed; soft ventral, sternal cuticle wrinkled; appendages of trunk region not preserved (arrows point to their insertions). Scale bar 30 μm.

2   Same specimen as in 5, Pls. 20:5; 21:2, 3. Posterior view of trunk; anus recognizable as a T-shaped slit (an) within membranous field, dorsally with faintly-developed 'supra-anal flap'; furcal margin now with primary and secondary row of furcal spines (pfsp, sfsp; i vcp = ventrocaudal processes). Scale bar 30 μm.

3   UB 73 (same specimen as in Pl. 32:2). Close-up of right furcal ramus and incipient ventrocaudal process (i vcp); spines broken off distal to their sockets (fsp); on ventral side of ramal margin the number of pits has increased in accordance with the spines (po). Scale bar 10 μm.

4   UB 74 (same specimen as in 7). Close-up of masticatory spine (msp) of mandibular basipod; tip split into at least two spinules; two of the accompanying pectinate setae (ps) preserved with their subordinate setules which appear thicker than those on the coxal surface (stl cox) and on the somewhat depressed posterolateral side of the labrum (stl la); this type of pectinate setae with increasing distance between the setules toward the tip with a tuft of setules is different from that on the

endites of the posterior appendages and may have served for different function (see Pls. 15, 16 and Fig. 35H); masticatory spine also covered with setules, recognizable as small prickles. Scale bar 5 μm.

5   Same specimen as in 2. Arrows point to setules on exopodal setae of 2nd antenna and mandible; both endopods (en a2, md) with terminal segment reduced to a small hump which carried the apical seta originally (see also Pls. 18:5, 21:3). Scale bar 10 μm.

6   Same specimen as in Pl. 21:4. Enditic spines of the antennal coxa (cox a2) reaching along the labral side (la) toward the mouth and approaching the mandibular gnathobase (gn md); median one of the 3 spines visible appears to be split distally; (gns = socket of gnathobasic seta). Scale bar 5 μm.

7   Same specimen as in 4. Detail of enditic armature of the maxillae; setae very thin and gently tapering toward their tip (see Fig. 4 of same plate). Scale bar 10 μm.

8   Same specimen as in Pl. 21:6. Setae and spines of proximal maxillulary endite; note the different size and furnishment with setules and their orientation; compared to later stages these setae were likely not yet adapted for filtration; on upper left: short spine with more rounded tip; arrows point to spherical structures common in the *orsten* material (simply artificial?; see also Müller & Walossek 1985b, Pl. 16:7). Scale bar 3 μm.

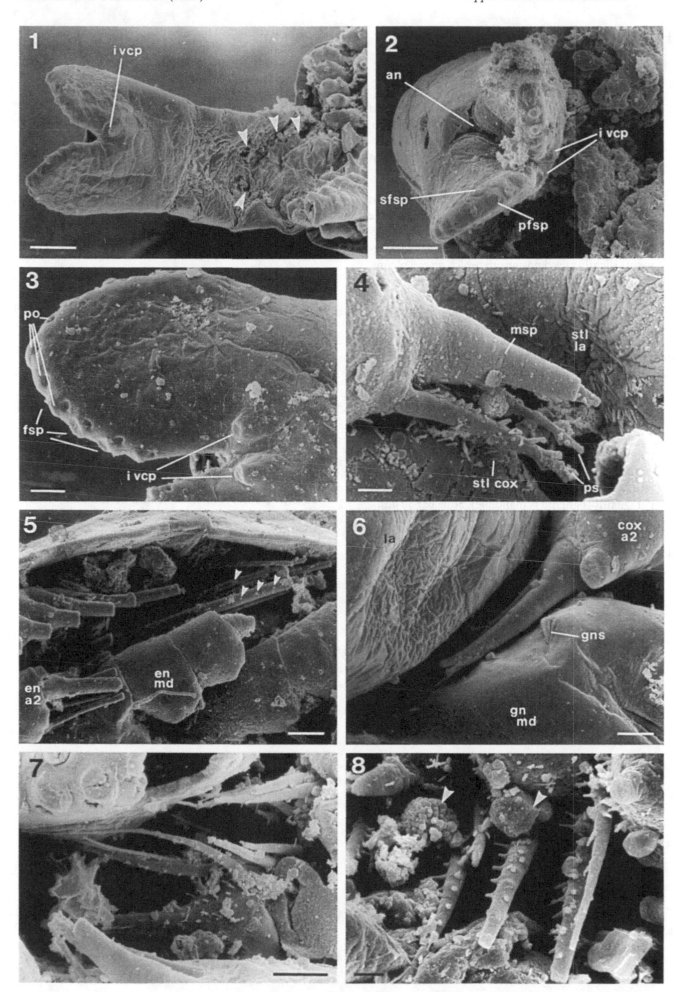

Plate 23

1: Stage TS5iB; 2–7: Stage TS5B

1   UB 75. Ventrolateral view of specimen with distorted head and slightly collapsed trunk (a2 = 2nd antenna; la = labrum; md = mandible; mx1, 2 = maxillae; stn = sternum; ths1–5i = thoracomeres; i vcp = incipient ventrocaudal process). Scale bar 100 µm.

2   UB 76 (same specimen as in 6). Nearly lateral view; shield (cs) broken off posteriorly, rendering visible the dorsal surfaces of maxillulary to anterior thoracic segments (mx1–ths1); ventral margins of 2nd to 4th segments slightly raised posteriorly which gives a pleura-like appearance to these; furcal rami (fr) not jointed, slightly dorsally directed (a2 = 2nd antenna; abd = abdomen; ex md = exopod of mandible; ths5 = 5th thoracomere). Scale bar 15 µm.

3   UB 77. Ventral view of somewhat deformed specimen with complete 2nd antennae (a2; except setation); labrum distorted distally; posterior appendages preserved with their proximal parts; region between 2nd thoracomere and abdomen deformed due to twisting; 1st maxilla with few but differentiated endites (mx1; pe, end2–3) as compared to the more equally designed endites of the posterior limbs (mx2, thp1, 2); sternum (stn) with deeply incised food groove and bulging paragnaths (pgn; md = mandible). Scale bar 50 µm.

4, 5: UB 78

4   Lateral view of incomplete specimen, seen slightly from posterior, showing the wing-like extended poste-rolateral corners of the shield (cs) and concave posterior margin (compare with *Bredocaris* in Müller & Walossek 1988b, Pl. 3:7, 8); inner lamella (il) exposed; trunk collapsed and wrinkled, abdomen and furca not preserved. Scale bar 50 µm.

5   Close-up of mandibular coxae (palp) and gnathobasic seta broken off; gns; plpf); distal end of labrum collapsed, rendering visible the curved inner edges of the gnathobases; anterior part of cutting edge with fine spinules, posterior part with few acute spinules; straight anterior margin of gnathobases fits into excavation at posterolateral edge of labrum (arrow); posterior tooth (pt) in line with the posterior margin of the gnathobase (mx1 = 1st maxilla). Scale bar 30 µm.

6   Same specimen as in 2. Close-up of complete gnathobasic seta (gns), lying between gnathobase and labrum; lower left: setules on right paragnath. Scale bar 10 µm.

7   UB 79 (same specimen as in Pl. 24:3). Similar view as in 5, but more from posterior; labrum with pits or knoblets (arrows) on its posterior side (see also Pls. 5:7, 9:7, 34:4); posterolateral sides of labrum slightly depressed, flanked by the antennal endites (cox, bas a2); mandibular coxae (cox md) oriented anteriorly, with their sharply angled gnathobases (gn) approaching the labrum; some of the marginal spinules still recognizable (spl, pt), those of the anterior part of the cutting edge now in a double row; membrane of coxal joint widely stretched due to anterior flexure of the limb; setal arrangement around median spine of maxillary endites well-recognizable (mx2; stn fgr = sternal food groove; mx1 = 1st maxilla; pgn = deformed paragnaths). Scale bar 30 µm.

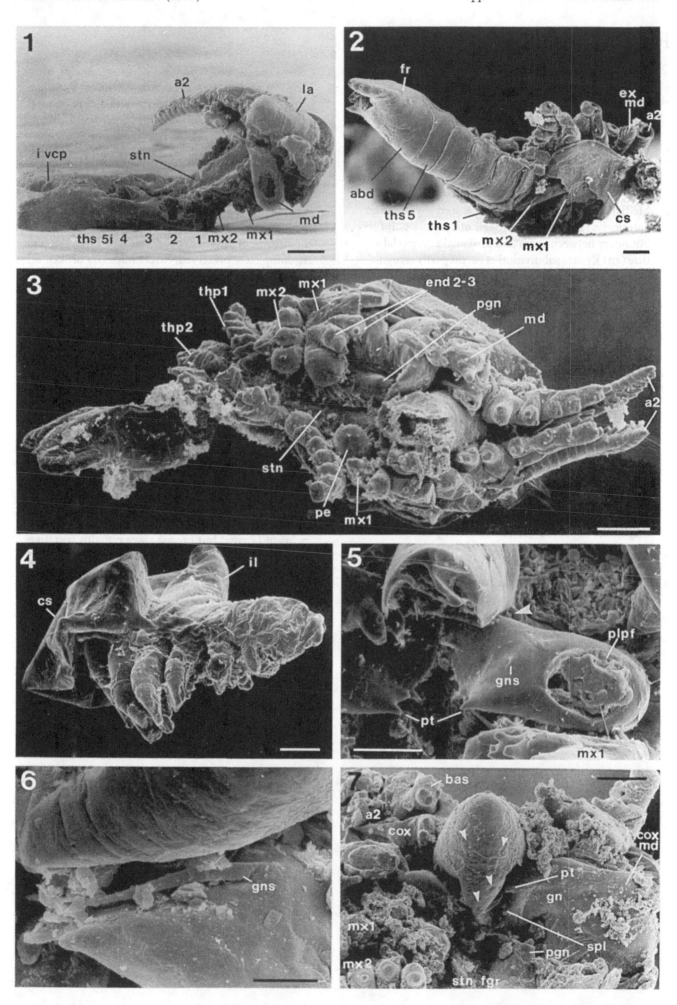

Plate 24

1–3: Stage TS5B continued; 4, 5 Stage TS7iB; 6–8: Stage TS8iB–9iB

1, 2: Same specimen as in 2, Pl. 23:2, 6

1   Ventral view of partly preserved right appendages (a2, md, mx1, 2, thp1, 2, 3 rud, 4 rud); (la = labrum; stn = sternum with deep median furrow). Scale bar 30 μm.

2   Median view of maxillae (mx1, 2), seen slightly from anteriorly; rami (en, ex) broken off; arrows point to boundary between corm and proximal endopodal article (en1); outer subdivision of corms partly recognizable, at least in the better sclerotized distal part; 4 endites in the 1st maxilla and 6 in the 2nd; proximal endites of 1st thoracopod (thp1) horn-like drawn out posteromedially (pe = proximal endite). Scale bar 30 μm.

3   Same specimen as in Pl. 23:7. Sternal food groove shallow due to collapsing of the body, except between the paragnaths (pgn); sternum (stn) with typical short, curved rows of setules; sternite of 2nd maxilla seemingly not yet coalesced (st mx2), consisting of two plates with a groove with two short slits medially (arrow); pores (po) are located on the sternitic plates close to the groove, a further pore is positioned just between the sternites of maxillary segment and 1st thoracomere (ths1; gn = gnathobase; la = labrum; pe mx1 = proximal endite of 1st maxilla). Scale bar 30 μm.

4–5: UB 80

4   Lateral view of distorted specimen; posterolateral corners of deformed shield raised and exposing the inner lamella; maxillary segment seemingly not coalesced (mx2); trunk laterally compressed anteriorly but flattened dorsoventrally posteriorly (saf = supra-anal flap; ths1–7i = thoracomeres). Scale bar 100 μm.

5   Incipient ventrocaudal processes with major spine (sp, broken), a shorter spine medially and a pore (arrow) corresponding to the major spine. Scale bar 3 μm.

6   UB 81 (same specimen as in Pls. 25:2–8). Lateral view of laterally deformed specimen, probably of stage TS9i; anterior head region distorted; shield fairly complete save for its anterolateral margin, which exposes the appendages; trunk sharply flexed against the body; appendages being successively more anteriorly oriented; some of the slender paddle-shaped exopods (ex) are completely preserved; furcal rami (fr) broken off distally; last thoracomere incipient (arrow). Scale bar 100 μm.

7   UB 645 (paratype, same specimen as in Pls. 25:1; 26:1, 2, possibly of stage TS8i rather than of TS9i). Ventral view of partly preserved specimen, seen slightly from posterior; shield margins (cs) somewhat rolled inward; appendages (md, mx1, 2, thp1–5), broken off distally save for the exopods of the 3rd to 5th thoracopods; median food path covered by foreign particles; on right: mandibular coxae, right one still carrying the basipod (bas) and its masticatory spine; left basipod broken off, exposing the palp foramen which is smaller than in preceding stages (ths7? = probable 7th thoracomere). Scale bar 30 μm.

8   UB 82. Ventral view (specimen lost); last segment incipient (arrow) but precise stage remains tentative; trunk flexed ventrally; anterior of head and distal parts of appendages distorted, also furcal rami incomplete; proximal endites of maxillae (pe mx1, 2) still with their anteriorly curved proximal row of setae; anterior thoracomeres seem to have short pleura-like lateral margins, but this may rather resulting from deformation (a1, 2 = antennae; cs = shield; fr = furcal rami; md = mandible; plpf = palp foramen; saf = supra-anal flap; thp1–3 = thoracopods; vcp = ventrocaudal process). Not to scale.

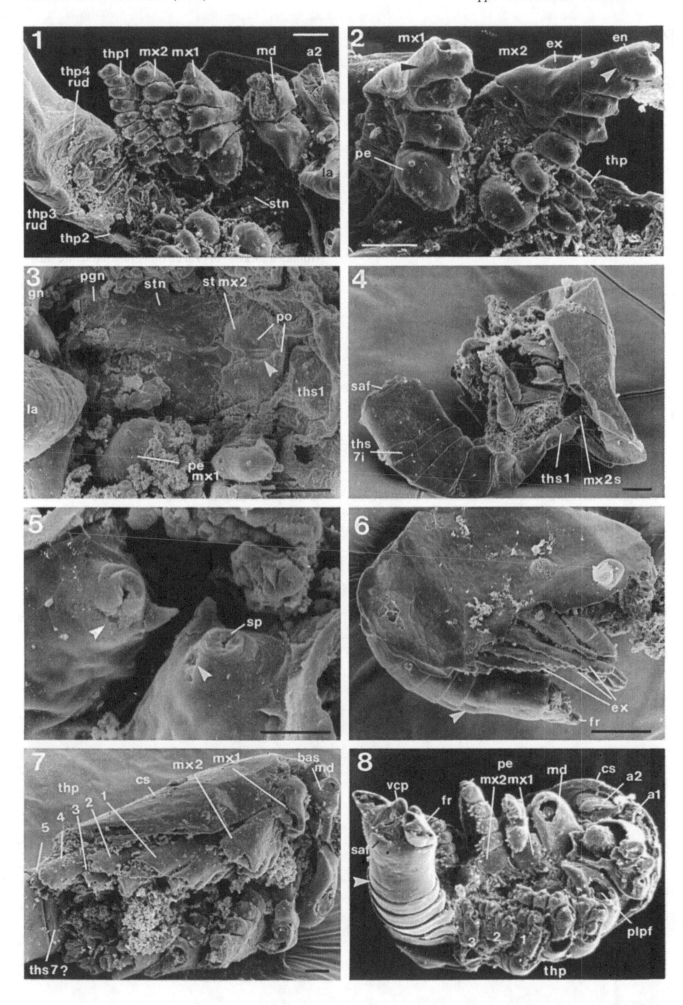

Plate 25

Stage TS8iB–9iB continued

1 Same specimen as in Pl. 24:7. Close-up of mandibular coxa with huge, blade-like gnathobase; surface slightly concave; posterior part of cutting edge with a few strong tooth-like spinules, anterior part with thinner acute spinules; gnathobasic seta not preserved, a small knob indicating its original position (gns; la = labrum; msp bas = basipodal masticatory spine; plpf = palp foramen; pt = posterior tooth). Scale bar 30 µm.

2–8: Same specimen as in Pl. 24:8

2 View of postmaxillulary appendages; some of their exopods still completely preserved; small knobs on exopodal margins represent sockets of original setation; lateral subdivision of corms recognizable only in 2nd maxilla, but effaced in the other limbs due to poor preservation (abd = abdomen; bas = basipod; fr = broken furcal ramus; saf = deformed supra-anal flap). Scale bar 30 µm.

3 Probable left 2nd maxilla with distal endites of corm and enditic surfaces of proximal endopodal articles; enditic surfaces with anterior group of bipectinate setae (1), median group of setae and/or spines (2) and curved posterior row of bipectinate setae (3; some complete ones on proximal endite visible); armature progressively less developed from proximal to distal (see also Pls. 26:3; 29:1 and Fig. 37). Scale bar 30 µm.

4 Detail of pectinate type of setae of the posterior rows; note the changing orientation of the comb rows of setules around the endites (arrows; sp = thinner, spine-like seta of median group; stl = setules on enditic surface). Scale bar 10 µm.

5 Distal ends of two exopods viewed from medially; ramal surface concave posteriorly, probably as it was already during life; two of the rigid marginal setae are partly preserved (arrows point to nodes originally having born setules), all other setae are either not preserved (rounded sockets) or broken off (straight fracture surfaces). Scale bar 10 µm.

6 High magnification of setal socket (seta broken off) with corona of tiny acute denticles (den). Scale bar 3 µm.

7 Posterior view of ventrocaudal processes; some of the rigid stout spines are still preserved, also setules or denticles on the surfaces of processes and spines; pores (po) appear to be partly closed by a flap. Scale bar 10 µm.

8 Close-up of spines on the ventrocaudal processes seen in dorsal view; spines with combs of denticles on their surfaces, more denticles are positioned in the vicinity of the spines. Scale bar 3 µm.

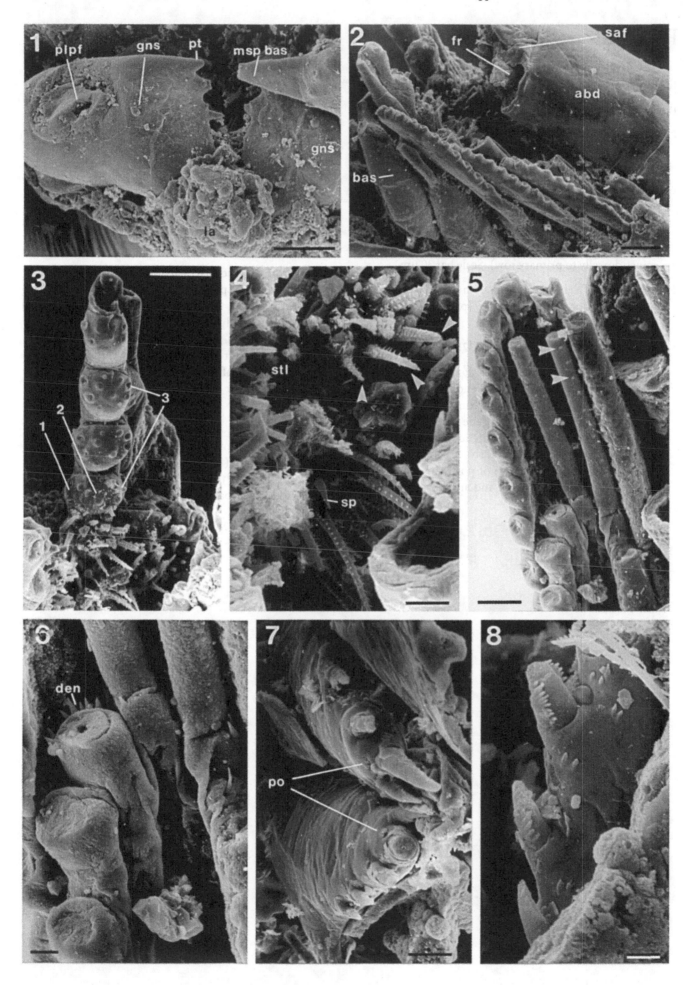

Plate 26

1, 2: Stage TS8iB–9iB continued; 3–5: TS10B

1, 2: Same specimen as in Pls. 24:7; 25:1.

1   Furcal rami and ventrocaudal processes at posterior end of abdomen; pores (po) occur on the articulated and slightly dorsally oriented furcal rami; their joint (j) is fixed medially which limited the range of dorsal movements; processes now with about 5 marginal spines and 3 pores; note the incision between the processes which proceeds anteriorly during ontogenesis (see Pls. 12:7; 15:1; sfsp = spine of secondary row of spines dorsally to primary row). Scale bar 30 μm.

2   Dorsal view of right furcal ramus with secondary row of spines (sfsp) dorsal to primary row (pfsp); no pores on dorsal surface. Scale bar 15 μm.

3–5: UB 771 (same specimen as in Pl. 32:3)

3   Ventral view of stretched fragment; right 1st maxilla preserved with its large proximal endite (pe) and the elongate 2nd one (end2 mx1); proximal endite of 2nd maxilla (mx2) larger than that of the thoracopods but similar to these in all other aspects; endites of corm with double row of anterior setae (1a, 1b), few median setae or spines (2) and a semicircle of posterior setae (3); appendages decreasing in size and armature progres-

sively rearward, 6th one with bifid endites only, similar to early stages of maxillulary development (see, e.g., Pls. 5:1, 3; 6:1); 7th and 8th limbs not preserved, but 9th pair present with its incipient rami (thp9 rud); 10th thoracomere apodous and almost ring-shaped save for the soft ventral side; abdomen (abd) incised mid-axially; ventrocaudal processes and furca not preserved; sternite of 2nd maxilla fused with sternum (stn); thoracic sternites composed of two plates (am = pliable membrane between trunk sternites). Scale bar 50 μm.

4   View from a medioproximal direction onto major row (3) of setae on the proximal endites (pe) of the maxillae (mx1, 2); in both this row (see also preceding figure) runs from dorsally around the posterior edge almost below the endite towards its anterior side where it meets the anterior set of setae (1); setae and spines on bulging surface of endites not preserved (fgr = food groove). Scale bar 30 μm.

5   Anterior surfaces of 2nd and 3rd thoracopods (thp2, 3); exopods (ex) arising from steeply sloping outer surface of the corms, the latter being subdivided into 3 portions laterally (fu = furrow); most likely 8–9 endites maximally (6 in the 2nd maxilla; en = insertion of endopod). Scale bar 30 μm.

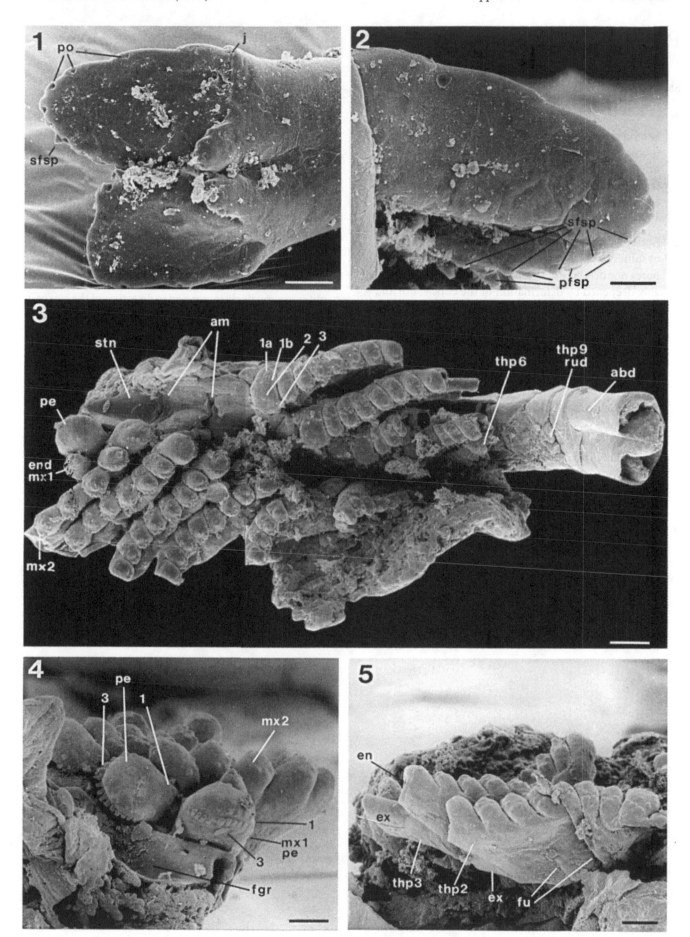

Plate 27

Stage TS11B

1, 2: UB 83

1   Lateral view of fragment lacking head, limbs and furca; thoracomeres shrivelled (specimen thus much larger originally); trunk end sharply angled against the anterior portion (abd = abdomen; app = remnants of thoracopods; vcp = ventrocaudal process). Scale bar 100 μm.

2   Abdomen with long ventrocaudal processes; number of spines has increased as the processes have enlarged; pores not distinctive; sharp median furrow (fu) probably deepened by deformation of the cuticle; 8th and 9th thoracomeres with remains of rudimentary appendages (app rud); last segment partly broken apart, exposing the interior filled with coarse phosphatic matter. Scale bar 30 μm.

3, 4: UB 84

3   Lateral view of specimen distorted in particular in the head region (C); cuticle of posterior portion of trunk seemingly rubbed off (arrow) exposing a non-structured filling (fi); ball-shaped structure attached to one of the thoracomeres is probably artificial (abd = abdomen; cs = shield; vcp = ventrocaudal process). Scale bar 100 μm.

4   Ventral view of this fragmentary specimen showing peculiar preservation: anterior appendages and labrum broken off rendering visible the bottom (ceiling in morphological terms) of the atrium oris (ao) anteriorly to esophagus (eso) running inwardly (stn fgr = sternal food groove). Scale bar 30 μm.

5–7: UB 85

5   Median view of thoracopods and inter-limb spaces which form the 'sucking chambers' of the filter apparatus (compare with Figs. 33 and 38); limbs inserting abaxially and being flattened in antero-posterior direction; their corms being convex anteriorly and concave posteriorly; endites pointing posteriorly and elongated asymmetrically elongated rearward; orientation of endites progressively increasing in direction of the axis of the limb (distal ends of endopods broken off); size and setation of endites changing from proximal to distal as well as from the anterior to the posterior limbs; sternal food groove less developed in the posterior part of trunk. Scale bar 30 μm.

6   Anterior view of anteriormost thoracopod; arrows point to furrows subdividing the corm laterally; pliable proximal shaft is clearly separate from the more firmly sclerotized distal part of the corm (sh; ex = fragments of exopods). Scale bar 30 μm.

7   View of outer edges of the large limb corms; subdivision of these (arrows) becoming indistinct in direction of the endites; joint (j) of exopod (ex) feebly demarcated anteriorly (but see Pl. 29:6 for posterior side; en = endopod, broken off distally). Scale bar 30 μm.

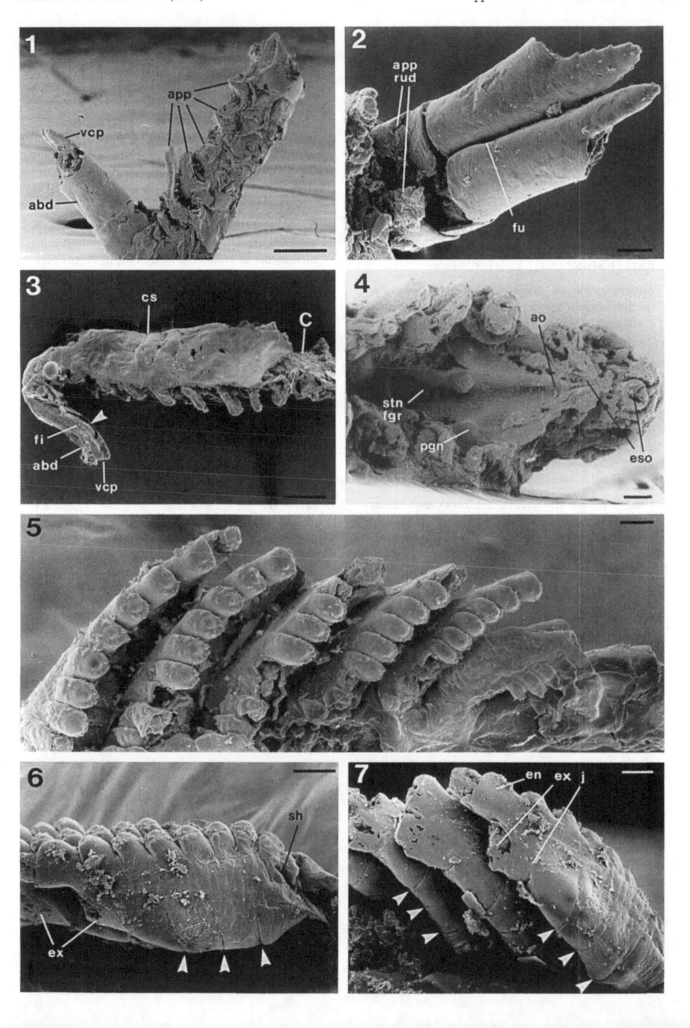

Plate 28

Stage TS12B, UB 644 (holotype; same specimen as in Pl. 32:4, see also Müller 1983, his Fig. 7A–C)

1 Lateral view (picture rotated into morphological orientation of animal); shield long, covering much of the trunk (abd = abdomen). Scale bar 100 μm.

2 High magnification of gnathobasal cutting edge, subdivided into anterior pars molaris (broadened and concave area) and pars incisivus with rigid tooth-like spines; anterior margin of cutting edge bearing thinner spinules (spl), concave area with setules (stl; pt = insertion of posterior tooth). Scale bar 10 μm.

3 Anterior view of shield, showing its roof-shape (slightly laterally compressed); anterior margin widely V-shaped opening ventrally and gently curving posteriorly reaching the deepest level behind the mandibles (md); arrow points to somewhat distorted hump on anterior surface of labrum (la). Scale bar 100 μm.

4 Anteroventral view; forehead region with eyes (eye) and antennae (a1, 2) distorted; all anterior structures seemingly reduced in size, at least relative to the other details; note the foliate habit of the postmandibular limbs, enhanced by shrinkage; margin of shield rolled inward (md = mandible). Scale bar 100 μm.

5 View of mouth area (a2, la, md, mx1); mandibular coxal bodies rounded laterally and with small palp foramen (plpf); gnathobase widening towards the cutting edge, surface concave; posterior margin thickened and slightly angled against rest of gnathobase (see also

Fig. 2E); insertion of gnathobasic seta not identified; both gnathobases seem to be rather symmetrical; distal end of labrum not preserved; large proximal endite (pe) of 1st maxilla provided with numerous setae and spines; elongate 2nd endite (end2) much smaller. Scale bar 30 μm.

6 Ventral view of proximal endites of maxillae (mx1, 2) and anterior thoracopods (thp1, 2); due to collapse prior to embedding the limbs are very thin in antero-posterior aspect save for the endites, demonstrating their nature as limbs sustained in life by turgor pressure; proximal endite of 2nd maxilla designed as in the thoracopods. Scale bar 30 μm.

7 Close-up of row of setae at lower margin of a proximal endite of a thoracopod; setae feathered with setules but much denser (1 per μm) than on true pectinate setae (1 per 2 μm) and slightly irregularly, indicating that they did not serve for filtration but as brushes transporting nutrient particles anteriorly mechanically. Scale bar 3 μm.

8 Ventral view, giving an impression of the mode of food intake: particles were transported through the narrow food path between the thoracopods towards the bulbous maxillulary proximal endites which passed them over to the gnathobases (abd = abdomen; a1, 2 = antennae; la = labrum; m = mouth tunnel; md = mandible; mx1, 2 = maxillae; thp1–6 = thoracopods; ths8–12 = posterior thoracomeres). Scale bar 100 μm.

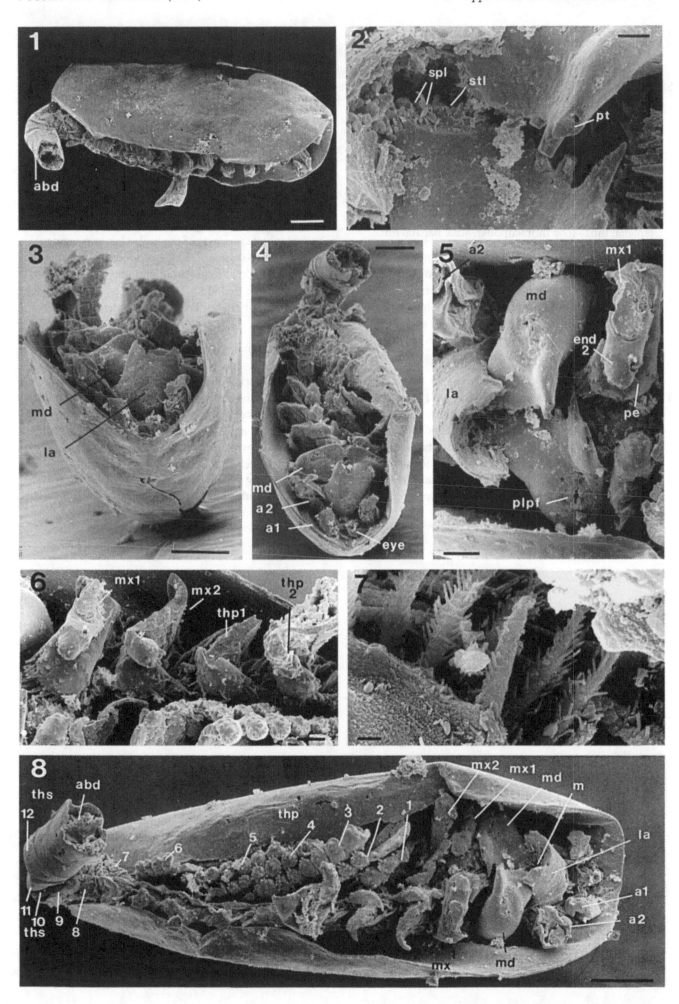

Plate 29

Stage TS13B

1–3: UB 86

1   Ventral view; limbs very thin in the middle due to collapse but outer and inner edges broader due to better sclerotization; arrow points to boundary between fragmented head (C) and thorax; food groove (fgr) deeply recessed due to collapse; sternites lying immediately on the shield (abd = abdomen; il = inner lamella; thp1–10 = thoracopods). Scale bar 100 µm.

2   Close-up of proximal two thoracopodal endites, with anterior group of setae (1a, b), spines or setae on median surface (2), and the U-shaped posterior row of bipectinate setae (3) running from dorsally along the posterior margin and slightly on the lower side anteriorly; it seems as if some of the setae or spines had nested in sockets and thus were articulated (arrows). Scale bar 10 µm.

3   Median view of left thoracopods (thp4–11), showing arrangement and orientation of the enditic setae; posterior end of trunk with limbs; 10th limb not definitely developed and comparable to earliest stages of 1st maxilla (see Pls. 5:1, 3; 6:1); 11th limb rudimentary, pliable, and with some paired spines medially indicating the future endites (arrow; abd = abdomen; en = endopod; ex = exopod). Scale bar 30 µm.

4, 5: UB 87

4   View of right side of peculiar specimen with distorted head (C), part of the shield (cs), curved trunk and cylindrical abdomen (abd) lacking the furca; anterior limbs distorted and somewhat dislocated; some of the posterior limbs are well preserved, even retaining some of their exopodal setae; orientation of limbs preserved as if fixed while beating in rhythm; segments of anterior trunk region with bipartite swellings dorsal to the pliable shafts (arrows) of the limbs; last two segments almost ring-shaped (en = endopod; ex = exopod; mx2 = dislocated fragment of 2nd maxilla; thp1–10 = thoracopods; ths1–13 = thoracomeres; lines indicate connections between limbs and segments). Scale bar 100 µm.

5   Posterior thoracopods of left series with their distinctly subdivided corms and complete exopods overlapping each other; orientation of exopods is predicted by their insertion at the steeply sloping outer edge of the corm; note the effacement of the anterior part of the exopod joint (j), while the posterior part is distinct to permit a wider back swing; depressions between thoracomeres may have been caused by limb muscles (ms = muscle scars); arrow points to boundary between elongated corm and endopod (en). Scale bar 30 µm.

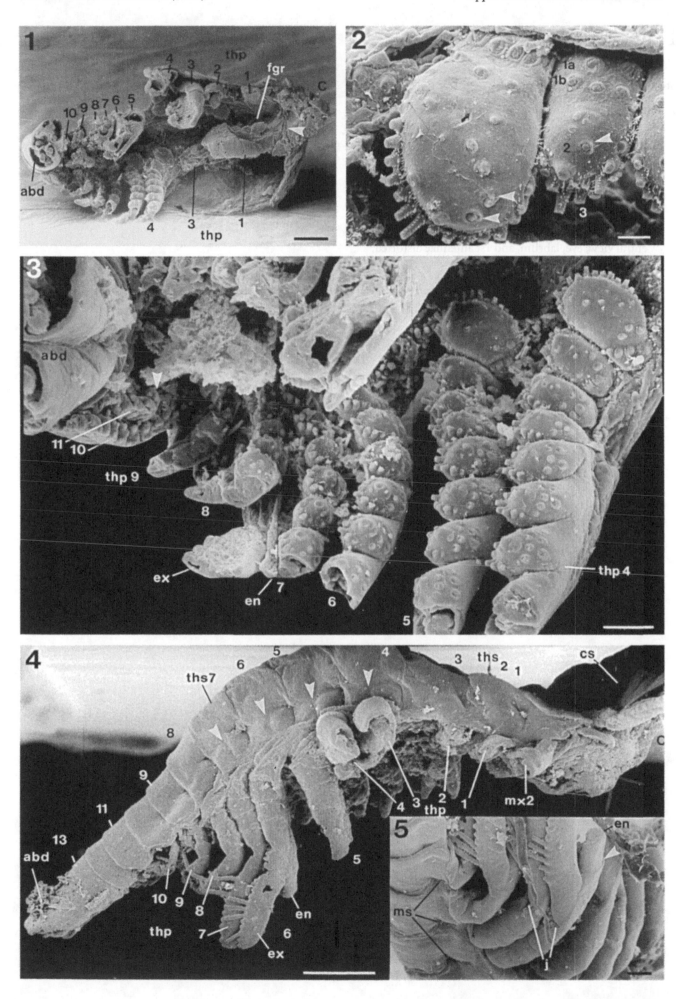

Plate 30

Stereo photographs (not scaled)

1   UB 9, stage L2A (same specimen as in Pl. 2:5, 9, 10). Ventral view; eye and 1st antenna not preserved; 2nd antenna and mandible directed ventrolaterally; ventral flexure of hind body is a common type of preservation of *Rehbachiella* larvae, possibly a typical life position of these (see also Müller & Walossek 1988b for *Bredocaris*, e.g., Pl. 12:1, 2, 4, 5).

2   UB 28, stage L5A (same specimen as in Pls. 6:9; 7:2). Ventral view; eye, labrum, and 1st antennae distorted; appendages posteriorly oriented.

3   UB 37, stage TS2A (same specimen as in Pls. 8:7, 8; 9:1–3). Ventral view; lobes of compound eye and mid-ventral lobe much inflated; trunk curved ventrally and twisted.

4   UB 44, stage TS5A (same specimen as in Pls. 10:5; 11:4, 5). Ventral view of body fragment; inflation may be caused either by gas production due to decomposition or hypo-salinity of the surrounding medium at the time of burial.

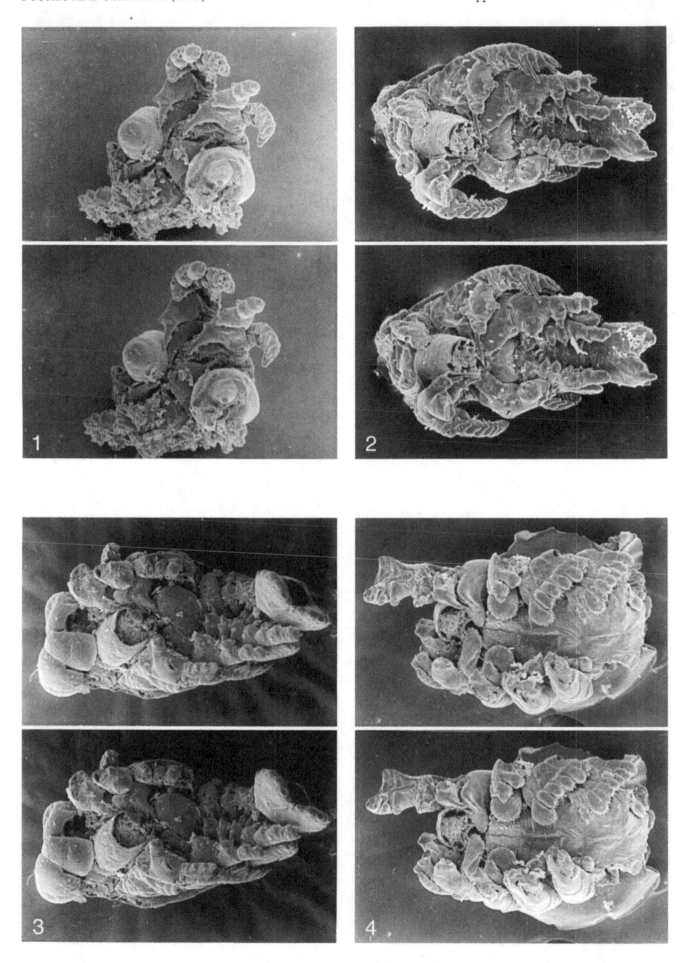

Plate 31

Stereo photographs continued

1 UB 51, stage TS7A (same specimen as in Pl. 13:1–4, 8). Ventral view; trunk broken at 5th thoracomere.

2, 3: UB 55, stage TS10A (same specimen as in Pls. 15; 16:1–7)

2 Lateral view of stretched specimen lacking shield and posterior limbs of right set of thoracopods.

3 View of posterior part of filter apparatus, with numerous pectinate setae still preserved.

4 UB 58, stage TS13A (same specimen as in Pl. 17:2). Ventral view of largest specimen at hand assigned to *Rehbachiella*; posterior body broken off.

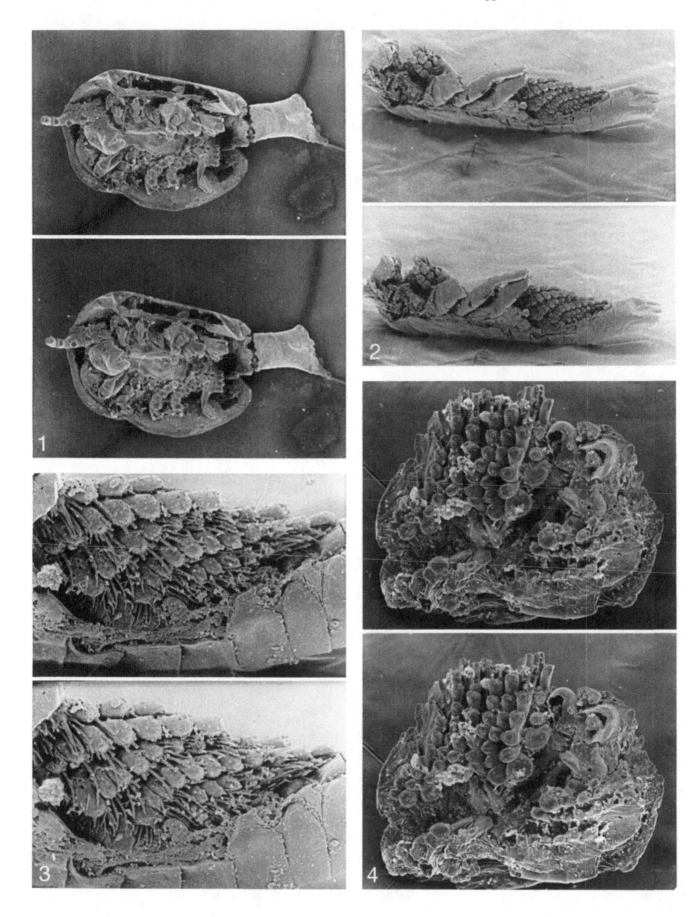

Plate 32

Stereo photographs continued

1   UB 63, stage TS3B (same specimen as in Pl. 19:4–7). Ventral view of specimen with ventrolaterally directed appendages and sharply ventrally flexed trunk; anus somewhat protruded.

2   UB 73, stage TS4B (same specimen as in Pl. 22:3). Ventral view of slightly curved specimen.

3   UB 771, stage TS10B (same specimen as in Pl. 26:5–7). Ventral view of fragment with laterally stretched, somewhat dislocated thoracopods.

4   UB 644, stage TS12B, holotype (same specimen as in Pl. 28). Ventral view of the largest specimen with the cephalic shield preserved.

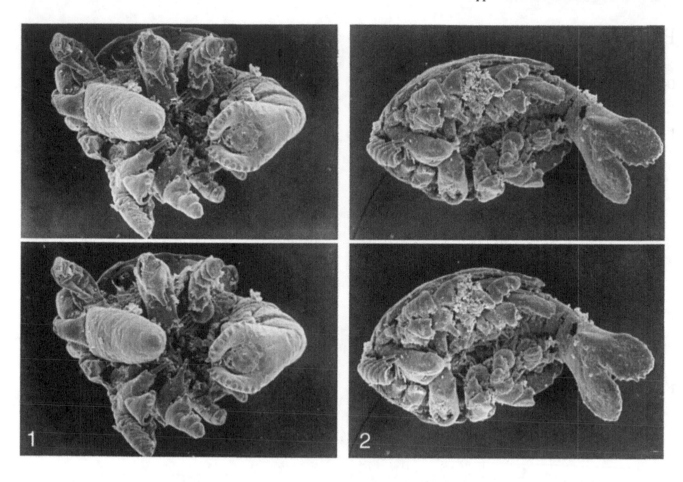

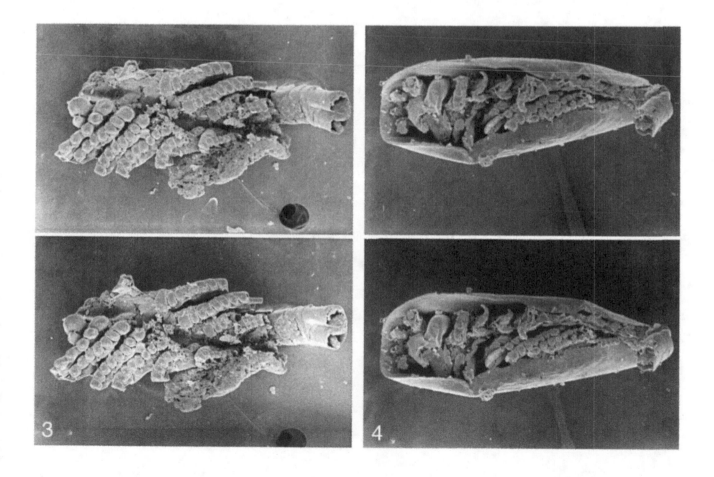

Plate 33

Unassignable specimens

1, 2: UB 88–91 (same specimens as in Pl. 34:1–4)

1   At least 4 specimens aggregated together, probably all of about stage TS4 (A and B!); of (1), in the upper right, only part of the distorted shield is recognizable; of (2), in the centre, the shield and the thoracomeres are exposed; (3) on left is a ventrally flexed and distorted specimen but showing many details (see also Pl. 34); (3) may belong to series B, stage TS4, because the 2nd endopodal article of 2nd antenna is subdivided (arrow) and the setation and annulation of the antennal exopod are clearly correlated with one another; number of setae on maxillulary exopod, number of furcal spines, and lack of secondary furcal spines are further indicators; (4) is concealed by phosphatic matter (see next Fig.; a1, 2 = antennae; cs = shield; fr = furcal rami; la = labrum; mx1, 2 = maxillae). Scale bar 50 μm.

2   View of opposite side; specimens numbered as in Fig. 1 (abd = abdomen; cs = shield; fr = furcal ramus; La = labrum). Scale bar 50 μm.

3, 4: UB 92

3   Median view of large fragment consisting of two limbs; specimen tentatively assigned to *Rehbachiella*, comprising the distal portion of the corms and parts of the rami; shape of enditic surfaces changes from being transversely elongated in the proximal endites to axially stretched in the endopodal articles; arrow points to boundary between endopod (en) and corm (co = corm; ex = exopod; see also Fig. 14B). Scale bar 100 μm.

4   Close up view of two endites with almost triangular surface; anterior setae arranged in two axial rows (1a, b), the inner one joining the U-shaped posterior row of pectinate setae (3); median surface little elevated and with few setae or spines (2); surface covered with numerous setules (stl); note the different sizes of the pectinate setae and the changing orientation of their setules (see also Fig. 14A). Scale bar 30 μm.

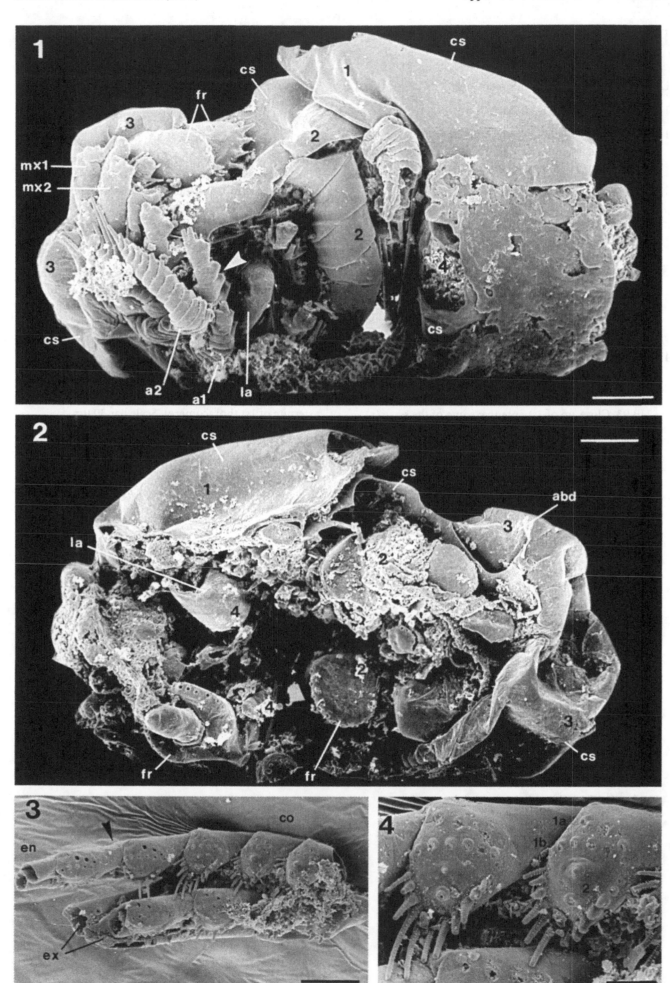

Plate 34

Unassignable specimens continued

1–4: same specimen as in Pl. 33:1, 2

1   View of distal segments of 1st antenna (a1) and endopods of 2nd antenna (a2) and mandible (md) of specimen (3); arrows point to apical articles of these appendages (end2 = 2nd endopodal article; ex = exopod). Scale bar 30 μm.

2   Ventrolateral view of furcal ramus; marginal spines broken off; arrows point to pits ventral to the spines, which may either have covered pores or have born hairs originally (den = denticles). Scale bar 10 μm.

3   Expanded anal membrane or extruded hind gut between the furcal rami; furcal spines broken off, leaving holes in the cuticle. Scale bar 30 μm.

4   Posterior view of labrum; arrows point to tubercles of probable sensory function (see also Pls. 5:7; 9:7; 23:7). Scale bar 10 μm.

5   UB 93. Close-up of 1st antenna; some of the setae (s) recognizable by their preserved sockets (den = denticles on annules of 1st antenna; see also Müller & Walossek 1985b for Skaracarida and 1988b for *Bredocaris*). Scale bar 15 μm.

6   UB 94. Lateral view of distorted head fragment; finely folded limb bases of 2nd antenna, 2nd maxilla (a2, mx2, thp1) collapsed in contrast to the strong mandibular coxa (md) and the slightly better sclerotized 1st maxilla (mx1; arrow on outer side; a1 = 1st antenna; la = labrum). Scale bar 30 μm.

7, 8: UB 95

7   Peculiar distortion of labrum (la) and surrounding appendages renders visible the ceiling (bottom) of the atrium oris (ao; md = mandible; pgn = paragnath). Scale bar 15 μm.

8   View of the sternal surface of this badly distorted specimen; thoracic sternal plates crushed against the cephalon, anterior plate with pore (po) medially (la = labrum; md = mandible; pgn = paragnath; st ths1 = sternite of 1st thoracomere). Scale bar 30 μm.

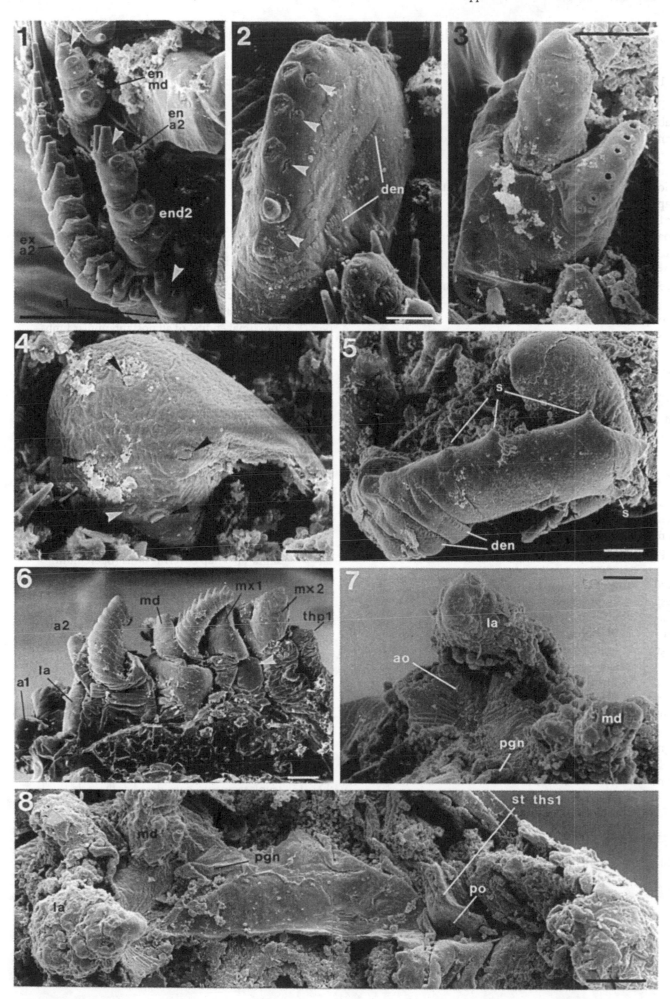

List of abbreviations with explanations

Terms mainly after Kaestner (1967), Moore & McCormick (1969) and McLaughlin (1980).

abd	abdomen (including the non-somitic telson)
abdl	length of abdomen
am	arthrodial membrane covering joints
an	anus
anf	anal field, membranous area around anus
ao	atrium oris, funnel-shaped mouth opening
app	unidentified appendage
a1, 2	first and second antennae
bas	basipod, distal portion of limb corm carrying the rami
bs	brush-like seta or spine
ce	presumed compound eye composed of two ovate blisters
co	corm of limb
cox	portion of limb corm proximal to basipod (enlarged 'proximal endite')
cs	cephalic shield formed by all pre-maxillary segments
csp	comb spine of more distal endites
dcsp	dorsocaudal spine dorsal to anus (only in early larvae)
den	denticles, often as fringe on ring-like ramal segments, also on lateral and outer sides of appendages
ds1–6	developmental stages 1–6 of limbs
en	endopod, inner ramus
end	endites, setiferous lobes of limb corm
epi	epipod(ite), outgrowth of outer side of limb corm
eso	esophagus
esp	enditic spine, often setulate distally
ex	exopod, outer ramus
fi	phosphatic internal filling of body cavity
fgr	food groove, formed by invaginated thoracic sternites (cephalic part = 'paragnath channel')
fr	furcal or caudal ramus
fsp	furcal spines, marginal armature of furcal rami
fu	furrow
g	gut, digestive tract, intestine
gn	gnathobase, blade-like median process of mandibular coxa
gns	gnathobasic seta; setulate seta on surface of gnathobase (note: 'Gnathobasen-Seta' of some authors refers to the armature of the coxal endite 'naupliar process' of the 2nd antenna)
h	height
hl	head length (distance between 1st antenna and 2nd maxilla)
i	incipient, also used for 1st step in thoracomere formation
il	inner lamella, pliable cuticle below the shield
j	joint, articulation
l	length
la	labrum
lo	lobes or blisters of presumed compound eye
L1–4	1st to 4th larval stages (1st one is a true 'orthonauplius')
m	mouth
md	mandible
mvl	midventral lobe (between lobes of presumed compound eye)
msp	masticatory spine of mandibular basipod
mx1, 2	first (maxillula) and second maxillae (maxilla)
mx1, 2s	segments of the maxillae
ne	naupliar eye
no	'neck organ'
pe	'proximal endite' of all postantennular limbs
pgn	paragnaths, pair of outgrowths of mandibular sternite
plp	palp, distal part of mandible comprising basipod and rami
plpf	palp foramen, insertion area of basipod
po	pore or pit
ps	pectinate seta, with regular row(s) of setules
pfsp	primary row of spines on margin of furcal ramus
pt	posterior tooth, posterior spinule at inner margin of mandibular gnathobase
rud	rudimentary, larval-shaped (not fully-developed, in contrast to vestigial; see also incipient)
s(tl)	seta (setule, minute bristle)
saf	supra-anal flap covering anus (operculum)
sec sh	secondary shield of Conchostraca after metamorphosis to pre-bivalved stage ('heilophora')
sfsp	secondary row of spines dorsally to primary row
sh	used for shaft-like proximal part of appendages
sp(l)	spine (spinule, short spine)
ST	specimen number (not illustrated ones)
st	sternite, sternal bar between postoral appendages
stn	sternum, formed by fusion of sternites of postoral segments
thl	length of thorax (distance between mx2 and anus)
thp1–12	trunk legs, considered as thoracopods
ths1–13	trunk segments, considered as thoracic
tl	total length
tr	larval trunk prior to delineation of segments, hind body
TS1i–13	postnaupliar stages (stages of 'thoracic phase')
vcp	pair of processes at ventrocaudal margin of telson
w	width